MOLECULAR TOPOLOGY

MOLECULAR TOPOLOGY

M DIUDEA, I. GUTMAN AND L. JANTSCHI

Nova Science Publishers, Inc.
Huntington, New York

Senior Editors: Susan Boriotti and Donna Dennis
Office Manager: Annette Hellinger
Graphics: Wanda Serrano
Information Editor: Tatiana Shohov
Book Production: Cathy DeGregory, Lynette Van Helden and Jennifer Vogt
Circulation: Ave Maria Gonzalez, Ron Hedges, Andre Tillman

Library of Congress Cataloging-in-Publication Data
Available upon request.

Includes bibliographical references and index.
ISBN 1-56072-957-0

M. Diudea, I. Gutman and J. Lorentz

CONTENTS

PREFACE

Many, yet not all, chemical substances consist of molecules. The fact that molecules have a "structure" is known since the middle of the XIX century. Since then, one of the principal goals of chemistry is to establish (causal) relations between the chemical and physical (experimentally observable and measurable) properties of substance and the structure of the corresponding molecules. Countless results along these lines have been obtained, and their presentation comprise significant parts of textbooks of organic, inorganic and physical chemistry, not to mention treatises on theoretical chemistry.

The vast majority of such "chemical rules" are qualitative in nature. A trivial example: if the molecule possesses a -COOH group then the corresponding chemical compound (usually, but not always) exhibits an acidic behavior.

A century-long tendency in chemistry is to go a step further and to find quantitative relations of the same kind. Here, however, one encounters a major problem. Molecular structure (to simplify: the features expressed by means of structural formulas) is a non-numerical notion. The measured physico-chemical properties of substances are quantities that are expressed by numbers (plus units, plus experimental errors). Hence, to find a relation between molecular structure and any physico-chemical property, one must somehow transform the information contained in the molecular structure into a number (or, more generally, into a sequence of numbers). Nobody knows how to make this transformation or these transformations.

At this moment there is no theory that could serve as a reliable guide for solving this problem. There have been many many many attempts in this direction. One group of them uses so-called topological indices. A topological index is a quantity that is somehow calculated from the molecular graph and for which we believe (or, sometimes, are able to demonstrate) that it reflects relevant structural features of the underlying molecule. This book is aimed at giving a reasonably comprehensive survey of the present, *fin de siècle*, state-of-the-art of the theory and practice of topological indices.

Some twenty years ago there were a dozen or so topological indices, only few of them with noteworthy chemical applications. Nowadays, their number increased enormously. The readers of this book are warned that in Chapter 7 the number of distinct topological indices will exceed 10,000. An alternative title of our book could be "Topological Indices - A Jungle Guide". There are two nasty, but inevitable questions: Is there any need for topological indices? Is there any real benefit for chemistry (or to generalize: for mankind) from the usage of topological indices?

Some twenty years ago these authors would certainly offer "yes" as answers, but would have a hard time to convince the less gullible part of the chemical community. Nowadays, the answers are still "yes", but their justification is much easier.

The applications of topological indices reached a level when they are directly used for designing pharmacologically valuable compounds. Let the titles of some recently published papers speak for themselves: *Quantitative Structure-Activity Relationship Studies on Local Anesthetics* [S.P. Gupta, *Chem. Rev.* **1991**, *91*, 1109-1119]; *Structure-Activity Study of Antiviral 5- Vinylpyrimidine Nucleoside Analogs Using Wiener's Topological Index* [S. Mendiratta, A. K. Madan, *J. Chem. Inf. Comput. Sci.* **1994**, *34*, 867-871]; *Structure-Activity Study on Antiulcer Agents Using Wiener's Topological Index and Molecular Topological Index* [A. Goel & A. K. Madan, *J. Chem. Inf. Comput. Sci.* **1995**, *35*, 504-509]; *Modelling Antileukemic Activity of Carboquinones with Electrotopological State and Chi Indices* [J. D. Gough & L. H. Hall, *J. Chem. Inf. Comput. Sci.* **1999**, *39*, 356-361]. Of all recent successes made by the aid of topological indices we mention just one. The paper G. Grassy, B. Calas, A. Yasri, R. Lahana, J. Woo, S. Iyer, M. Kaczorek, R. Floc'h, & R.Buelow, *Computer Assisted Rational Design of Immunosuppressive Compounds*, [*Nature Biotechnol.* **1998**, *16*, 748-752] reports on a search for peptides possessing immunosuppressive activity. They used 27 structure-descriptors, of which 12 topological indices. From a combinatorial library of about 280,000 compounds they selected 26 peptides for which high activity was predicted. Five of them were actually synthesized and tested experimentally. The most potent of these showed an immunosuppressive activity approximately 100 times higher than the lead compound.

One may suspect that in pharmaceutical companies many analogous researches have been (and are currently being) undertaken, with even better results, but - understandably - are not publicized.

* * *

Returning to topological indices: They, of course, are not the miraculous philosopher's stone of our times. They are far from other powerful tools of theoretical chemistry (such as thermodynamics or quantum mechanics). They, however, offer a meager hope to connect structure with properties, and to do this in a quantitative manner. They, perhaps, deserve the attention of a limited group of chemists. They, perhaps, deserve that every chemist should know a bit about them. They, perhaps, deserve to be mentioned in (undergraduate) courses of organic, physical and pharmacological chemistry.

* * *

Although each author contributed to the entire book, Chapters 1, 2, 4, 6 and 8 were written by M.V.D., Chapters 3 and 5 by I.G. and Chapters 7 and 9 by L.J. Each author takes responsibility only for the materials outlined in the chapters written by himself.

Cluj and Kragujevac, Fall 1999

Mircea V. Diudea   Ivan Gutman   Lorentz Jantschi

Chapter 1

INTRODUCTION TO MOLECULAR TOPOLOGY

Graph theory applied in the study of molecular structures represents an interdisciplinary science, called chemical graph theory or molecular topology. By using tools taken from the graph theory, set theory and statistics it attempts to identify structural features involved in structure-property activity relationships. The partitioning of a molecular property and recombining its fragmental values by additive models is one of its main tasks. Topological characterization of chemical structures allows the classification of molecules and modeling unknown structures with desired properties.

Before detailing the specific questions of molecular topology, some basic definitions[1] in graph theory are needed.

1.1 GRAPHS

A *graph,* $G = G(V, E)$ is a pair of two sets: $V = V(G)$, a finite nonempty set of N points (i.e. vertices) and $E = E(G)$, the set of Q unordered pairs of distinct points of V. Each pair of points (v_i, v_j) (or simply (i,j)) is a line (i.e. edge), $e_{i,j}$, of G if and only if $(i,j) \in E(G)$. In a graph, N equals the cardinality, $|V|$, of the set V while Q is identical to $|E|$. A graph with N points and Q lines is called a (N, Q) graph (i.e. a graph of order N and dimension Q). Two vertices are *adjacent* if they are joined by an edge. If two distinct edges are incident with a common vertex then they are *adjacent edges*. The angle between edges as well as the edge length are disregarded. The term *graph* was introduced by Sylvester.[2]

There is a variety of graphs, some of them being mentioned below.

A *directed graph* or *digraph* consists of a finite nonempty set V of points along with a collection of ordered pairs of distinct points. The elements of E are *directed lines* or *arcs*.[1]

In a *multigraph* two points may be joined by more than one line. Figure 1.1. shows the three types of graphs above mentioned.

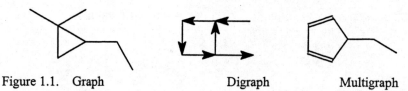

Figure 1.1.   Graph   Digraph   Multigraph

A *path graph*, *P*, is an unbranched chain. A *tree*, *T*, is a branched structure. A *star* is a set of vertices joined by a common vertex; it is denoted by $S_{N'}$, with $N' = N-1$. A *cycle*, *C*, is a chain which starts and ends in one and the same vertex. (Figure 1.2).

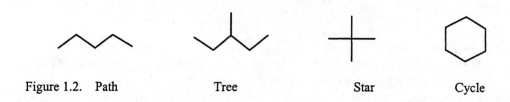

Figure 1.2.   Path   Tree   Star   Cycle

A *complete graph*, K_N, is the graph with any two vertices adjacent. The number of edges in a complete graph is $N(N-1)/2$. In Figure 1.3, complete graphs with $N = 1$ to 5 are presented

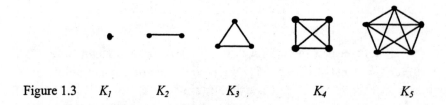

Figure 1.3 K_1 K_2 K_3 K_4 K_5

A *bigraph* (i.e. *bipartite graph*) is a graph whose vertex set V can be partitioned into two disjoint subsets: $V_1 \cup V_2 = V$; $V_1 \cap V_2 = \emptyset$ such that any edge $(i,j) \in E(G)$ joins V_1 with V_2.[1,3] A graph is bipartite if and only if all its cycles are even.[4]

If any vertex $i \in V_1$ is adjacent to any vertex $j \in V_2$ then G is a *complete bipartite graph* and is symbolized by $K_{m,n}$, with $m = |V_1|$ and $n = |V_2|$ A star is a complete bigraph $K_{1,n}$. It is obvious that $K_{m,n}$ has mn edges. Figure 1.4 presents some bigraphs.

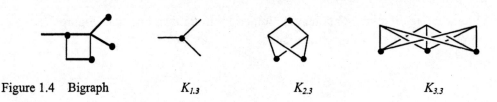

Figure 1.4   Bigraph $K_{1,3}$ $K_{2,3}$ $K_{3,3}$

A *rooted graph* is a graph in which heteroatoms or carbons with an unshared electron are specified[5,6] (Figure 1.5).

Figure 1.5. Rooted graphs

A *homeomorph of a graph G* is a graph resulted by inserting vertices of degree 2 (Figure 1.6)[3]

Figure 1.6. Homeomorphs of tehrahedron

A *planar graph* is a graph which can be drawn in the plane so that any two edges intersect to each other at most by their endpoints.[7] The regions defined by a plane graph are called *faces*, F, the unbounded region being the *exterior face*[1] (e.g. f_4 in Figure 1.7). For any spherical polyhedron with $|V|$ vertices, $|E|$ edges and $|F|$ faces the Euler formula[8] is true: $|V| - |E| + |F| = 2$. A graph is planar if and only if it has no subgraphs homeomorphic to K_5 or $K_{3,3}$ (Kuratowski's theorem).[9]

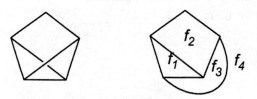

Figure 1.7. A planar graph and its faces

The *line graph*, $L(G)$, of a graph G, is constructed such that its points represent lines of G and two points of $L(G)$ are adjacent if the corresponding lines of G are incident to a common point.[1] Figure 1.8 illustrates this derivative of a graph (see also Sect. 8.2).

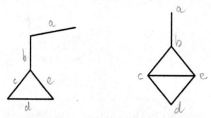

Figure 1.8. A graph and its line graph

The *complementary graph* of a graph $G = (V, E)$ is a graph $\overline{G} = (V, \overline{E})$, having the same set of vertices but joined with edges if and only if they were not present in G. The degree of each vertex in $\overline{G}$ equals the difference between the vertex degree in the complete graph K_N and the corresponding vertex in G.[7] (Figure 1.9).

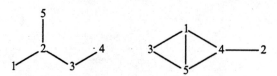

Figure 1.9. A graph and its complement

A graph G is *labeled*, $G(Lb)$, when its points are distinguished (e.g. by their numbers) from those of the corresponding abstract graph.[10] There exists $N!$ possibilities of numbering a graph of order N, $G(Lb_i)$; $i = 1,2,...N!$

Two graphs $G=(V, E)$ and $G_1=(V_1, E_1)$ are *isomorphic* (written $G \cong G_1$) if there exists a function $f: V \to V_1$ which obeys the conditions:[7,11,12]

(1) f is a bijection (one-to-one and onto)

(2) for all vertices $i, j \in V$; $(i,j) \in E \leftrightarrow (f(i), f(j)) \in E_1$.

The function f is called an isomorphism.

If f is the permutation operation, then there exists a permutation for which $G(Lb)$ and $G_1(Lb)$ coincide (see Figure 1.10 - see also Sect. 8.1).

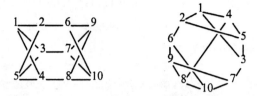

Figure 1.10. Two isomorphic graphs

A *subgraph* of a graph G is a graph $G_1 = (V_1, E_1)$ having $V_1 \subset V$ and $E_1 \subset E$ (Figure 1.11.).

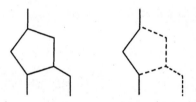

Figure 1.11. A graph and one of its subgraphs

A *spanning subgraph* is a subgraph $G_1 = (V, E_1)$ containing all the vertices of G but $E_1 \subset E$ (Figure 1.12.).

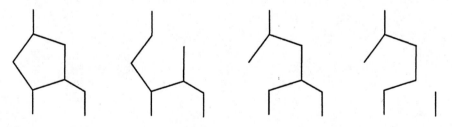

Figure 1.12. A graph and some of its spanning subgraphs

1.2. WALKS

A *walk* is a finite string, $w_{1,n} = (v_i)_{1 \le i \le n}$, $v_i \in V(G)$ such that any pair $(v_{i-1}, v_i) \in E(G)$, $i = \overline{2,...,n}$. Revisiting of vertices and edges is allowed.[1,3,13] The walk is *closed* if $v_1 = v_n$ and is *open* otherwise. When closed, it is also called *self-returning* walk. The *set of all walks* in G is denoted by $\widetilde{W}(G)$. The *length* of a walk $w_{1,n} = (v_i)_{1 \le i \le n}$ equals the number of occurrences of edges in it.

The concept of walk is very extended. If no other conditions are imposed, the walk is called a *random* walk. Additional conditions specify various kinds of walk.[14,15]

A *trail* (i.e. *Eulerian walk*) is a walk having all its edges distinct. Revisiting of vertices is allowed.

A *path* (i.e. *self-avoiding* walk) is a finite string, $p_{1,n} = (v_i)_{1 \le i \le n}$, $v_i \in V(G)$ such that any pair $(v_{i-1}, v_i) \in E(G)$, $i = \overline{2,...,n}$ and $v_i \ne v_j$, $(v_{i-1}, v_i) \ne (v_{j-1}, v_j)$ for any $1 \le i < j \le n$. Revisiting of vertices and edges, as well as branching is prohibited. The *set of all paths* in G is denoted by $P(G)$.

A graph is *connected* if every pair of vertices is joined by a path. A maximal connected subgraph of G is called a *component*. A disconnected graph has at least two components.[1]

A *terminal path*, $tp_{1,n} = (v_i)_{1 \le i \le n}$, $v_i \in V(G)$, is the path involving a walk $w = v_1$, $v_2,...,v_n$, v_k, that *is no more a path* in G, for any $v_k \in V(G)$ such that $(v_n, v_k) \in E$.

A closed path is a *cycle* (i.e. *circuit*). The *girth* of a graph, $g(G)$, is the length of a shortest cycle (if any) in G. The *circumference*, $c(G)$ is the length of a longest cycle.[1] A cycle is both a self-returning and a self-avoiding walk. A n-membered cycle includes n terminal paths in it.

A path is *Hamiltonian* if $n = |V|$. In other words, a Hamiltonian path visits once all the vertices in G. If such a path is a closed one, then it is a *Hamiltonian circuit*. Figure 1.13 illustrates each type of the above discussed walks.

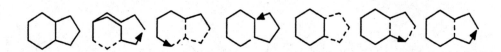

Figure 1.13.   Closed walk   path   trail   cycle   Hamiltonian Hamiltonian
                                                                      path   circuit

The *distance*, d_{ij}, between two vertices v_i and v_j is the length of a *shortest* path joining them, if any : $d_{ij} = min\ l(p_{ij})$; otherwise $d_{ij} = \infty$. A shortest path is often called a *geodesic*. The *eccentricity* of a vertex i, ecc_i, in a connected graph is the maximum distance between i and any vertex j of G: $ecc_i = max\ d_{ij}$. The *radius* of a graph, $r(G)$, is the minimum eccentricity among all vertices i in G: $r(G) = min\ ecc_i = min\ max\ d_{ij}$.

Conversely, the *diameter* of a graph, $d(G)$, is the maximum eccentricity in G: $d(G) = max\ ecc_i = max\ max\ d_{ij}$. The *set of all distances* (i.e. geodesics) in G is denoted by $D(G)$.

The *detour*, δ_{ij}, between two vertices v_i and v_j is the length of a *longest* path joining these vertices, if any : $\delta_{ij} = max\ l(p_{ij})$; otherwise $\delta_{ij} = \infty$. The *set of all detours* (i.e. longest paths) in G is denoted by $\Delta(G)$.

In a connected graph, the distance and the detour are *metrics*, that is, for all vertices v_i, v_j and v_k,

1. $m_{ij} \geq 0$, with $m_{ij} = 0$ if and only if $v_i = v_j$.
2. $m_{ij} = m_{ji}$
3. $m_{ij} + m_{ik} \geq m_{jk}$

When $l(p_{ij})$ is expressed in number of edges, the distance is called *topological distance*; when it is measured in *meters* or submultiples:(nm, pm) it is a *metric distance*. Table 1.1 illustrates the two types of distances.

Table 1.1. Topological and Metric Distances

Chemical Compound	Topological Distance	Metric Distance (pm)
$CH_3 - CH_3$	1	154
$CH_2 = CH_2$	1	134
$CH \equiv CH$	1	121

An *invariant* of a graph is a graph theoretical property, which is preserved by isomorphism.[1] In other words, it remains unchanged, irrespective of the numbering or pictorial representation of G.

The *degree, deg v_i*, (i.e. *valency*, sometimes denoted by k or δ) of a vertex v_i in G is the number of edges incident in v_i.[1] Since any edge has two endpoints, it contributes twice to the sum of degrees of vertices in G, such that $\sum_i deg\ v_i = 2Q$, a result which was the first theorem of graph theory (Euler, 1736).[1] In a (N, Q) graph, $0 \leq deg\ v_i \leq N-1$, for any vertex v_i. If all vertices have the same degree, k, the graph is called *k-regular*; otherwise it is *irregular* (Figure 1.14). The 4-regular graph in Figure 1.14 is both an Eulerian and Hamiltonian graph.

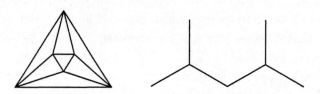

Figure 1.14. A regular and an irregular graph

1.3 CHEMICAL GRAPHS

A *chemical graph* is a model of a chemical system, used to characterize the interactions among its components: atoms, bonds, groups of atoms or molecules. A structural formula of a chemical compound can be represented by a *molecular graph*, its vertices being atoms and edges corresponding to covalent bonds.

Usually hydrogen atoms are not depicted in which case we speak of *hydrogen depleted molecular graphs*. (Figure 1.15).

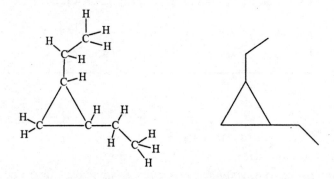

Figure 1.15. A molecular graph and its hydrogen depleted representation

The heavy atoms different from carbon (i.e. heteroatoms) can be represented, as shown in Figure 1.5. Similarly, a transform of a molecule (e.g. a chemical reaction) can be visualized by a *reaction graph*, whose vertices are chemical species and edges reaction pathways. Within this book, only molecular graphs are considered.

REFERENCES

1. Harary, F. *Graph Theory*, Addison - Wesley, Reading, M.A., 1969.
2. 2. Sylvester, J. J. On an Application of the New Atomic Theory to the Graphical Representation   of the Invariants and Covariants of Binary Quantics - With Three Appendices, *Am. J. Math.* **1874**, *1*, 64-90.
3. Trinajstić, N. *Chemical Graph Theory* , CRC Press, Inc., Boca Raton, Florida, 1983.
4. Kőnig, D. Theorie der endlichen und unendlichen Graphen. Leipzig, 1936.
5. Reprinted Chelsea, New York, 1950.
6. Kier, L. B.; Hall, L.H. Molecular Connectivity in Chemistry and Drug Research,
7. Acad. Press, New York, San Francisco, London, 1976.
8. Balaban, A. T.; Filip, P. Computer Program for Topological Index J (Average Distance
9. Sum Connectivity), Commun. Math. Comput. Chem. MATCH, **1984**, 16, 163-190.
10. Ionescu, T. *Graphs-Applications*, (in Romanian), Ed. Ped. Bucharest, 1973.
11. Euler, L. Solutio Problematis ad Geometriam Situs Pertinentis. *Comment. Acad. Sci. I. Petropolitanae* **1736**, *8*, 128-140.
12. Kuratowski, K. Sur le Problème des Courbes Gauches en Topologie, *Fund. Math.*
13. **1930**, *15*, 271-283.
14. 10. Klin, M.H.; Zefirov, N.S. Group Theoretical Approach to the Investigation of
15. Reaction Graphs for Highly Degenerate Rearrangements of Chemical
16. Compounds. II. Fundamental Concepts. *Commun. Math. Comput.*
17. *Chem. (MATCH)*, **1991**, *26*, 171-190.
18. 11. Razinger, M.; Balasubramanian, K.; Munk, M. E. Graph Automorphism Perception
19. Algorithms in Computer-Enhanced Structure Elucidation, *J. Chem. Inf. Comput. Sci.*
20. **1993**, *33*, 197-201.
21. Read, R. C.; Corneil, D. C. The Graph Isomorphism Disease, *J. Graph Theory,*
22. **1977**, *1*, 339-363.
23. Cayley, E. On the Mathematical Theory of Isomers, *Philos. Mag.* **1874**, *67*, 444-446.
24. Berge, C. *Graph Theory and Its Applications* (in Romanian), Ed. Tehnică,
25. Bucharest, 1969.
26. Randić, M. ; Woodworth, W. L.; Graovac, A. Unusual Random Walks,
27. Int. J. Quant. Chem. **1983**, 24, 435-452.

Chapter 2

TOPOLOGICAL MATICES

A molecular graph can be represented by: a sequence of numbers, a polynomial, a single number or a matrix.[1] These representations are aimed to be unique, for a given structure. Topological matrices can be accepted as a rational basis for designing topological indices.[2] The main types of matrix descriptors are listed and illustrated in the following.

2.1. ADJACENCY MATRIX

Since early 1874, Sylvester[3] has associated to an organic molecule a matrix $A(G)$. This is a square table, of dimensions NxN, whose entries are defined as:

$$[\mathbf{A}]_{ij} = \begin{cases} 1 & \text{if} & i \neq j \text{ and} & (i,j) \in E(G) \\ 0 & \text{if} & i = j \text{ or} & (i,j) \notin E(G) \end{cases} \tag{2.1}$$

$A(G)$ characterizes a graph up to isomorphism. It allows the reconstruction of the graph. $A(G)$ is symmetric vs. its main diagonal, so that the transpose $\mathbf{A}^T(G)$ leaves $A(G)$ unchanged:

$$\mathbf{A}^T(G) = \mathbf{A}(G) \tag{2.2}$$

Figure 2.1 illustrates the adjacency matrix for the graph $G_{2.1}$, and its powers, $\mathbf{A}^e$, till $e = 3$. Note that the entries $[\mathbf{A}^e]_{ij}$ represent *walks* of length e, $^e w$,[4] whereas the diagonal entries, $[\mathbf{A}^e]_{ii}$ count *self returning walks* (or closed walks), $^e srw$. The sum of the i-th row, RS, or of the i-th columns, CS of the entries in $\mathbf{A}^e$ equals the number of walks (of length e) starting from the vertex i. It is called the walk degree, $^e w_i$; for $e = 1$, one retrieves the classical vertex degree, $deg_i = {}^1 w_i$.

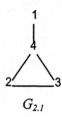

$G_{2.1}$

	A	$^1w_i = deg_i$	A²	^2w_i	A³	^3w_i
1	0 0 0 1	1	1 1 1 0	3	0 1 1 3	5
2	0 0 1 1	2	1 2 1 1	5	1 2 3 4	10
3	0 1 0 1	2	1 1 2 1	5	1 3 2 4	10
4	1 1 1 0	3	0 1 1 3	5	3 4 4 2	13

Figure 2.1. Adjacency matrices for the graph $G_{2.1}$.

If multibonds are taken into account, a variant of $A(G)$, denoted $C(G)$, (the connectivity matrix) can be written:

$$[\mathbf{C}]_{ij} = \begin{cases} b_{ij} & \text{if} \quad i \neq j \text{ and } (i,j) \in E(G) \\ 0 & \text{if} \quad i = j \text{ or } (i,j) \notin E(G) \end{cases} \tag{2.3}$$

where b_{ij} is the conventional bond order: 1; 2; 3; 1.5 for simple, double, triple and aromatic bonds, respectively.

In its general form, the walk degree, can be defined as:

$$^e w_i = \sum_j [\mathbf{C}^e]_{ij} \tag{2.4}$$

The raising at a power e, of a square matrix, can be eluded by applying the algorithm of Diudea, Topan and Graovac.[5] It evaluates a (topological) property of a vertex i, by iterative summation of the first neighbors contributions. The algorithm, called ^eW_M, is defined as:

$$\mathbf{M} + {}^e\mathbf{W} = {}^e\mathbf{W}_M \tag{2.5}$$

$$[{}^{e+1}\mathbf{W}_M]_{ii} = \sum_j ([\mathbf{M}]_{ij}[{}^e\mathbf{W}_M]_{jj}); \qquad [{}^0\mathbf{W}_M]_{jj} = 1 \tag{2.6}$$

$$[{}^{e+1}\mathbf{W}_M]_{ij} = [{}^e\mathbf{W}_M]_{ij} = [\mathbf{M}]_{ij} \tag{2.7}$$

where $\mathbf{M}$ is any square matrix and $^e\mathbf{W}$ is the diagonal matrix of walk degrees. The diagonal elements, $[^e\mathbf{W}_M]_{ii}$ equal the RS_i of $\mathbf{M}^e$, or in other words, they are walk degrees, $^e w_{M,i}$ (weighted by the property collected by $\mathbf{M}$):[5]

$$[^e\mathbf{W}_M]_{ii} = \sum_j [\mathbf{M}^e]_{ij} = {}^e w_{M,i} \tag{2.8}$$

The half sum of the local invariants, $^e w_{M,i}$, in a graph, defines a global invariant, called the *walk number*, $^e W_M$:

$$^e W_M = {}^e W_M(G) = \frac{1}{2}\sum_i {}^e W_{M,i} \tag{2.9}$$

When $\mathbf{M} = \mathbf{A}$; $\mathbf{C}$, the quantity $^e W_M$ (or simply $^e W$) represents the so called *molecular walk count*;[6] when $\mathbf{M} = \mathbf{D}$, (i.e. the distance matrix - see below) then $^e W_M$ equals the Wiener number of rank e (see Chap. Topological Indices).

The sum of diagonal elements in a square matrix is called *trace*, $Tr(\mathbf{M}^e)$:

$$Tr(\mathbf{M}^e) = \sum_i [\mathbf{M}^e]_{ii} \tag{2.10}$$

The half sum of diagonal elements offers a global invariant, $^e SRW_M$ (*Self Returning Walk* number):

$$^e SRW_M = \frac{1}{2}\sum_i [\mathbf{M}^e]_{ii} = MOM(\mathbf{M}^e) \tag{2.11}$$

which equals the *moment* of order e of the matrix $\mathbf{M}$, $MOM(\mathbf{M}^e)$. When $\mathbf{M} = \mathbf{A}$, the elements $[\mathbf{A}^e]_{ii}$ count both self returning walks and circuits of length e. $MOM(\mathbf{A}^e)$ is related to the spectral properties of molecular graphs (e.g. the energy of molecular orbitals).[7]

Figure 2.2 illustrates the *graphical evaluation* of $^e w_i$ and $^e W$ numbers, by using weighted graphs $G\{^e w_i\}$.

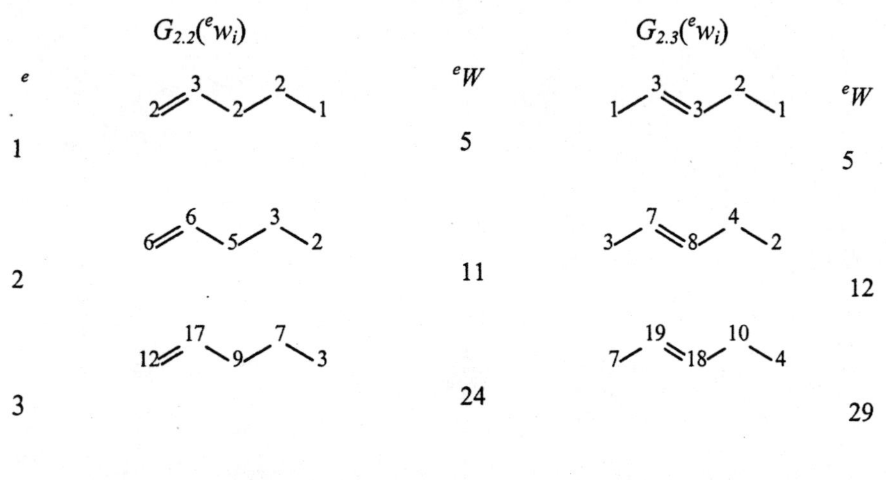

Figure 2.2. Graphical evaluation of $^e w_i$ and $^e W$: $e = 1\text{-}3$.

M. Diudea, I. Gutman and J. Lorentz

For indicating the edge adjacency, the **EA** matrix is used. The edge adjacency can be obtained from the line graph, $L(G)$ (see Sect. 8.2). When a relation between vertices and edges is needed, the incidency matrix, **VEA**, can be constructed (Figure 2.3).[4, 8]

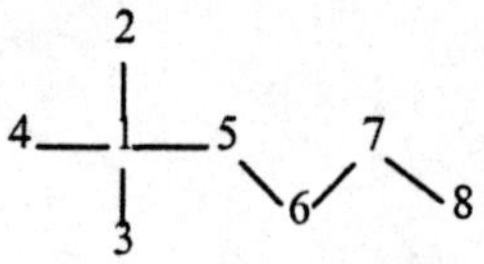

$$G_{2.4}$$

$$\textbf{EA}(G_{2.4}) \qquad\qquad \textbf{VEA}(G_{2.4})$$

	12	13	14	15	56	67	78
12	0	1	1	1	0	0	0
13	1	0	1	1	0	0	0
14	1	1	0	1	0	0	0
15	1	1	1	0	1	0	0
56	0	0	0	1	0	1	0
67	0	0	0	0	1	0	1
78	0	0	0	0	0	1	0

	12	13	14	15	56	67	78
1	1	1	1	1	0	0	0
2	1	0	0	0	0	0	0
3	0	1	0	0	0	0	0
4	0	0	1	0	0	0	0
5	0	0	0	1	1	0	0
6	0	0	0	0	1	1	0
7	0	0	0	0	0	1	1
8	0	0	0	0	0	0	1

Figure 2.3. Matrices **EA** and **VEA** for the graph $G_{2.4}$:

2.2. LAPLACIAN MATRIX

The Laplacian matrix is defined as:[9-14]

$$\mathbf{La}(G) = \mathbf{DEG}(G) - \mathbf{A}(G) \tag{2.12}$$

where $\mathbf{DEG}$ is the diagonal matrix of vertex degrees and $\mathbf{A}$ is the adjacency matrix. In multigraphs, $\mathbf{A}$ is changed by $\mathbf{C}$ (connectivity) matrix. For the graph $G_{2.5}$ (3-methyl-heptan), the Laplacian is shown in Figure 2.4.

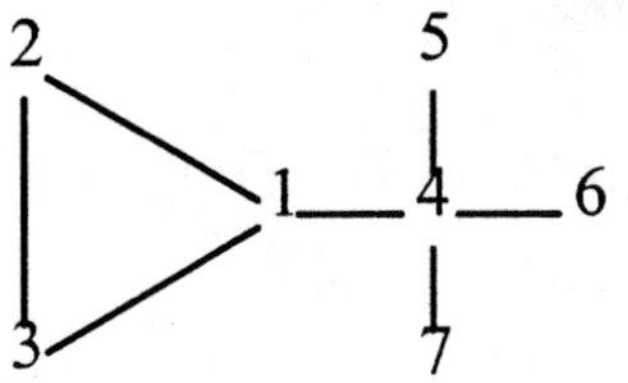

$$\mathbf{La}\,(G_{2.5}) = \begin{array}{c} 1 \\ 2 \\ 3 \\ 4 \\ 5 \\ 6 \\ 7 \end{array}\begin{bmatrix} 3 & -1 & -1 & -1 & 0 & 0 & 0 \\ -1 & 2 & -1 & 0 & 0 & 0 & 0 \\ -1 & -1 & 2 & 0 & 0 & 0 & 0 \\ -1 & 0 & 0 & 4 & -1 & -1 & -1 \\ 0 & 0 & 0 & -1 & 1 & 0 & 0 \\ 0 & 0 & 0 & -1 & 0 & 1 & 0 \\ 0 & 0 & 0 & -1 & 0 & 0 & 1 \end{bmatrix}$$

$$G_{2.5}$$

Spectrum of eigenvalues:

λ_1	λ_2	λ_3	λ_4	λ_5	λ_6	λ_7
0	0.3983	1.0000	1.0000	3.0000	3.3399	5.2618

$$t(G_{2.5}) = 3 \; ; \; Q(G_{2.5}) = 7$$

Figure 2.4. The Laplacian matrix of the graph $G_{2.5}$.

The Laplacian matrix is also referred to as the Kirchhoff matrix.[12,15,16] It is involved in the matrix-tree theorem.[17] Thus, the *number of spanning trees, t(G)*, in a cycle-containing structure, is given by:

$$t(G) = \big| \, det \, ([\mathbf{La}]_{ij}) \, \big| \tag{2.13}$$

Where $[\mathbf{La}]_{ij}$ is a submatrix of $\mathbf{La}$, from which the row i and column j were deleted. The number $t(G)$ can also be calculated from the spectrum of eigenvalues, λ_i, of the Laplacian, by relation [12]

$$t(G) = (1/N)\prod_{i=2}^{N} \lambda_i \qquad (2.14)$$

The Laplacian spectrum can be used for calculating the Wiener number[9, 13] and represents a source of other graph invariants (see Chap. Wiener- Type Indices). For example, the number of edges, Q, in a graph can be calculated by[14]

$$Q = \frac{1}{2}\sum_{i=2}^{N} \lambda_i = (1/2)Tr(\mathrm{La}) \qquad (2.15)$$

2.3 DISTANCE MATRIX

Distance Matrix $\mathbf{D}(G)$, was introduced in 1969 by Harary.[4] It is a square symmetric table, of dimension $N x N$, whose entries are defined as:

$$[\mathbf{D}]_{ij} = \begin{cases} N_{e,(i,j)}; (i,j) \in D(G), & if\ i \neq j \\ 0 & if\ i = j \end{cases} \qquad (2.16)$$

where $N_{e,(i,j)} = d_{ij}$, the topological distance between i and j. The matrix $\mathbf{D}$, (denoted hereafter $\mathbf{D}_e$ by reasons that will become clear in the following), for the graph $G_{2.6}$. is illustrated in Figure 2.5. The $RS(\mathbf{D}_e)_i$ denotes the distance from the vertex i to all N -1 vertices in graph.

The entries $[\mathbf{D}_e]_{ij}$ are defined by:

$$[\mathbf{D}_e]_{ij} = e : [A^e]_{ij} \neq [A^{e-1}]_{ij}; \quad e = 1, 2, ...d(G) \qquad (2.17)$$

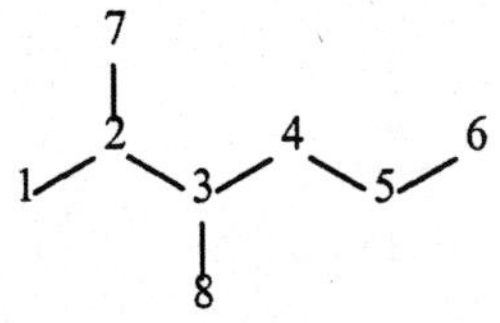

$G_{2.6}$

$\mathbf{D}_e\,(G_{2.6})$:

	1	2	3	4	5	6	7	8	RS_i
1	0	1	2	3	4	5	2	3	20
2	1	0	1	2	3	4	1	2	14
3	2	1	0	1	2	3	2	1	12
4	3	2	1	0	1	2	3	2	14
5	4	3	2	1	0	1	4	3	18
6	5	4	3	2	1	0	5	4	24
7	2	1	2	3	4	5	0	3	20
8	3	2	1	2	3	4	3	0	18

Figure 2.5. Distance matrix for the graph $G_{2.6}$

$\mathbf{D}_e$ matrix can be built up by calculating the boolean powers A^e; $e \in [1, d(G)]$, where $A = I + A$, with I being the unity matrix and $d(G)$ the diameter of graph. The procedure is illustrated for the graph $G_{2.4}$ in Figure 2.6.

By applying the ^eW_M algorithm (eqs 2.5-2.7) on $\mathbf{D}_e$ results in $^eW_{De}$ numbers, which are Wiener numbers of rank e^{18} (see Chap. Topological Indices). The diagonal entries in the matrix $(\mathbf{D}_e)^e$ represent degrees of the self returning walks, $^esrw_{D,i}$, weighted by distance.

Figure 2.7 illustrates the *graphical evaluation* of $^ew_{D,i}$ quantities, by using the weighted graph $G\{^eW_{D,i}\}$. Note that the matrix $\mathbf{D}_e$ can be considered as the *connectivity* matrix of a complete graph, K_N (having the same number, N, of vertices as the initial graph) with the weight (i.e. multiplicity) of edges equaling the distance d_{ij}.

$\mathbf{D}_e\,(G_{2.4})$

	1	2	3	4	5	6	7	8
1	0	1	1	1	1	2	3	4
2	1	0	2	2	2	3	4	5
3	1	2	0	2	2	3	4	5
4	1	2	2	0	2	3	4	5
5	1	2	2	2	0	1	2	3
6	2	3	3	3	1	0	1	2
7	3	4	4	4	2	1	0	1
8	4	5	5	5	3	2	1	0

$A^1(G_{2.4})$

	1	2	3	4	5	6	7	8
1	1	1	1	1	1	0	0	0
2	1	1	0	0	0	0	0	0
3	1	0	1	0	0	0	0	0
4	1	0	0	1	0	0	0	0
5	1	0	0	0	1	1	0	0
6	0	0	0	0	1	1	1	0
7	0	0	0	0	0	1	1	1
8	0	0	0	0	0	0	1	1

$A^2(G_{2.4})$

	1	2	3	4	5	6	7	8
1	1	1	1	1	1	1	0	1
2	1	1	1	1	1	0	0	0
3	1	1	1	1	1	0	0	0
4	1	1	1	1	1	0	0	1
5	1	1	1	1	1	1	1	0
6	1	0	0	0	1	1	1	1
7	0	0	0	0	1	1	1	1
8	0	0	0	0	0	1	1	1

$A^3(G_{2.4})$

	1	2	3	4	5	6	7	8
1	1	1	1	1	1	1	1	0
2	1	1	1	1	1	1	0	0
3	1	1	1	1	1	1	0	0
4	1	1	1	1	1	1	0	0
5	1	1	1	1	1	1	1	1
6	1	1	1	1	1	1	1	1
7	1	0	0	0	1	1	1	1
8	0	0	0	0	1	1	1	1

$A^4(G_{2.4})$

	1	2	3	4	5	6	7	8
1	1	1	1	1	1	1	1	1
2	1	1	1	1	1	1	1	0
3	1	1	1	1	1	1	1	0
4	1	1	1	1	1	1	1	0
5	1	1	1	1	1	1	1	1
6	1	1	1	1	1	1	1	1
7	1	1	1	1	1	1	1	1
8	1	0	0	0	1	1	1	1

$A^5(G_{2.4})$

	1	2	3	4	5	6	7	8
1	1	1	1	1	1	1	1	1
2	1	1	1	1	1	1	1	1
3	1	1	1	1	1	1	1	1
4	1	1	1	1	1	1	1	1
5	1	1	1	1	1	1	1	1
6	1	1	1	1	1	1	1	1
7	1	1	1	1	1	1	1	1
8	1	1	1	1	1	1	1	1

Figure 2.6. The construction of $\mathbf{D}_e\,(G_{2.4})$ by using boolean powers, A^e for $G_{2.4}$.

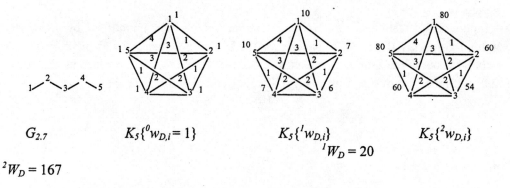

$G_{2.7}$ $K_5\{^0w_{D,i}=1\}$ $K_5\{^1w_{D,i}\}$ $K_5\{^2w_{D,i}\}$

$^1W_D = 20$

$^2W_D = 167$

Figure 2.7. Graphical evaluation of the numbers $^ew_{D,i}$ and ^eW_D.

2.4. DETOUR MATRIX

In cycle-containing graphs, when the shortest path (i.e. geodesic) is replaced by the longest path between two vertices i and j, the maximum path matrix, or the *detour matrix*, Δ_e, can be constructed[19,20]

$$[\Delta_e]_{ij} = \begin{cases} N_{e,(i,j)}; \ (i,j) \in \Delta(G), & \text{if } i \neq j \\ 0 & \text{if } i = j \end{cases} \quad (2.18)$$

Figure 2.8. illustrates this matrix for 1-Ethyl-2-methyl-cyclopropane, $G_{2.8}$.

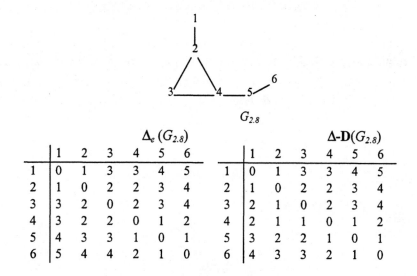

$$G_{2.8}$$

$\Delta_e\,(G_{2.8})$							$\Delta\text{-D}(G_{2.8})$					
	1	2	3	4	5	6		1	2	3	4	5	6
1	0	1	3	3	4	5	1	0	1	3	3	4	5
2	1	0	2	2	3	4	2	1	0	2	2	3	4
3	3	2	0	2	3	4	3	2	1	0	2	3	4
4	3	2	2	0	1	2	4	2	1	1	0	1	2
5	4	3	3	1	0	1	5	3	2	2	1	0	1
6	5	4	4	2	1	0	6	4	3	3	2	1	0

Figure 2.8. Detour, Δ_e, and detour-distance, $\Delta\text{-D}$, matrices for the graph $G_{2.8}$.

The two types of paths, the shortest and the largest ones, can be combined in one and the same square matrix, $\Delta\text{-D}$, (originally called *Maximum minimum Path*, **MmP**,[20] whose entries are defined as:

$$[\Delta\text{-D}]_{ij} = \begin{cases} N_{e,(i,j)}; \ (i,j) \in \Delta(G), & \text{if } i < j \\ N_{e,(i,j)}; \ (i,j) \in D(G), & \text{if } i > j \\ 0 & \text{if } i = j \end{cases} \quad (2.19)$$

It is easily seen that the upper triangle is identical to that in the matrix Δ_e while the lower triangle coincides to that in the D_e matrix.

2.5. 3D - DISTANCE MATRICES

When one considers the genuine distances between atoms (i.e. the distances measured through space), one obtains the *geometric matrix*, $\mathbf{G}$.[21, 22] When the distances refers to the vertices of a graph embedded on a graphite or a diamond lattice, we speak of *topographic matrix*, $\mathbf{T}$.[23] It is exemplified in Figure 2.9, for *cis-* ($G_{2.9a}$) and *trans-*butadiene ($G_{2.9b}$).

Distance /Distance matrix,[24] $\mathbf{D} / \mathbf{D}$, reports ratios of the geometric distance (i.e. measured through space, for a graph embedded in a 2D or a 3D grid) to graph distances (i.e. measured through bonds). It is also exemplified in Figure 2.9.

2.6. COMBINATORIAL MATRICES

Recently, two path-defined matrices have been proposed: the *distance-path*,[18] $\mathbf{D}_p$, and the *detour-path*,[25] Δ_p, (see also[26]) whose elements are combinatorially calculated from the classical distance (i.e. *distance-edge*), $\mathbf{D}_e$, and detour (i.e. *detour-edge*), Δ_e, matrices

$$[\mathbf{D}_p]_{ij} = \begin{cases} N_{p,(i,j)}; (i,j) \in D(G), & if \ i \neq j \\ 0 & if \ i = j \end{cases} \tag{2.20}$$

$$[\Delta_p]_{ij} = \begin{cases} N_{p,(i,j)}; (i,j) \in \Delta(G), & if \ i \neq j \\ 0 & if \ i = j \end{cases} \tag{2.21}$$

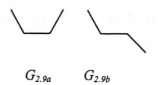

$$G_{2.9a} \qquad G_{2.9b}$$

$$\mathbf{T}(G_{2.9a})$$

$$\begin{bmatrix} 0 & 1 & \sqrt{3} & 2 \\ 1 & 0 & 1 & \sqrt{3} \\ \sqrt{3} & 1 & 0 & 1 \\ 2 & \sqrt{3} & 1 & 0 \end{bmatrix}$$

$$\mathbf{T}(G_{2.9b})$$

$$\begin{bmatrix} 0 & 1 & \sqrt{3} & \sqrt{7} \\ 1 & 0 & 1 & \sqrt{3} \\ \sqrt{3} & 1 & 0 & 1 \\ \sqrt{7} & \sqrt{3} & 1 & 0 \end{bmatrix}$$

$$\mathbf{D} / \mathbf{D}(G_{2.9a}))$$

$$\begin{bmatrix} 0 & 1 & \sqrt{3}/2 & 2/3 \\ 1 & 0 & 1 & \sqrt{3}/2 \\ \sqrt{3}/2 & 1 & 0 & 1 \\ 2/3 & \sqrt{3}/2 & 1 & 0 \end{bmatrix}$$

$$\mathbf{D} / \mathbf{D}(G_{2.9b})$$

$$\begin{bmatrix} 0 & 1 & \sqrt{3}/2 & \sqrt{7}/3 \\ 1 & 0 & 1 & \sqrt{3}/2 \\ \sqrt{3}/2 & 1 & 0 & 1 \\ \sqrt{7}/3 & \sqrt{3}/2 & 1 & 0 \end{bmatrix}$$

Figure 2.9. (3D) - Distance matrices

$$N_{p,(i,j)} = \binom{[\mathbf{M}_e]_{ij}+1}{2} = \{([\mathbf{M}_e]_{ij})^2 + [\mathbf{M}_e]_{ij}\}/2 , \quad \mathbf{M} = \mathbf{D}; \Delta \tag{2.22}$$

$N_{p,(i,j)}$ represents the number of all internal paths[27] of length $1 \le |p| \le |(i,j)|$ included in the path (i,j).

Matrices $\mathbf{D}_p$ and Δ_p allow the direct calculation of the *hyper-Wiener*, *WW*, and *hyper-detour* , *ww*, indices, respectively. (see Chap. Topological Indices). Matrix $\mathbf{D}_p$, like $\mathbf{D}_e$, allows the immediate reconstruction of the original graph: the adjacency is given by the entries [1] (see Figure 2.10).

2.7. WIENER MATRICES

Randić proposed a square matrix, denominated Wiener matrix,[28, 29] $\mathbf{W}$, and exploited it as a source of structural invariants, useful in *QSPR/QSAR*. For trees, the non-diagonal entries in such a matrix are defined as:

$$[\mathbf{W}_{e/p}]_{ij} = N_{i,e/p}\, N_{j,e/p} \tag{2.23}$$

where N_i and N_j denote the number of vertices lying on the two sides of the edge/path, e/p (having i and j as endpoints). The diagonal entries are zero.

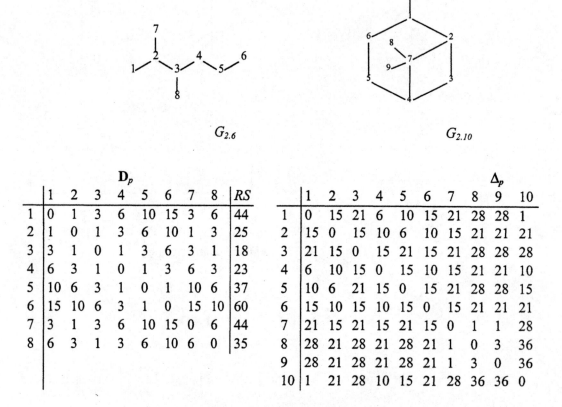

$G_{2.6}$　　　　　　　　　　　　　　　　　　$G_{2.10}$

$\mathbf{D}_p$

	1	2	3	4	5	6	7	8	RS
1	0	1	3	6	10	15	3	6	44
2	1	0	1	3	6	10	1	3	25
3	3	1	0	1	3	6	3	1	18
4	6	3	1	0	1	3	6	3	23
5	10	6	3	1	0	1	10	6	37
6	15	10	6	3	1	0	15	10	60
7	3	1	3	6	10	15	0	6	44
8	6	3	1	3	6	10	6	0	35

Δ_p

	1	2	3	4	5	6	7	8	9	10
1	0	15	21	6	10	15	21	28	28	1
2	15	0	15	10	6	10	15	21	21	21
3	21	15	0	15	21	15	21	28	28	28
4	6	10	15	0	15	10	15	21	21	10
5	10	6	21	15	0	15	21	28	28	15
6	15	10	15	10	15	0	15	21	21	21
7	21	15	21	15	21	15	0	1	1	28
8	28	21	28	21	28	21	1	0	3	36
9	28	21	28	21	28	21	1	3	0	36
10	1	21	28	10	15	21	28	36	36	0

Figure 2.10. Combinatorial matrices

Eq. 2.23 defines just the *edge/path contributions* to a global index: it is the *Wiener* number,[30] W, when defined on edge, (i.e. $(i,j)\in E(G)$) and *hyper-Wiener* number,[31] WW, when defined on path (i.e. $(i,j)\in P(G)$) - see Chap. Topological Indices).

Wiener matrices are illustrated in Figure 2.10, for the graph $G_{2.10}$. $\mathbf{W}_e$ is an adjacency matrix weighted by the number of *external* paths which include a given edge, e. Note that any topological index defined on edge, can be written as a weighted adjacency matrix.

$\mathbf{W}_p$ allows the reconstruction of the original graph according to the Randić conjecture:[29] "take a single line in $\mathbf{W}_p$ at once. Identify the largest entry $[\mathbf{W}_p]_{ij}$ in that line and replace it by 1. After the completion of all lines, make the matrix symmetric. Thus results in the matrix **A**, from which the reconstruction is trivial".

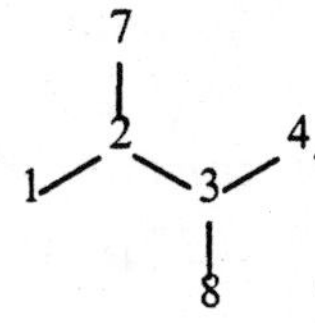

$$G_{2.6}$$

	W_e										**W_p**							
	1	2	3	4	5	6	7	8	RS_i	1	2	3	4	5	6	7	8	RS_i
1	0	7	0	0	0	0	0	0	7	0	7	5	3	2	1	1	1	20
2	7	0	15	0	0	0	7	0	29	7	0	15	9	6	3	7	3	50
3	0	15	0	15	0	0	0	7	37	5	15	0	15	10	5	5	7	62
4	0	0	15	0	12	0	0	0	27	3	9	15	0	12	6	3	3	51
5	0	0	0	12	0	7	0	0	19	2	6	10	12	0	7	2	2	41
6	0	0	0	0	7	0	0	0	7	1	3	5	6	7	0	1	1	24
7	0	7	0	0	0	0	0	0	7	1	7	5	3	2	1	0	1	20
8	0	0	7	0	0	0	0	0	7	1	3	7	3	2	1	1	0	18

Figure 2.11. Wiener matrices for the graph $G_{2.6}$

2.8. SZEGED MATRICES

Since the Wiener matrix is not defined in cyclic structures, (see eq 2.23) Gutman[32] has changed the meaning of N_i and N_j as follows:

$$n_{i,e/p} = | \{v | v \in V(G); \ d_{iv} < d_{jv}\} | \quad (2.24)$$

$$n_{j,e/p} = | \{v | v \in V(G); \ d_{jv} < d_{iv}\} | \quad (2.25)$$

Thus, $n_{i,e/p}$ and $n_{j,e/p}$ denote the cardinality of the sets of vertices lying closer to i and to j, respectively; *vertices equidistant to i and j are not counted*. These quantities are the ground for the novel invariant, called the *Szeged index*[32-38] (see Chap. 5).

Consequently, eq 2.23 can be re-written as:

$$[\mathbf{SZD}_{e/p}]_{ij} = n_{i,e/p} \ n_{j,e/p} \quad (2.26)$$

where $[\mathbf{SZD}_{e/p}]_{ij}$ are the non-diagonal entries in the new matrices, called the *Szeged-Distance matrices*,[33] edge-defined (i.e. $(i,j) \in E(G)$), $\mathbf{SZD}_e$, or path-defined, (i.e. $(i,j) \in P(G)$), $\mathbf{SZD}_p$. The diagonal entries in these matrices are zero. Figure 2.12 illustrates the Szeged-Distance matrices for the graphs $G_{2.6}$ (acyclic) and $G_{2.12}$ (cyclic).

Usually, a path (i,j) is characterized by its endpoints i and j, namely, by their associated numbers. In Wiener matrices a path is characterized by the numbers N_i and N_j (see above). Now, let renounce to the characterization of j and build up a square matrix whose entries look at a single endpoint, i. According to this principle, referred to as *the principle of unsymmetric characterization of a path*,[39-41] a new matrix, called *the unsymmetric Szeged matrix*, **USZ** , was constructed. The entries $[\mathbf{UM}]_{ij}$, $\mathbf{M} = \mathbf{SZD}$ (*Szeged-Distance* - eq 2.28) and **SZΔ** (*Szeged-Detour* - eq 2.29), are defined as:

$$[\mathbf{UM}]_{ij} = n_{i,(i,j)} \tag{2.27}$$

$$n_{i,(i,j)} = \left| \left\{ v \middle| v \in V(G); d_{iv} < d_{jv} \right\} \right| \tag{2.28}$$

$$n_{i,(i,j)} = \left| \left\{ v \middle| v \in V(G); \Delta_{iv} < \Delta_{jv} \right\} \right| \tag{2.29}$$

The diagonal entries in these matrices are zero. Note that the symbol $n_{i,(i,j)}$ recall the path (i,j) but the quantity given by eq 2.28 is identical to $n_{i,e/p}$, eq 2.24. Figure 2.13 illustrates the unsymmetric Szeged matrices for the graph $G_{2.13}$

These matrices can be symmetrized by the procedure

$$\mathbf{SM_p} = \mathbf{UM} \bullet (\mathbf{UM})^T \tag{2.30}$$

$$\mathbf{SM_e} = \mathbf{SM_p} \bullet \mathbf{A} \tag{2.31}$$

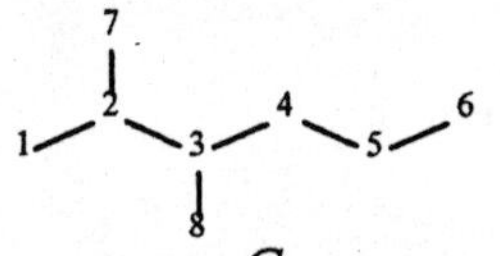

$G_{2.6}$

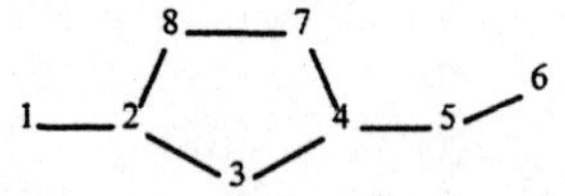

$G_{2.12}$

SZD$_e$($G_{2.6}$)

	1	2	3	4	5	6	7	8	RS_i
1	0	7	0	0	0	0	0	0	7
2	7	0	15	0	0	0	7	0	29
3	0	15	0	15	0	0	0	7	37
4	0	0	15	0	12	0	0	0	27
5	0	0	0	12	0	7	0	0	19
6	0	0	0	0	7	0	0	0	7
7	0	7	0	0	0	0	0	0	7
8	0	0	7	0	0	0	0	0	7

SZD$_e$($G_{2.12}$)

	1	2	3	4	5	6	7	8
1	0	7	0	0	0	0	0	0
2	7	0	12	0	0	0	0	6
3	0	12	0	12	0	0	0	0
4	0	0	12	0	12	0	8	0
5	0	0	0	12	0	7	0	0
6	0	0	0	0	7	0	0	0
7	0	0	0	8	0	0	0	12
8	0	6	0	0	0	0	12	0

SZD$_p$($G_{2.6}$)

	1	2	3	4	5	6	7	8	RS_i
1	0	7	5	15	9	15	1	15	67
2	7	0	15	9	15	10	7	3	66
3	5	15	0	15	10	12	5	7	69
4	15	9	15	0	12	6	15	3	75
5	9	15	10	12	0	7	9	15	87
6	15	10	12	6	7	0	15	10	85
7	1	7	5	15	9	15	0	15	67
8	15	3	7	3	15	10	15	0	68

SZD$_p$($G_{2.12}$)

	1	2	3	4	5	6	7	8
1	0	7	5	10	12	12	10	5
2	7	0	12	8	12	10	12	6
3	5	12	0	12	8	12	6	8
4	10	8	12	0	12	6	8	12
5	12	12	8	12	0	7	8	12
6	12	10	12	6	7	0	12	10
7	10	12	6	8	8	12	0	12
8	5	6	8	12	12	10	12	0

Figure 2.12. Szeged-distance matrices

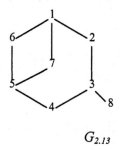

$G_{2.13}$

USZD($G_{2.13}$)

	1	2	3	4	5	6	7	8
1	0	6	2	6	2	7	7	6
2	2	0	3	2	2	3	3	3
3	2	5	0	5	2	7	7	7
4	2	2	3	0	2	3	3	3
5	2	6	2	6	0	7	7	6
6	1	1	1	1	1	0	1	1
7	1	1	1	1	1	1	0	1
8	2	2	1	2	2	2	2	0

USZΔ($G_{2.13}$)

	1	2	3	4	5	6	7	8
1	0	4	3	4	2	5	5	5
2	4	0	4	2	4	3	3	4
3	3	4	0	4	3	4	4	7
4	4	2	4	0	4	3	3	4
5	2	4	3	4	0	5	5	5
6	3	2	4	2	3	0	1	4
7	3	2	4	2	3	1	0	4
8	3	1	1	1	3	2	2	0

Figure 2.13. Unsymmetric Szeged matrices for the graph $G_{2.13}$

where $\mathbf{A}$ is the adjacency matrix. The symbol • indicates the Hadamard (pairwise) matrix product[42] (i.e. $[\mathbf{M_a} \bullet \mathbf{M_b}]_{ij} = [\mathbf{M_a}]_{ij} [\mathbf{M_b}]_{ij}$).

For the symmetric matrices, the letter $\mathbf{S}$ is usually missing.

Two indices are calculated[33, 40, 43] on the Szeged matrices, $\mathbf{M}$, $\mathbf{M} = \mathbf{SZD}$; $\mathbf{SZΔ}$

$$IE(M) = \sum_e [\mathbf{M}_e]_{ij} = IE2(UM) = \sum_e [\mathbf{UM}]_{ij}[\mathbf{UM}]_{ji} \tag{2.32}$$

$$IP(M) = \sum_p [\mathbf{M}_p]_{ij} = IP2(UM) = \sum_p [\mathbf{UM}]_{ij}[\mathbf{UM}]_{ji} \tag{2.33}$$

where summation goes over all edges, e, (resulting an *index*) and over all paths, p, (resulting a *hyper-index*)[33] respectively. The symbol varies by the operator used and by the type of matrix: symmetric or unsymmetric (see Sect. 6.1). It is obvious that $I(M\} = I2(UM)$. Note that $IE(SZD)$ means the *classical Szeged index*, symbolized Sz by Gutman.

2.9. Path Matrix P

Randić[44] defined the entries in the **P** matrix as the quotient between the number of paths P' in a subgraph, $G' = G\text{-}(i,j)$, (resulted by cutting the edge (i,j) from the graph G), to the number of paths P in G

$$[\mathbf{P}]_{ij} = \begin{cases} P'_{ij}/P & if \;\; i \neq j \;\; and \;\; (i,j) \in E(G) \\ 0 & if \;\; i = j \;\; or \;\; (i,j) \notin E(G) \end{cases} \tag{2.34}$$

When the subgraph $G\text{-}(i,j)$ is disconnected, then the contributions for each component are added. This matrix is illustrated in Figure 2.14. The index calculated on this matrix is called the P'/P index. By a similar procedure, Randić et al.[45] defined the *graphical bond order* related to a certain graph invariant (see Chap. Topological Indices).

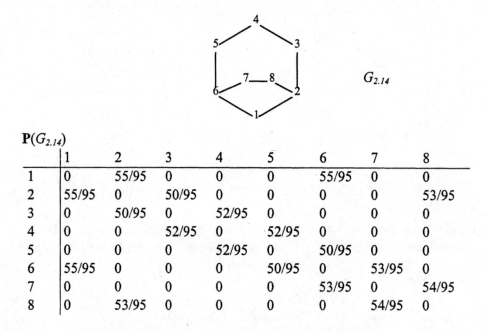

$G_{2.14}$

P($G_{2.14}$)

	1	2	3	4	5	6	7	8
1	0	55/95	0	0	0	55/95	0	0
2	55/95	0	50/95	0	0	0	0	53/95
3	0	50/95	0	52/95	0	0	0	0
4	0	0	52/95	0	52/95	0	0	0
5	0	0	0	52/95	0	50/95	0	0
6	55/95	0	0	0	50/95	0	53/95	0
7	0	0	0	0	0	53/95	0	54/95
8	0	53/95	0	0	0	0	54/95	0

Figure 2.14. Path matrix for the graph $G_{2.14}$

2.10. Hosoya Matrix

Randić[46] introduced the *Hosoya matrix* by an analogue cutting procedure. He calculated the Hosoya number,[47] Z, on the spanning subgraph $G\text{-}(i,j)$ of a tree

$$[\mathbf{Z}]_{ij} = \begin{cases} Z(G-(i,j)); & (i,j) \in P(G) \quad if\, i \neq j \\ 0 & if\, i = j \end{cases} \tag{2.35}$$

The Z number counts the modes of selecting k edges in a graph such that they are non-adjacent to each other (i.e. the number of k-matching of G - see Chap. Topological Indices). The matrix is illustrated in Figure 2.5 for the graph $G_{2.15}$.

The Z matrix and the path numbers, calculated on it, were further generalized for cycle-containing graphs as well as for edge-weighted molecular graphs.[48,49]

2.11. CLUJ MATRICES

The non-diagonal entries, $[\mathbf{UM}]_{ij}$, $\mathbf{M} = \mathbf{CJD}$ (Cluj-Distance) or $\mathbf{CJ\Delta}$ (Cluj-Detour), are defined as

$$[\mathbf{UM}]_{ij} = \max_{k=1,2,...} \left| V_{i,(i,j)_k} \right| \tag{2.36}$$

$$V_{i,(i,j)_k} = \left\{ v \; \middle| \; v \in V(G); d_{iv} < d_{jv}; \; (i,v)_h \cap (i,j)_k = \{i\}; \; (i,j)_k \in D(G) \, or \, \Delta(G) \right\}; h,k = 1,2,... \tag{2.37}$$

where $\left| V_{i,(i,j)_k} \right|$ is the cardinality of the set $V_{i,(i,j)_k}$, which is taken as the maximum over all paths $(i,j)_k$. $D(G)$ and $\Delta(G)$ are the sets of distances (i.e. geodesics) and detours (i.e. elongations), respectively.

The set $V_{i,(i,j)_k}$ consists of vertices, v, lying *closer* to the vertex i (condition $d_{iv} < d_{jv}$). This variant of Cluj matrices is called[51] *at least one path external to the path* (i,j), since at least one of the paths $(v,i)_h$ must be *external* with respect to the path $(i,j)_k$: $(i,v)_h \cap (i,j)_k = \{i\}$. In cycle-containing structures, more than one path $(i,j)_k$ may exist, thus supplying various sets $V_{i,(i,j)_k}$. By definition, the (ij)- entry in the Cluj matrices is taken as $max\,|V_{i,(i,j)_k}|$. The diagonal entries are zero. For paths $(i,v)_h$ no other restriction is imposed. The above definitions hold for any connected graph. The Cluj matrices are square arrays, of dimension NxN, usually unsymmetric (excepting some symmetric regular graphs). They can be symmetrized cf. eqs 2.30 and 2.31. Figure 2.16 illustrates these matrices for the graphs.

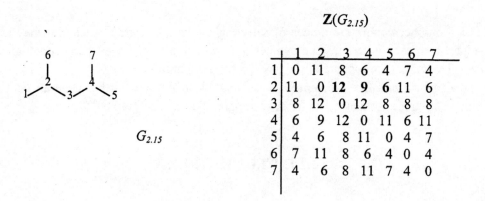

$$\mathbf{Z}(G_{2.15})$$

	1	2	3	4	5	6	7
1	0	11	8	6	4	7	4
2	11	0	**12**	**9**	6	11	6
3	8	12	0	12	8	8	8
4	6	9	12	0	11	6	11
5	4	6	8	11	0	4	7
6	7	11	8	6	4	0	4
7	4	6	8	11	7	4	0

$G_{2.15}$

(i,j)	$G\text{-}(i,j)$	Non-adjacent two edge selections ($k = 2$)		$Z(G\text{-}(i,j))$
(2,3)				$1(0)+5(1)+6(2)=12$
(2,4)				$1(0)+4(1)+4(2)=9$
(2,5)				$1(0)+3(1)+2(2)=6$

Figure 2.15. Construction of Hosoya matrix, **Z**, for the graph $G_{2.15}$

2.11.1. CJ Matrices

The *unsymmetric Cluj matrix*, **UCJ** , has been recently proposed by Diudea.[39-41,43,50] It is defined by using either the *distance* or the *detour* concept:

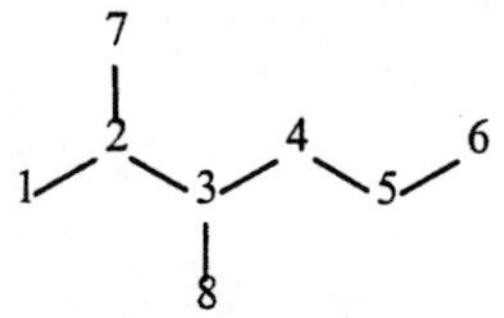

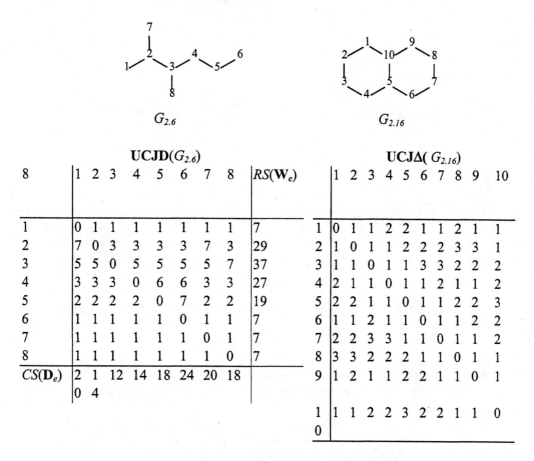

$G_{2.6}$

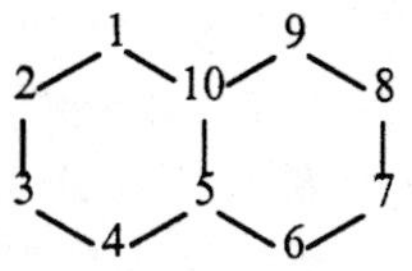

$G_{2.16}$

UCJD($G_{2.6}$)

8	1	2	3	4	5	6	7	8	*RS*(**W**$_e$)
1	0	1	1	1	1	1	1	1	7
2	7	0	3	3	3	3	7	3	29
3	5	5	0	5	5	5	5	7	37
4	3	3	3	0	6	6	3	3	27
5	2	2	2	2	0	7	2	2	19
6	1	1	1	1	1	0	1	1	7
7	1	1	1	1	1	1	0	1	7
8	1	1	1	1	1	1	1	0	7
CS(**D**$_e$)	20	14	12	14	18	24	20	18	

UCJΔ($G_{2.16}$)

	1	2	3	4	5	6	7	8	9	10
1	0	1	1	2	2	1	1	2	1	1
2	1	0	1	1	2	2	2	3	3	1
3	1	1	0	1	1	3	3	2	2	2
4	2	1	1	0	1	1	2	1	1	2
5	2	2	1	1	0	1	1	2	2	3
6	1	1	2	1	1	0	1	1	2	2
7	2	2	3	3	1	1	0	1	1	2
8	3	3	2	2	2	1	1	0	1	1
9	1	2	1	1	2	2	1	1	0	1
10	1	1	2	2	3	2	2	1	1	0

Figure 2.16. Unsymmetric Cluj matrices for the graphs $G_{2.6}$ and $G_{2.16}$.

It is obvious that, in trees, **UCJD** is identical to **UCJΔ**, due to the uniqueness of the path joining a pair of vertices (*i,j*).

In trees, **UCJD** matrix shows an interesting property:

$$RS(\mathbf{UCJD}) = RS(\mathbf{W}_e) \tag{2.38}$$

$$CS(\mathbf{UCJD}) = CS(\mathbf{D}_e) \tag{2.39}$$

Thus, **UCJD** contains some information included in both $\mathbf{W}_e$ and $\mathbf{D}_e$ matrices. The half sum of entries in all the three matrices equal the Wiener index (see Chap. Topological Indices):

$$IP(UCJD) = (1/2)\sum_i \sum_j [\mathbf{UCJD}]_{ij} = (1/2)\sum_i \sum_j [\mathbf{W}_e]_{ij} = (1/2)\sum_i \sum_j [\mathbf{D}_e]_{ij} = W \quad (2.40)$$

Note that the operator $IP(M)$, (meaning the half sum of entries in a square matrix), as well as $IE2(M)$ and $IP2(M)$ (see eqs 2.32 and 2.33) may be calculated both for symmetric and unsymmetric matrices. When the last two operators are calculated on a symmetric matrix, the terms of sum represent squared entries in that matrix. This is the reason for the number 2 in the symbol of these operators. Only in trees, and only for Cluj distance indices,

$$IE2(UM) = IP(UM) \qquad (2.41)$$

2.11.2. CF Matrices

It happens that $V_{i,(i,j)_k}$ be sets of disconnected vertices. This fact is undesirable when molecular graphs (which are always connected graphs) are investigated. If $V_{i,(i,j)_k}$ real (connected) chemical fragments are wanted, the *Cluj fragmental matrices*[52] are defined. In this version, the sets $V_{i,(i,j)_k}$ are defined as

$$V_{i,(i,j)_k} = \{v | v \in V(G_p); \; G_p = G - (i,j)_k; \, d_{iv}(G_p) < d_{jv}(G_p); \; (i,j)_k \in D(G) \; or \; \Delta(G)\} \qquad (2.42)$$

where $d_{iv}(G_p)$ and $d_{jv}(G_p)$ are the topological distances between a vertex v and vertices i and j, respectively, in the spanning subgraph G_p resulted by cutting the path $p = (i,j)_k$ (except its endpoints) from G.

The set $V_{i,(i,j)_k}$ consists now of vertices lying *closer* to the vertex i in G_p. This version is called[50] *all paths external to the path* (i,j) since all paths $(i,v)_h$, $h = 1,2,...$ (see eq 2.37) are *external* with respect to the path $(i,j)_k$, because the last path was already cut off. The diagonal entries are zero.

When $(i,j)_k \in D(G)$, then *Cluj Fragmental Distance matrix*, **CFD**, is defined; in case $(i,j)_k \in \Delta(G)$, the matrix is *Cluj Fragmental Detour*, **CFΔ**. The entries $[UM]_{ij}$, M = **CFD ; CFΔ** represent connected subgraphs, both in G_p and G.

Theorem 2.1.

For any $i, j \in V(G)$, and for any path joining i and j, $p_{i,j} \in P_{i,j}(G)$, the *Cluj Fragment*, $CF_i(G_p) \equiv V_{i,(i,j)_k}$ (cf. eq. 2.42), is a fragment (i.e. connected subgraph).

Demonstration:

Let $v \in CF_i(G_p)$, involving $d_{iv}(G_p) < d_{jv}(G_p)$ (Szeged-Cluj criterion). It follows that d_{iv} is finite and a shortest path joining i and v, $p_{iv} \in P_{iv}(G_p)$ may exist (for simplicity G_p is hereafter missing).

For any vertex k lying on that path, $k \in p_{iv}$, we *have to prove* that $d_{ik} < d_{jk}$ and (cf. criterion) $k \in CF_i$ and CF_i is connected.

From $k \in p_{iv}$, it follows that there exists a path joining k and i, $p_{ik} \in P_{ik}$, such that $p_{ik} \subseteq p_{iv}$ and a path joining k and v, $p_{kv} \in P_{kv}$ with $p_{kv} \subseteq p_{iv}$. It is immediate that $p_{iv} = p_{ik} \cup p_{kv}$. Since p_{iv} is a geodesic it follows that it is a sum of geodesics. Thus, we can write $d_{iv} = d_{ik} + d_{kv}$.

Case 1: d_{jv} is finite (G_p is connected). There exists a path p_{jv}, which is the shortest path joining v and j such that $d_{jv} \le d_{jk} + d_{kv}$ (d is a metric) and, from hypothesis,

$$d_{ik} + d_{kv} = d_{iv} < d_{jv} \le d_{jk} + d_{kv} \tag{2.43}$$

following that $d_{ik} < d_{jk}$, $k \in CF_i$ and CF_i is connected.

Case 2: d_{jv} is infinite (G_p is disconnected). There is no path p_{jv}, to join j and v. The following relations hold

$$d_{ik} = d_{iv} - d_{kv} < \infty - d_{kv} < \infty \tag{2.44}$$

$$d_{jk} \ge d_{jv} - d_{kv} \ge \infty - d_{kv} = \infty \tag{2.45}$$

It is immediate that $d_{ik} < d_{jk}$, $k \in CF_i$ and CF_i is connected.

The Cluj matrices, $\mathbf{UCJD}_p$, $\mathbf{UCFD}_p$, $\mathbf{UCJ\Delta}_p$ and $\mathbf{UCF\Delta}_p$, for the graph $G_{2.17}$ are illustrated in Figure 2.17 along with the corresponding fragmentation. A disconnected subgraph, CJD_i is herein encountered.

In acyclic structures, $\mathbf{CJD}_e = \mathbf{CFD}_e = \mathbf{SZD}_e = \mathbf{W}_e$ and $\mathbf{CJD}_p = \mathbf{CFD}_p = \mathbf{W}_p$. In cyclic graphs, $\mathbf{CJD}_e = \mathbf{CFD}_e = \mathbf{SZD}_e$ while $\mathbf{CJD}_p \ne \mathbf{CFD}_p \ne \mathbf{SZD}_p$, $\mathbf{CJ\Delta}_p \ne \mathbf{CF\Delta}_p \ne \mathbf{SZ\Delta}_p$ and $\mathbf{W}_{e/p}$ are not defined.

Relationships between the indices constructed on these matrices will be discussed in the Chap. Cluj Indices.

An interesting property is shown by the detour-based matrices: $\mathbf{CJ\Delta}_p$ and $\mathbf{CF\Delta}_p$. Let consider the vertices 8 (of degree 1) and 5 (of degree 2) in $G_{2.17}$, Figure 2.17. The vertex 8 is an *external* vertex (with a terminal path ending in it) while the vertex 5 is an *internal* one (usually a terminal path not ending in it). An external vertex, like 8, shows all its entries in the Cluj matrices equal to 1 (see Figure 2.17). The same entries are shown by the internal vertex 5. This unusual property is called *the internal ending of all detours* joining a vertex i and the remaining vertices in G. Such a vertex is called an *internal endpoint*.[50] There exist graphs with all the vertices internal endpoints and their detours are *Hamiltonian paths* now. This kind of graph we call the *full Hamiltonian detour graph*, *FHΔ* (see Chap. 8).

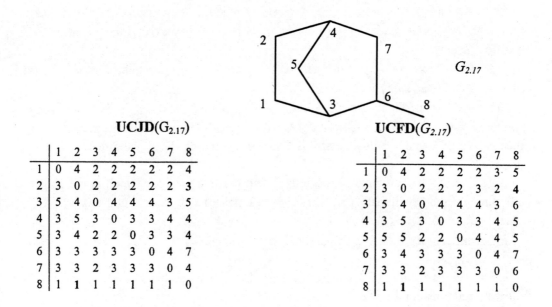

UCJD(G$_{2.17}$)

	1	2	3	4	5	6	7	8
1	0	4	2	2	2	2	2	4
2	3	0	2	2	2	2	2	3
3	5	4	0	4	4	4	3	5
4	3	5	3	0	3	3	4	4
5	3	4	2	2	0	3	3	4
6	3	3	3	3	3	0	4	7
7	3	3	2	3	3	3	0	4
8	1	1	1	1	1	1	1	0

UCFD(G$_{2.17}$)

	1	2	3	4	5	6	7	8
1	0	4	2	2	2	2	3	5
2	3	0	2	2	2	3	2	4
3	5	4	0	4	4	4	3	6
4	3	5	3	0	3	3	4	5
5	5	5	2	2	0	4	4	5
6	3	4	3	3	3	0	4	7
7	3	3	2	3	3	3	0	6
8	1	1	1	1	1	1	1	0

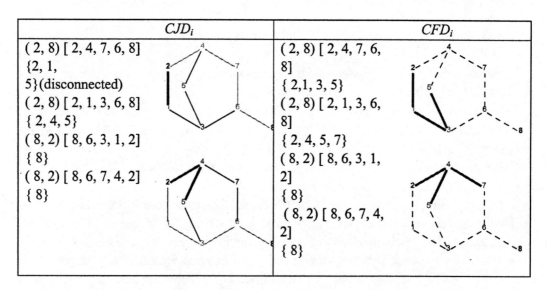

CJD_i	CFD_i
(2, 8) [2, 4, 7, 6, 8] {2, 1, 5}(disconnected) (2, 8) [2, 1, 3, 6, 8] { 2, 4, 5} (8, 2) [8, 6, 3, 1, 2] { 8} (8, 2) [8, 6, 7, 4, 2] { 8}	(2, 8) [2, 4, 7, 6, 8] { 2,1, 3, 5} (2, 8) [2, 1, 3, 6, 8] { 2, 4, 5, 7} (8, 2) [8, 6, 3, 1, 2] { 8} (8, 2) [8, 6, 7, 4, 2] { 8}

Figure 2.17. Unsymmetric Cluj matrices and fragmentation for the graph $G_{2.17}$.

$UCJ\Delta(G_{2.17})$

	1	2	3	4	5	6	7	8
1	0	1	1	1	1	2	1	2
2	1	0	1	1	1	1	2	1
3	2	2	0	3	3	2	2	2
4	2	2	2	0	2	2	2	3
5	1	1	1	1	0	1	1	1
6	3	2	2	2	2	0	2	7
7	1	3	1	1	1	1	0	1
8	1	1	1	1	1	1	1	0

$UCF\Delta(G_{2.17})$

	1	2	3	4	5	6	7	8
1	0	1	1	1	1	2	1	2
2	1	0	1	1	1	1	2	1
3	2	2	0	3	4	2	2	2
4	2	2	2	0	4	2	2	3
5	1	1	1	1	0	1	1	1
6	3	2	2	2	2	0	2	7
7	1	3	1	1	1	1	0	1
8	1	1	1	1	1	1	1	0

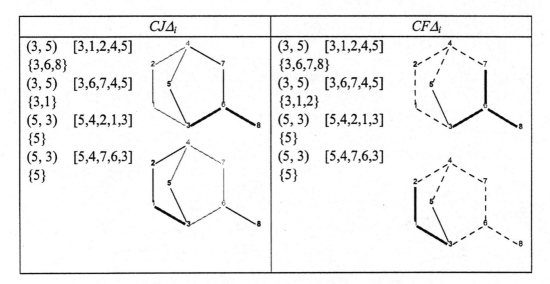

$CJ\Delta_i$		$CF\Delta_i$	
(3, 5) [3,1,2,4,5] {3,6,8}		(3, 5) [3,1,2,4,5] {3,6,7,8}	
(3, 5) [3,6,7,4,5] {3,1}		(3, 5) [3,6,7,4,5] {3,1,2}	
(5, 3) [5,4,2,1,3] {5}		(5, 3) [5,4,2,1,3] {5}	
(5, 3) [5,4,7,6,3] {5}		(5, 3) [5,4,7,6,3] {5}	

Figure 2.17. (continued) Unsymmetric Cluj matrices and fragmentation for the graph $G_{2.17}$.

2.12. DISTANCE EXTENDED MATRICES

Tratch *et al.*[53] have proposed an *extended distance matrix*, $\mathbf{E}$, whose entries are the product of the entries in the $\mathbf{D}_e$ matrix and a multiplier, m_{ij}, which is the number of paths in the graph of which path (i,j) is a subgraph. In acyclic structures, it equals the entries in the Wiener matrix $\mathbf{W}_p$, so that $\mathbf{E}$ is further referred to as $\mathbf{D\_W}_p$ matrix

$$[\mathbf{D\_W}_p]_{ij} = [\mathbf{D}_e]_{ij} \, m_{ij} = [\mathbf{D}_e]_{ij} \, [\mathbf{W}_p]_{ij} = d_{ij} N_i N_j \qquad (2.46)$$

where d_{ij} is the topological distance between i and j and N_i, N_j have the same meaning as in case of the Wiener matrix (see above). The $\mathbf{D\_W}_p$ matrix is just the

Hadamard product[42] of the $\mathbf{D}_e$ and $\mathbf{W}_p$ matrices. The half sum of its entries gives an expanded Wiener number.[28,53] Figure 2.18 illustrates this matrix for the graph $G_{2.6}$.

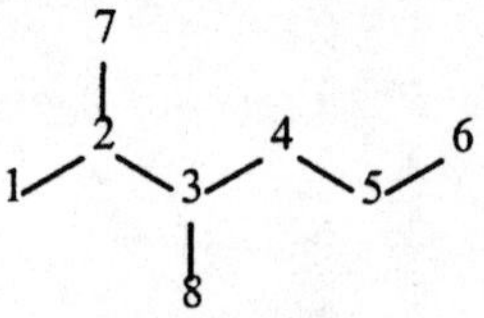

$G_{2.6}$

D_W$_p$($G_{2.6}$)

	1	2	3	4	5	6	7	8	RS_i
1	0	7	10	9	8	5	2	3	44
2	7	0	15	18	18	12	7	6	83
3	10	15	0	15	20	15	10	7	92
4	9	18	15	0	12	12	9	6	81
5	8	18	20	12	0	7	8	6	79
6	5	12	15	12	7	0	5	4	60
7	2	7	10	9	8	5	0	3	44
8	3	6	7	6	6	4	3	0	35
CS_i	44	83	92	81	79	60	44	35	

D_UCJD($G_{2.6}$)

	1	2	3	4	5	6	7	8	$RS_i(\mathbf{W}_p)$
1	0	1	2	3	4	5	2	3	20
2	7	0	3	6	9	12	7	6	50
3	10	5	0	5	10	15	10	7	62
4	9	6	3	0	6	12	9	6	51
5	8	6	4	2	0	7	8	6	41
6	5	4	3	2	1	0	5	4	24
7	2	1	2	3	4	5	0	3	20
8	3	2	1	2	3	4	3	0	18
$CS_i(\mathbf{D}_p)$	44	25	18	23	37	60	44	35	

Figure 2.18. Distance-extended matrices, for the graph $G_{2.6}$.

Similarly, Diudea[39] has performed the Hadamard product $\mathbf{D}_e \bullet \mathbf{UCJD}$

$$[\mathbf{D\_UCJD}]_{ij} = [\mathbf{D}_e]_{ij} [\mathbf{UCJD}]_{ij} = d_{ij} N_{i, (ij)} \tag{2.47}$$

This matrix (illustrated in Figure 2.18 for the graph $G_{2.6}$) shows, in trees, the equalities

$$CS(\mathbf{D\_UCJD}) = CS(\mathbf{D}_p) \tag{2.48}$$

$$RS(\mathbf{D\_UCJD}) = RS(\mathbf{W}_p) \tag{2.49}$$

Thus, $IP(D\_UCJD)$ calculates the hyper-Wiener index (as the half sum of its entries). The $\mathbf{D\_UCJD}$ matrix is a direct proof of the finding[27] that the *sum of all internal paths* (given by $\mathbf{D}_p$) *equals the sum of all external paths* (given by $\mathbf{W}_p$) *with respect to all pairs* (*i,j*) *in a graph*. The matrix $\mathbf{D\_UCJD}$ offers a new definition of the hyper-Wiener number (see Chap. Topological Indices and eq 2.47). Various other combinations: **D_M** or **Δ_M, M** being a symmetric or unsymmetric square matrix, were performed in trees or in cycle-containing graphs, by means of the CLUJ software program.

Similarly, a *3D*-extension[39] (e.g. by using the geometric matrix, **G**) allows the construction of various *3D-distance extended matrices*, such as **G_UCJD** (see Figure 2.19). They can offer *3D- sensitive* indices.

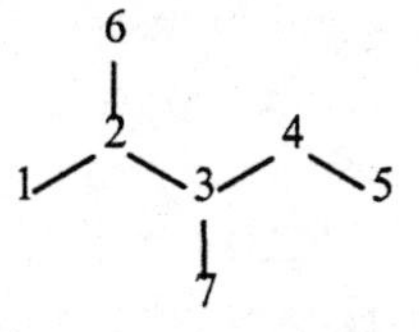

$G_{2.18}$

UCJD($G_{2.18}$)

	1	2	3	4	5	6	7
1	0	1	1	1	1	1	1
2	6	0	3	3	3	6	3
3	4	4	0	5	5	4	6
4	2	2	2	0	6	2	2
5	1	1	1	1	0	1	1
6	1	1	1	1	1	0	1
7	1	1	1	1	1	1	0

G($G_{2.18}$)

	1	2	3	4	5	6	7
1	0.0000	1.5414	2.5709	3.9411	4.5163	2.5178	3.0305
2	1.5414	0.0000	1.5543	2.5634	3.0891	1.5388	2.5821
3	2.5709	1.5543	0.0000	1.5468	2.5852	2.5930	1.5395
4	3.9411	2.5634	1.5468	0.0000	1.5364	3.0398	2.5461
5	4.5163	3.0891	2.5852	1.5364	0.0000	3.6199	3.9326
6	2.5178	1.5388	2.5930	3.0398	3.6199	0.0000	3.2366
7	3.0305	2.5821	1.5395	2.5461	3.9326	3.2366	0.0000

G_UCJD($G_{2.18}$)

	1	2	3	4	5	6	7
1	0.0000	1.5414	2.5709	3.9411	4.5163	2.5178	3.0305
2	9.2484	0.0000	4.6629	7.6902	9.2673	9.2328	7.7463
3	10.2836	6.2172	0.0000	7.7340	12.9260	10.3720	9.2370
4	7.8822	5.1268	3.0936	0.0000	9.2184	6.0796	5.0922
5	4.5163	3.0891	2.5852	1.5364	0.0000	3.6199	3.9326
6	2.5178	1.5388	2.5930	3.0398	3.6199	0.0000	3.2366
7	3.0305	2.5821	1.5395	2.5461	3.9326	3.2366	0.0000

Figure 2.19. *3D*-Distance-extended Cluj matrix for the graph $G_{2.18}$.

2.13. RECIPROCAL MATRICES

In chemical graph theory, the distance matrix accounts for the *through bond* interactions of atoms in molecules. However, these interactions decrease as the distance between atoms increases. This reason lead to the introduction, in 1993, by the group of

Balaban[54] and Trinajstić,[55] respectively, of the *reciprocal distance* matrix, RD_e. The entries in this matrix are defined by

$$[RD_e]_{ij} = 1 / [D_e]_{ij} \tag{2.50}$$

RD_e matrix allows the calculation of a Wiener number analogue, called the *Harary index*[55] (see Chap. Topological Indices), in the honor of Frank Harary.

Since topological matrices are considered *natural* sources in deriving graph descriptors,[2,28,29] some other matrices having entries as *reciprocal* (topological) *property* : $[RM]_{ij} = 1/[M]_{ij}$; $M = W_{e/p}$, D_p, USZD and UCJD have been recently proposed by Diudea,[56] as a ground for new Harary-type indices (see Chap. Topological Indices). Figure 2.20. illustrates some reciprocal property matrices, for the graph $G_{2.8}$.

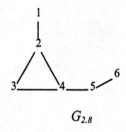

$G_{2.8}$

$RD_e(G_{2.8})$

	1	2	3	4	5	6
1	0	1	1/2	1/2	1/3	1/4
2	1	0	1	1	1/2	1/3
3	1/2	1	0	1	1/2	1/3
4	1/2	1	1	0	1	1/2
5	1/3	1/2	1/2	1	0	1
6	1/4	1/3	1/3	1/2	1	0

$RD_p(G_{2.8})$

	1	2	3	4	5	6
1	0	1	1/3	1/3	1/6	1/10
2	1	0	1	1	1/3	1/6
3	1/3	1	0	1	1/3	1/6
4	1/3	1	1	0	1	1/3
5	1/6	1/3	1/3	1	0	1
6	1/10	1/6	1/6	1/3	1	0

$RUSZD(G_{2.8})$

	1	2	3	4	5	6
1	0	1	1	1	1/2	1/3
2	1/5	0	1/2	1/2	1/3	1/4
3	1/4	1	0	1	1/3	1/4
4	1/4	1/3	1/3	0	1/4	1/4
5	1/3	1/2	1/2	1/2	0	1/5
6	1/2	1/2	1/2	1	1	0

$RUCJD(G_{2.8})$

	1	2	3	4	5	6
1	0	1	1	1	1	1
2	1/5	0	1/2	1/2	1/3	1/3
3	1/4	1	0	1	1/3	1/3
4	1/4	1/3	1/3	0	1/4	1/4
5	1/2	1/2	1/2	1/2	0	1/5
6	1	1	1	1	1	0

Figure 2.20. Reciprocal matrices for the graph $G_{2.8}$.

2.14. WALK MATRICES

Diudea [96Diu1] has recently proposed the *walk matrix*,[18,57] $W_{(M1,M2,M3)}$, constructed by the principle of the single endpoint characterization of a path,[18,58]

$$[\mathbf{W}_{(M_1,M_2,M_3)}]_{ij} = {}^{[M_2]_{ij}}W_{M_1,i}[\mathbf{M}_3]_{ij} = [RS((\mathbf{M}_1)^{[M_2]_{ij}})]_i[\mathbf{M}_3]_{ij} \qquad (2.51)$$

where $W_{M1,i}$ is the walk degree, of elongation $[\mathbf{M}_2]_{ij}$, of the vertex i, weighted by the property collected in matrix $\mathbf{M}_1$ (i.e., the ith row sum of the matrix $\mathbf{M}_1$, raised to power $[\mathbf{M}_2]_{ij}$). The diagonal entries are zero. It is a square, (in general) non-symmetric matrix. This matrix, that mixes three square matrices, is a true matrix operator (see below).

Let, first, $(\mathbf{M}_1, \mathbf{M}_2, \mathbf{M}_3)$ be $(\mathbf{M}_1, 1, 1)$, where 1 is the matrix with the off-diagonal elements equal to 1. In this case, the (i,j)-elements of matrix $\mathbf{W}_{(M1,1,1,)}$ will be

$$[\mathbf{W}_{(M_1,1,1)}]_{ij} = [RS(\mathbf{M}_1)]_i = W_{M_1,i} \qquad (2.52)$$

Next, consider the combination $(\mathbf{M}_1, 1, \mathbf{M}_3)$; the corresponding walk matrix can be expressed as the Hadamard product

$$\mathbf{W}_{(M_1,1,M_3)} = \mathbf{W}_{(M_1,1,1)} \bullet \mathbf{M}_3 \qquad (2.53)$$

Examples are given in Figure 2.21 for the Graph $G_{2.18}$, in case: $\mathbf{M}_1 = \mathbf{A}$ and $\mathbf{M}_3 = \mathbf{D}_e$.

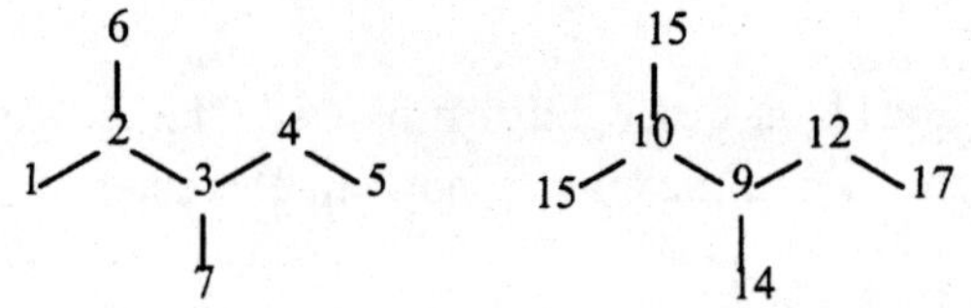

$G_{2.18}$ $G_{2.18}\{RS(\mathbf{D}_e)_i\}$

$\mathbf{A}(G_{2.18})$

	1	2	3	4	5	6	7	RS_i
1	0	1	0	0	0	0	0	1
2	1	0	1	0	0	1	0	3
3	0	1	0	1	0	0	1	3
4	0	0	1	0	1	0	0	2
5	0	0	0	1	0	0	0	1
6	0	1	0	0	0	0	0	1
7	0	0	1	0	0	0	0	1

$\mathbf{D}_e(G_{2.18})$

	1	2	3	4	5	6	7	RSi
1	0	1	2	3	4	2	3	15
2	1	0	1	2	3	1	2	10
3	2	1	0	1	2	2	1	9
4	3	2	1	0	1	3	2	12
5	4	3	2	1	0	4	3	17
6	2	1	2	3	4	0	3	15
7	3	2	1	2	3	3	0	14

$\mathbf{W}_{(A,1,1)}$ $(G_{2.18})$

	1	2	3	4	5	6	7	RS_i
1	0	1	1	1	1	1	1	6
2	3	0	3	3	3	3	3	18
3	3	3	0	3	3	3	3	18
4	2	2	2	0	2	2	2	12
5	1	1	1	1	0	1	1	6
6	1	1	1	1	1	0	1	6
7	1	1	1	1	1	1	0	6

$\mathbf{W}_{(De,1,1)}(G_{2.18})$

	1	2	3	4	5	6	7	RS_i
1	0	15	15	15	15	15	15	90
2	10	0	10	10	10	10	10	60
3	9	9	0	9	9	9	9	54
4	12	12	12	0	12	12	12	72
5	17	17	17	17	0	17	17	102
6	15	15	15	15	15	0	15	90
7	14	14	14	14	14	14	0	84

$\mathbf{W}_{(A,1,De)} = \mathbf{W}_{(A,1,1)} \bullet \mathbf{D}_e$

	1	2	3	4	5	6	7	$deg_i RS(\mathbf{D}_e)_i$
1	0	1	2	3	4	2	3	15
2	3	0	3	6	9	3	6	30
3	6	3	0	3	6	6	3	27
4	6	4	2	0	2	6	4	24
5	4	3	2	1	0	4	3	17
6	2	1	2	3	4	0	3	15
7	3	2	1	2	3	3	0	14
$CS(\mathbf{AD}_e)_i$	24	14	12	18	28	24	22	142

$\mathbf{W}_{(De,1,A)} = \mathbf{W}_{(De,1,1)} \bullet A$

	1	2	3	4	5	6	7	$RS(\mathbf{D}_e)_i deg_i$
1	0	15	0	0	0	0	0	15
2	10	0	10	0	0	10	0	30
3	0	9	0	9	0	0	9	27
4	0	0	12	0	12	0	0	24
5	0	0	0	17	0	0	0	17
6	0	15	0	0	0	0	0	15
7	0	0	14	0	0	0	0	14
$RS(\mathbf{AD}_e)_i$	10	39	36	26	12	10	9	142

Figure 2.21. $\mathbf{W}_{(M1,M2,M3)}$ algebra for the graph $G_{2.18}$.

The sum of all entries in $\mathbf{W}_{(M1,1,M3)}$ can be obtained by

$$\mathbf{u}\mathbf{W}_{(M_1,1,M_3)}\mathbf{u}^T = \sum_i [RS(\mathbf{W}_{(M_1,1,M_3)})]_i = \mathbf{u}(\mathbf{M}_1\mathbf{M}_3)\mathbf{u}^T \qquad (2.54)$$

where $\mathbf{u}$ and $\mathbf{u}^T$ are the unit vector (of order N) and its transpose, respectively. The row sum vector in $\mathbf{W}_{(M1,1,M3)}$ can be achieved by the pairwise product of the row sums in $\mathbf{M}_1$ and $\mathbf{M}_3$, respectively

$$[RS(\mathbf{W}_{(M_1,1,M_3)})]_i = [RS(\mathbf{M}_1)]_i [RS(\mathbf{M}_3)]_i \qquad (2.55)$$

This vector represents a collection of pairwise products of local (topological) properties (encoded as corresponding row sums in $\mathbf{M}_1$ and $\mathbf{M}_3$ - see above). Eq 2.54 is a joint of the Cramer and Hadamard matrix algebra, by means of $\mathbf{W}_{(M1,1,M3)}$.

As walk numbers, eq 2.55 can be written

$$W_{\mathbf{W}_{(M_1,1,M_3)},i} = W_{M_1,i} W_{M_3,i} \qquad (2.56)$$

When $\mathbf{M}_1 = \mathbf{M}_3$, then eq 2.56 becomes

$$W_{\mathbf{W}_{(M,1,M)},i} = W_{M,i} W_{M,i} = {}^2 W_{M,i} \qquad (2.57)$$

and, by extension

$$W_{\mathbf{W}_{(M,n,M)},i} = {}^n W_{M,i} W_{M,i} = {}^{(n+1)} W_{M,i} \qquad (2.58)$$

where $\mathbf{n}$ is the matrix having entries $[\mathbf{n}]_{ij}$ and $^{(n+1)}W_{M,i}$ means a (weighted) walk number, of length $n+1$. As global walk numbers, eq 2.58 can be written

$$2W_{\mathbf{W}_{(M,n,M)}} = \mathbf{u}\mathbf{W}_{(M,n,M)}\mathbf{u}^T = \mathbf{u}((\mathbf{M})^n\mathbf{M})\mathbf{u}^T = 2^{(n+1)} W_M \qquad (2.59)$$

Eqs 2.54 and 2.59 prove that $\mathbf{W}_{(M1,M2,M3)}$ is a true matrix operator.

Figure 2.21 illustrates that the sum of entries in $\mathbf{W}_{(A,1,De)}$ equals that in $\mathbf{W}_{(De,1,A)}$. However, the two matrices are not identical. Only the vectors of their walk numbers (i.e. row sums) are identical. In this particular case, the walk numbers mean the local contributions to the *degree-distance* index of Dobrynin,[59] reinvented by Estrada,[60] or the non-trivial part of the Schultz index.[61] In walk number symbols, the local index can be written as

$$W_{\mathbf{W}_{(A,1,D_e)},i} = W_{A,i} W_{D_e,i} = W_{\mathbf{W}_{(D_e,1,A)},i} \qquad (2.60)$$

The twin unsymmetric walk matrices (having reversed sequence $\mathbf{M}_1,1,\mathbf{M}_3$) show, thus, common row sums but different column sums. However, the common point of these matrices is the Cramer product $\mathbf{AD}_e$ (or in general, $\mathbf{M}_1\mathbf{M}_3$):

$$CS(\mathbf{W}_{(A,1,De)}) = CS(\mathbf{AD}_e) \qquad (2.61)$$

$$CS(\mathbf{W}_{(De,1,A)}) = RS(\mathbf{AD}_e) \qquad (2.62)$$

A particular case of the walk matrix, $RW_{(A,De,1)}$, (see also Sect. 2.13) is identical to the *restricted random walk* matrix of Randić.[62]

2.15. SCHULTZ MATRICES

The Schultz matrices, $SCH(G)$ are related to the molecular topological index, *MTI*, or the Schultz index,[61] (see Chap. Topological Indices). Diudea and Randić[63] have extended the Schultz's definition by using a combination of three square matrices, one of them being obligatory the adjacency matrix

$$SCH_{(M_1,A,M_3)} = M_1(A + M_3) = M_1A + M_1M_3 \qquad (2.63)$$

It is easily seen that $SCH_{(A,A,De)}$ is the matrix on which the Schultz original index can be calculated. Analogue Schultz matrices, of sequence: (D_e,A,D_e), (RD_e,A,RD_e) and $(W_p A,W_p)$ have been proposed and the corresponding indices tested for correlating ability.[64-66]

A Schultz-extended matrix is related to the walk matrix by[57,63]

$$uSCH_{(M_1,A,M_3)}u^T = uW_{(M_1,1,(A+M_3))}u^T = uW_{(M_1,1,A)}u^T + uW_{(M_1,1,M_3)}u^T \quad (2.64)$$

When one of the square matrices are unsymmetric, the resulting Schultz matrix will also be unsymmetric. Matrices $W_{(M1,M2,M3)}$ involved in the calculation of $SCH_{(De,A,UCJD)}$, for the graph $G_{2.18}$ are illustrated in Figure 2.22. It can be seen that the sum of all entries in the walk matrix $W_{(D_e,1,(A+UCJD))}$ equals that in $SCH_{(De,A,UCJD)}$, calculated by Cramer algebra.

2.16. LAYER AND SEQUENCE MATRICES

Layer matrices have been proposed in connection to the sequences of walks: *DDS* (Distance Degree Sequence),[67-70] *PDS* (Path Degree Sequence),[71-74] and *WS* (Walk Sequence).[1] They are built up on the layer partitions in a graph.

A layer partition $G(i)$ with respect to the vertex i, in G, is defined as[5,70,75]

$$G(i) = \{G(v)_j, j \in [0, ecc_i] \quad and \quad v \in G(v)_j \Leftrightarrow d_{iv} = j\} \qquad (2.65)$$

where ecc_i is the eccentricity of i. Figure 2.23 illustrates the layer partitions for the graph $G_{2.19}$.

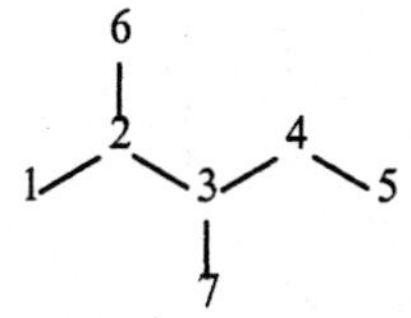

$G_{2.18}$

UCJD

	1	2	3	4	5	6	7	RS_i
1	0	1	1	1	1	1	1	6
2	6	0	3	3	3	6	3	24
3	4	4	0	5	5	4	6	28
4	2	2	2	0	6	2	2	16
5	1	1	1	1	0	1	1	6
6	1	1	1	1	1	0	1	6
7	1	1	1	1	1	1	0	6
								92

A + UCJD

	1	2	3	4	5	6	7	RS_i
1	0	2	1	1	1	1	1	7
2	7	0	4	3	3	7	3	27
3	4	5	0	6	5	4	7	31
4	2	2	3	0	7	2	2	18
5	1	1	1	2	0	1	1	7
6	1	2	1	1	1	0	1	7
7	1	1	2	1	1	1	0	7
								104

W$_{(De,1,A)}$

	1	2	3	4	5	6	7	RS_i
1	0	15	0	0	0	0	0	15
2	10	0	10	0	0	10	0	30
3	0	9	0	9	0	0	9	27
4	0	0	12	0	12	0	0	24
5	0	0	0	17	0	0	0	17
6	0	15	0	0	0	0	0	15
7	0	0	14	0	0	0	0	14
								142

W$_{(De,1,UCJD)}$

	1	2	3	4	5	6	7	RS_i
1	0	15	15	15	15	15	15	90
2	60	0	30	30	30	60	30	240
3	36	36	0	45	45	36	54	252
4	24	24	24	0	72	24	24	192
5	17	17	17	17	0	17	17	102
6	15	15	15	15	15	0	15	90
7	14	14	14	14	14	14	14	0
								1050

W$_{(De,1,(A+UCJD))}$

	1	2	3	4	5	6	7	RS_i
1	0	30	15	15	15	15	15	105
2	70	0	40	30	30	70	30	270
3	36	45	0	54	45	36	63	279
4	24	24	36	0	84	24	24	216
5	17	17	17	34	0	17	17	119
6	15	30	15	15	15	0	15	105
7	14	14	28	14	14	14	0	98
								1192

SCH$_{(De,A,UCJD)}$

	1	2	3	4	5	6	7	RS_i
1	30	27	25	28	39	28	29	206
2	14	18	15	16	23	14	16	116
3	14	13	15	12	15	14	11	94
4	24	20	19	22	19	24	20	148
5	38	31	29	32	37	38	33	238
6	28	27	25	28	39	30	29	206
7	28	24	23	24	31	28	26	184
								1192

Figure 2.22. Walk and Schultz matrices for the graph $G_{2.18}$.

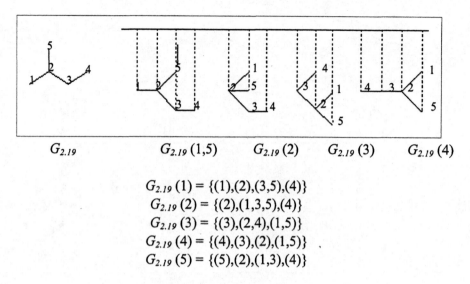

$G_{2.19}$ $G_{2.19}(1,5)$ $G_{2.19}(2)$ $G_{2.19}(3)$ $G_{2.19}(4)$

$$G_{2.19}(1) = \{(1),(2),(3,5),(4)\}$$
$$G_{2.19}(2) = \{(2),(1,3,5),(4)\}$$
$$G_{2.19}(3) = \{(3),(2,4),(1,5)\}$$
$$G_{2.19}(4) = \{(4),(3),(2),(1,5)\}$$
$$G_{2.19}(5) = \{(5),(2),(1,3),(4)\}$$

Figure 2.23. Layer partitions $G(i)$ for the graph $G_{2.19}$.

Let $G(v)_j$ be the jth layer of vertices v located at distance j, in the layer partition $G(i)$

$$G(v)_j = \{v \mid d_{iv} = j\} \tag{2.66}$$

The entries in a layer matrix, **LM,** collect the property $\mathbf{M}_v$ (topological or chemical) for all vertices v belonging to the layer $G(v)_j$

$$[\mathbf{LM}]_{ij} = \sum_{v \in G(v)_j} \mathbf{M}_v \tag{2.67}$$

The matrix **LM** can be written as

$$\mathbf{LM}\,(G) = \{\,[\mathbf{LM}]_{ij};\ i \in V(G);\ j \in [0, d(G)]\,\} \tag{2.68}$$

where $d(G)$ is the diameter of the graph. The dimensions of such a matrix are Nx $(d(G)+1)$. Figure 2.24 illustrates some layer matrices: **LC** (Layer of Cardinalities), **LDS** (Layer of Distance Sums) and $\mathbf{L}^e\mathbf{W}$ (Layer of Walk degrees, of length e), for the graph $G_{2.18}$.

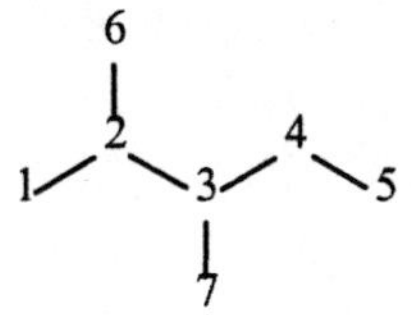

$G_{2.18}$

j		**LC**$(G_{2.18})$					**LDS**$(G_{2.18})$			
	0	1	2	3	4	0	1	2	3	4
1	1	1	2	2	1	15	10	24	26	17
2	1	3	2	1	0	10	39	26	17	0
3	1	3	3	0	0	9	36	47	0	0
4	1	2	2	2	0	12	26	24	30	0
5	1	1	1	2	2	17	12	9	24	30
6	1	1	2	2	1	15	10	24	26	17
7	1	1	2	3	0	14	9	22	47	0

$C(G_{2.18}) = 7$ $\qquad DS(G_{2.18}) = 92$

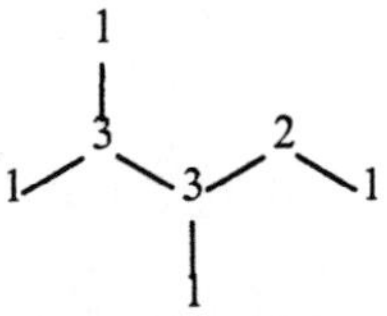

$G_{2.18}\{^{1}W_i\}$ 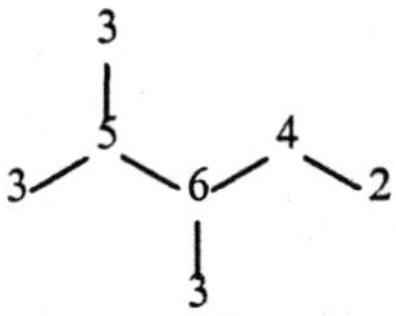 $\qquad G_{2.18}\{^{2}W_i\}$ 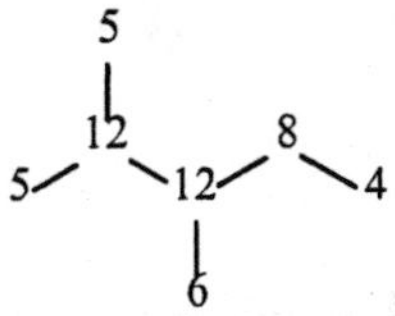 $\qquad G_{2.18}\{^{3}W_i\}$ 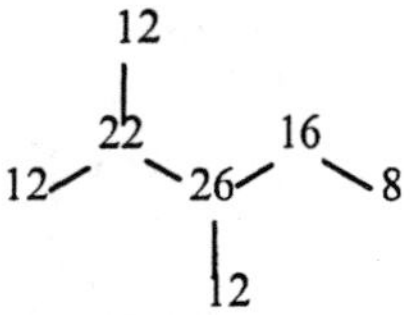 $\qquad G_{2.18}\{^{4}W_i\}$

j	**L^{1}W**$(G_{2.18})$					**L^{2}W**$(G_{2.18})$					**L^{3}W**$(G_{2.18})$					**L^{4}W**$(G_{2.18})$				
	0	1	2	3	4	0	1	2	3	4	0	1	2	3	4	0	1	2	3	4
1	1	3	4	3	1	3	5	9	7	2	5	12	17	14	4	12	22	38	28	8
2	3	5	3	1	0	5	12	7	2	0	12	22	14	4	0	22	50	28	8	0
3	3	6	3	0	0	6	12	8	0	0	12	26	14	0	0	26	50	32	0	0
4	2	4	4	2	0	4	8	8	6	2	8	16	18	10	0	16	34	34	24	0
5	1	2	3	4	2	2	4	6	8	6	4	8	12	18	10	8	16	26	34	24
6	1	3	4	3	1	3	5	9	7	2	5	12	17	14	4	12	22	38	28	8
7	1	3	5	3	0	3	6	9	8	0	6	12	20	14	0	12	26	38	32	0

$^{1}W(G_{2.18}) = 12$ $\qquad ^{2}W(G_{2.18}) = 26$ $\qquad ^{3}W(G_{2.18}) = 52$ $\qquad ^{4}W(G_{2.18}) = 108$

Figure 2.24. Layer matrices for the graph $G_{2.18}$.

Some properties of the **LM** matrices need to be pointed out:

(1) The sum of entries in any row equals the sum on the column $j = 0$ and equals the global property $M(G)$. When this property involves edges (e.g. a walk) the quantity $M(G)$ must be divided by 2 for being equivalent to the walk numbers, ^eW_M

$$\Sigma_j[\mathbf{LM}]_{ij} = \Sigma_i[\mathbf{LM}]_{io} = M(G) \tag{2.69}$$

(2) The entries in the column $j = 1$ of matrix $\mathbf{L}^e\mathbf{W}$ become the entries in the column $j = 0$ of the matrix $\mathbf{L}^{e+1}\mathbf{W}$

$$[\mathbf{L}^e\mathbf{W}]_{il} = [\mathbf{L}^{e+1}\mathbf{W}]_{io} \tag{2.70}$$

The above relation is valid for any graph, excepting the multigraphs. It represents the essence of the ^eW_M algorithm (see Sect. 2.1) and also of the Morgan algorithm.[76]

(3) The **LC** matrix (layer matrix of cardinalities) counts vertices lying on concentric layers/shells around each vertex $i \in V(G)$. Thus, the property $\mathbf{M}_v = 1$ (i.e. the cardinality) and:

$$\Sigma_j[\mathbf{LC}]_{ij} = \Sigma_i[\mathbf{LC}]_{io} = |V(G)| = N(G) \tag{2.71}$$

$$\Sigma_j[\mathbf{LC}]_{il} = 2|E(G)| = 2Q(G) \tag{2.72}$$

In fact this matrix follows just the layer partitions in G. The **LC** matrix can be viewed as a collection of *distance degree sequences*,[67-70] DDS_i (i.e. the number of vertices lying at the distance j form the vertex i - see below).

A *sequence matrix*,[70,75] **SM**, is defined as
$$[\mathbf{SM}]ij = \text{no. of } M \text{ of length } j \text{ traversing the vertex } i \tag{2.73}$$

where M stands for some topological quantities involving edges: (all) paths, shortest paths (i.e. distances), longest paths (i.e. detours), cycles, (see also Chap. 8) etc.

A global sequence of M, called the M sequence (i.e. *spectrum*), is derived from such matrices

$$MS_j = f\sum_i[\mathbf{SM}]ij; \quad j = 1, 2, ..., N \tag{2.74}$$

where f is 1/2 for path-type sequences and $1/j$ for the cycle sequence. Eq 2.74 provides the global sequence DDS, from the sequence matrix of distance degrees, **SDD**, which is similar to **LC**, excepting the column $j = 0$ and the zero-columns $j = d(G)+1$, $d(G)+2,...,d(G)+N$. Thus, the **LC** matrix is the joint point of the **LM** and **SM** matrices. Figure 2.25 illustrates some sequence matrices and their spectra for the graph $G_{2.20}$.

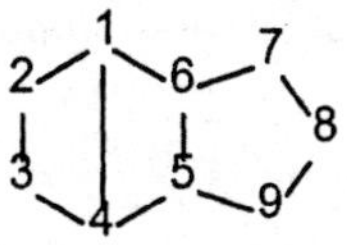

$G_{2.20}$

SAP($G_{2.20}$)

j	1	2	3	4	5	6	7	8
1	3	5	7	6	7	5	3	3
2	2	3	6	8	10	8	7	4
3	2	3	6	8	10	8	7	4
4	3	5	7	6	7	5	3	3
5	3	5	7	7	5	4	3	3
6	3	5	7	7	5	4	3	3
7	2	3	5	8	8	9	5	4
8	2	2	4	8	10	10	6	2
9	2	3	5	8	8	9	5	4

APS = 11.17.27.33.35.31.21.15

SCy($G_{2.20}$)

1-3	4	5	6	7	8	9
0	2	0	1	1	0	1
0	1	0	1	0	0	1
0	1	0	1	0	0	1
0	2	0	1	1	0	1
0	1	1	1	1	0	1
0	1	1	1	1	0	1
0	0	1	0	1	0	1
0	0	1	0	1	0	1
0	0	1	0	1	0	1

CyS = 0.0.0.2.1.1.1.0.1

SDD ($G_{2.20}$)

j	1	2	3	4	5-8
1	3	3	2	0	0
2	2	2	2	2	0
3	2	2	2	2	0
4	3	3	2	0	0
5	3	4	1	0	0
6	3	4	1	0	0
7	2	3	2	1	0
8	2	2	2	2	0
9	2	3	2	1	0

DDS = 11.13.8.4.0.0.0.0

SSP($G_{2.20}$)

1	2	3	4	5-8
3	5	3	0	0
2	3	4	4	0
2	3	4	4	0
3	5	3	0	0
3	5	3	0	0
3	5	3	0	0
2	3	3	3	0
2	2	2	2	0
2	3	3	3	0

SPS = 11.17.14.8.0.0.0.0

SΔD ($G_{2.20}$)

j	1-4	5	6	7	8
1	0	1	1	3	3
2	0	0	0	4	4
3	0	0	0	4	4
4	0	1	1	3	3
5	0	2	1	2	3
6	0	2	1	2	3
7	0	0	1	3	4
8	0	0	0	6	2
9	0	0	1	3	4

ΔDS = 0.0.0.0.3.3.15.15

SLP($G_{2.20}$)

1-4	5	6	7	8
0	1	1	3	3
0	0	0	7	4
0	0	0	7	4
0	1	1	3	3
0	2	1	3	3
0	2	1	3	3
0	0	1	5	4
0	0	0	6	2
0	0	1	5	4

LPS = 0.0.0.0.3.3.21.15

Figure 2.25. Sequence matrices and their spectra for the graph $G_{2.20}$.

It is easily seen that the spectra of *all paths*, *APS*, of *distance degrees*, *DDS*, of *shortest paths*, *SPS*, of *detour degrees*, *ΔDS*, and of *longest paths*, *LPS*, are different from each other in cycle-containing graphs. They are equal to each other in acyclic graphs, by virtue of the uniqueness of the path joining any two vertices.

Layer and sequence matrices can be represented in a *line form*.[70] For the graph $G_{2.18}$, the matrices **LC** and **SDD** can be written as:

$$\textbf{LC } (G_{2.18}) = \{ \ 1 \ (1,1,2,2,1) \ ; \ 2 \ (1,3,2,1) \ ; \ 3 \ (1,3,3) \ ; \ 4 \ (1,2,2,2); \ 5(1,1,1,2,2)$$
$$6(1,1,2,2,1); \ 7(1,1,2,3) \ \}$$
$$\textbf{SDD } (G_{2.18}) = \{ \ 1 \ (1,2,2,1) \ ; \ 2 \ (3,2,1) \ ; \ 3 \ (3,3) \ ; \ 4 \ (2,2,2); \ 5(1,1,2,2)$$
$$6(1,2,2,1); \ 7(1,2,3) \ \}$$

A *canonical* form can be written: the rows are ordered in decreasing length, (as non zero elements) and, at the same length, in lexicographic ordering.[70]

Layer and sequence matrices are used in studies of basic topological properties of the graphs as well as in calculating some topological indices (eg. indices of centrality and centrocomplexity - see Chap. Topological Indices).

* * *

Other matrices. Any topological index, defined on *edge*, can be written as weighted adjacency matrix.[8, 28, 31, 77-79] A *resistance distance matrix* was proposed by Klein et al. [80, 81] in connection with the electrical network theory. A *topological state matrix*, taking into account the paths and chemical identity of vertices was proposed by Hall and Kier.[82] A series of matrices, considering the heteroatoms and stereochemistry was proposed by Schultz et al.[83-91] as extensions of the molecular topological index (see Chap. Topological Indices).

REFERENCES

1. Randić, M.; Woodworth, W. L.; Graovac, A. Unusual Random Walks,
2. Int. J. Quantum Chem. **1983**, 24, 435-452.
3. Randić, M. Generalized Molecular Descriptors, *J. Math.Chem.* **1991**, *7*, 155-168.
4. Sylvester, J. J. On an Application of the New Atomic Theory to the Graphical Representation of the Invariants and Covariants of Binary Quantics - With Three Appendices, *Am. J. Math.* **1874**, *1*, 64-90.
5. Harary, F. *Graph Theory*, Addison-Wesley, Reading, Ma, 1969.
6. Diudea, M. V.; Topan, M.; Graovac, A. Layer Matrices of Walk Degrees,
7. J. Chem. Inf. Comput. Sci. **1994**, 34, 1071 -1078.
8. Rucker, G.; Rucker, C. Counts of all Walks as Atomic and Molecular Descriptors,
9. J. Chem. Inf. Comput. Sci. **1993**, 33, 683-695.
10. 7. Graovac, A.; Babić, D. The Evaluation of Quantum Chemical Indices by the Method
11. of Moments, Int. J. Quantum. Chem., Quantum Chem Symp. **1990**, 24, 251 - 262.

12. 8. Diudea, M. V.; Ivanciuc, O. *Molecular Topology*, Comprex Cluj, 1995 (in Romanian)
13. 9. Gutman, I.; Lee, S. L.; Chu, C. H.; Luo, Y. L. Chemical Applications of the Laplacian
14. Spectrum of Molecular Graphs: Studies of the Wiener Number, *Indian J. Chem.*
15. **1994**, *33A*, 603-608.
16. 10. Gutman, I.; Mohar, B. The Quasi-Wiener and the Kirchhoff Indices Coincide,
17. J. Chem. Inf. Comput. Sci. 1996, 36, 982-985.
18. 11. Ivanciuc, O. The Laplacian Polynomial of Molecular Graphs, *Rev. Roum. Chim.*
19. **1993**, *38*, 1499-1508.
20. Mohar, B. Laplacian Matrices of Graphs, *Prepr. Ser. Dept. Math. Univ. E. K. Ljubljana*, **1988**, *26*, 385-392; MATH/CHEM/COMP 1988, Ed. A.Graovac',(Stud. Phys. Theor. Chem. 63, Elsevier, Amsterdam, 1989), p. 1.
21. Mohar, B.; Babić, D.; Trinajstić, N. A Novel Definition of the Wiener Index for Trees
22. J. Chem. Inf. Comput. Sci. **1993**, 33, 153-154.
23. Trinajstić, N.; Babić, D.; Nikolić, S.; Plavšić, D.; Amić, D.; Mihalić, Z. The Laplacian Matrix in Chemistry, *J. Chem. Inf. Comput. Sci.* **1994**, *34*, 368-376.
24. Biggs, N. L.; Lloyd, E. K.; Wilsen, R. J. *Graph Theory 1736 - 1936* , Oxford
25. Univ.Press, Oxford 1976, pp.133-135.
26. Kirchhoff, G. Über die Auflösung der Gleichungen, auf welche man bei der Untersuchung der linearen Verteilung galvanischer Ströme geführt wird, *Ann. Phys. Chem.* **1847**, *72*, 497-508.
27. Cvetković, D. M.; Doob, M.; Sachs, H. *Spectra of Graphs*, Acad. Press, New York, 1980.
28. Diudea, M. V. Walk Numbers $^e W_M$: Wiener-Type Numbers of Higher Rank,
29. J. Chem. Inf. Comput. Sci. **1996**, 36, 535-540.
30. Amić, D.; Trinajstić, N. On the Detour Matrix, *Croat. Chem. Acta*, **1995**, *68*, 53-62.
31. Ivanciuc, O.; Balaban, A. T. Design of Topological Indices.Part. 8. Path Matrices
32. and Derived Molecular Graph Invariants, *Commun. Math. Comput. Chem. (MATCH)*,
33. **1994**, *30*, 141-152.
34. Crippen, G. M. A Novel Approach to Calculation of Conformation: Distance Geometry, *J. Comput. Phys.* **1977**, *24*, 96-107.
35. Diudea, M. V.; Horvath, D.; Graovac, A. 3D-Distance Matrices and Related
36. Topological Indices, *J. Chem. Inf. Comput. Sci.* **1995**, *35*, 129 -135.
37. Randić, M. Molecular Topographic Descriptors, *Studies Phys. Theor. Chem.* **1988**, *54*, 101-108.
38. Randić, M.; Kleiner, A. F.; De Alba, L. M. Distance-Distance Matrices for Graph
39. Embeded on 3-Dimensional Lattices, *J. Chem. Inf. Comput. Sci.* **1994**, *34*, 277-286.
40. Diudea, M. V.; Katona, G.; Lukovits, I.; Trinajstić, N. Detour and Cluj-Detour Indices, *Croat. Chem. Acta* **1998**, *71*, 459-471.
41. Lukovits, I.; Razinger, M. On Calculation of the Detour Index, *J. Chem. Inf. Comput. Sci.* **1997**, *37*, 283-286.

42.   Klein, D. J.; Lukovits, I.; Gutman, I. On the Definition of the Hyper-Wiener Index for Cycle-Containing Structures. *J. Chem. Inf. Comput. Sci.* **1995**, *35,* 50-52.
43.   Randić, M.; Guo, X.; Oxley, T.; Krishnapriyan, H. Wiener Matrix: Source of Novel Graph Invariants, *J. Chem. Inf. Comput. Sci.* **1993**, *33,* 700-716.
44.   Randić, M.; Guo, X.; Oxley, T.; Krishnapriyan, H.; Naylor, L. Wiener Matrix
45.   Invariants, *J. Chem. Inf. Comput. Sci.* **1994**, *34,* 361-367.
46.   Wiener, H. Structural Determination of Paraffin Boiling Point, *J. Amer. Chem. Soc.* **1947**, *69***,** 17-20.
47.   Randić, M. Novel Molecular Descriptor for Structure-Property Studies,
48.   Chem. Phys. Lett. **1993**, 211, 478-483.
49.   32. Gutman, I. A Formula for the Wiener Number of Trees and Its Extension to Graphs Containing Cycles, *Graph Theory Notes New York,* **1994**, *27,* 9-15.
50.   33. Diudea, M. V.; Minailiuc, O.; Katona, G.; Gutman, I. Szeged Matrices and Related Numbers, *Commun. Math. Comput. Chem.(MATCH),* **1997**, *35,* 129-143.
51.   34. Dobrynin, A. A.; Gutman, I. On a Graph Invariant Related to the Sum of all Distances in a Graph, *Publ. Inst. Math. (Beograd),* **1994**, *56,* 18-22.
52.   35. Dobrynin, A. A.; Gutman, I.; Dömötör, G. A Wiener-Type Graph Invariant for Some Bipartite Graphs, *Appl. Math. Lett.* **1995**, *8,* 57-62.
53.   36. Gutman, I.; Klavžar, S. An Algorithm for the Calculation of the Szeged Index of   Benzenoid Hydrocarbons, *J. Chem. Inf. Comput. Sci.* **1995**, *35,* 1011-1014.
54.   37. Khadikar, P. V.; Deshpande, N. V.; Kale, P. P.; Dobrynin, A. A.; Gutman, I.; Dömötör, G.   The Szeged Index and an Analogy with the Wiener Index, *J. Chem. Inf. Comput. Sci.* **1995**, *35,* 547-550.
55.   38. Gutman, I.; Dobrynin, A. A. The Szeged Index – A Success Story,
56.   Graph Theory Notes, N.Y., **1998**, 34, 37-44.
57.   39. Diudea, M. V. Cluj Matrix, CJ$_u$: Source of Various Graph Descriptors, *Commun. Math Comput. Chem. (MATCH),* **1997**, *35,* 169-183.
58.   40. Diudea, M. V.; Pârv, B.; Topan, M. I.   Derived Szeged and Cluj Indices,
59.   J. Serb. Chem. Soc. **1997**, *62***,** 267-276.
60.   41. Diudea, M. V. Cluj Matrix Invariants, *J. Chem. Inf. Comput. Sci.* **1997**, *37,* 300-305
61.   42. Horn, R. A.; Johnson, C. R. *Matrix Analysis*; Cambridge Univ. Press, Cambridge, **1985**.
62.   43. Diudea, M. V.; Gutman, I. *Croat. Chem. Acta,* **1998**, *71,* 21-51.
63.   44. Randic, M. Generalized Molecular Descriptors, *J. Math. Chem.* **1991**, *7,* 155-168.
64.   45. Randic, M.; Mihalic, Z.; Nikolic, S.; Trinajstic, N. Graphical Bond Orders: Novel
65.   Structural Descriptors, *J. Chem. Inf. Comput. Sci.* **1994**, *34,* 403-409.
66.   46. Randic, M. Hosoya Matrix - A Source of New molecular Descriptors, *Croat. Chem. Acta.* **1994**. *34.* 368-376.

67. 47. Hosoya, H. Topological Index. A Newly Proposed Quantity Characterizing the Topological Nature of Structural Isomers of Saturated Hydrocarbons. *Bull. Chem. Soc. Jpn.* **1971**, *44*, 2332-2339.

68. 48. Plavšić, D.; Šošić, M.; Landeka, I.; Gutman, I.; Graovac, A. On the Relation between the Path Numbers ^1Z and ^2Z and the Hosoya Z Index, *J. Chem. Inf. Comput. Sci.* **1996**, *36*, 1118-1122.

69. 49. Plavšić, D.; Šošić, M.; Đaković, Z.; Gutman, I.; Graovac, A. Extension of the Z Matrix to Cycle-containing and Edge-Weighted Molecular Graphs, *J. Chem. Inf. Comput. Sci.* **1997**, *37, 529-534.*

70. 50. Diudea, M. V.; Parv, B.; Gutman, I. Detour-Cluj Matrix and Derived Invariants,
71. J. Chem. Inf. Comput. Sci. **1997**, 37, 1101-1108.

72. 51. Gutman, I.; Diudea, M. V. Defining Cluj Matrices and Cluj Matrix Invariants,
73. J. Serb. Chem. Soc. **1998** 63, 497-504.

74. 52. Diudea, M. V.; Katona, G. Molecular Topology of Dendrimers,
75. in *Advances in Dendritic Macromolecules*, Ed. G.A. Newkome, JAI Press Inc.,
76. Stamford, Con. 1999, vol.4, pp. 135-201.

77. 53. Tratch, S. S.; Stankevich, M. I.; Zefirov, N. S. Combinatorial Models and Algorithms in Chemistry. The Expanded Wiener Number- a Novel Topological Index,
78. J. Comput. Chem. **1990**, 11, 899-908.

79. 54. Ivanciuc, O.; Balaban, T. S.; Balaban, A. T.; Reciprocal Distance Matrix, Related Local Vertex Invariants and Topological Indices, *J. Math Chem.* **1993**, *12*, 309-318.

80. 55. Plavšić, D.; Nikolić, S.; Trinajstić, N.; Mihalić, Z. On the Harary Index for the
81. Characterization of Chemical Graphs, *J. Math. Chem.* **1993**, *12*, 235-250.

82. 56. Diudea, M. V. Indices of Reciprocal Property or Harary Indices,
83. J. Chem. Inf. Comput. Sci. **1997**, 37, 292-299.

84. 57. Diudea, M. V. Valencies of Property, *Croat. Chem. Acta*, (in press)

85. 58. Diudea, M. V. Wiener and Hyper-Wiener Numbers in a Single Matrix, *J. Chem. Inf. Comput. Sci.* **1996**, *36*, 833-836

86. 59. Dobrynin, A. A.; Kochetova, A. A. Degree Distance of a Graph: A Degree Analogue of the Wiener Index, *J. Chem. Inf. Comput. Sci.* **1994**, *34*, 1082-1086.

87. 60. Estrada, E.; Rodriguez, L.; Gutiérrez, A. Matrix Algebraic Manipulation of Molecular Graphs. 1. Distance and Vertex-Adjacency Matrices, *Commun. Math.Comput. Chem. (MATCH)*, **1997**, *35*, 145-156.

88. 61. Schultz, H. P. Topological Organic Chemistry. 1. Graph Theory and Topological Indices of Alkanes. *J. Chem. Inf. Comput. Sci.* **1989**, *29*, 227-228.

89. 62. Randić, M. Restricted Random Walks on Graphs, *Theor. Chim. Acta,* **1995**, *92*, 97-106.

90. 63. Diudea, M. V.; Randić, M. Matrix Operator, $W_{(M1,M2,M3)}$ and Schultz-Type Numbers, *J. Chem. Inf. Comput. Sci.* **1997**, *37*, 1095-1100.

91. 64. Diudea, M. V. Novel Schultz Analogue Indices, *Commun. Math. Comput. Chem.*
92. *(MATCH)*, **1995**, *32*, 85-103.

93.   65. Diudea, M. V.; Pop, C. M. A Schultz-Type Index Based on the Wiener Matrix, *Indian J. Chem.* **1996**, *35A*, 257-261.

94.   66. Estrada, E.; Rodriguez, L. Matrix Algebraic Manipulation of Molecular Graphs.

95.   Harary- and MTI-Like Molecular Descriptors, *Commun. Math.Comput. Chem.*

96.   *(MATCH)*, **1997**, *35*,157-167.

97.   67. Balaban, A. T.; Diudea, M. V. Real Number Vertex Invariants: Regressive Distance Sums and Related Topological Indices, *J. Chem. Inf. Comput. Sci.* **1993**, *33*, 421-428.

98.   68. Diudea, M. V.; Minailiuc, O. M.; Balaban, A. T. Regressive Vertex Degrees (New Graph Invariants) and Derived Topological Indices, *J. Comput. Chem.* **1991**, *12*, 527-535.

99.   69. Dobrynin, A. Degeneracy of Some Matrix Invariants and Derived Topological Indices, *J. Math. Chem.* **1993**, *14*, 175 - 184.

100. 70. Skorobogatov, V. A.; Dobrynin, A. A. Metric Analysis of Graphs, *Commun. Math Comput. Chem. (MATCH)*, **1988**, *23*, 105-151.

101. 71. Bonchev, D.; Mekenyan, O.; Balaban, A. T. Iterative Procedure for the Generalized Graph Center in Polycyclic Graphs, *J. Chem. Inf. Comput. Sci.* **1989**, *29*, 91-97.

102. 72. Halberstam, F. Y.; Quintas, L.V. Distance and Path Degree Sequences for Cubic Graphs, Pace University, New York, **1982**, A Note on Table of Distance and Path Degree Sequences for Cubic Graphs, Pace University New York, **1982**.

103. 73. Randić, M. On Characterization of Molecular Branching, *J. Amer. Chem. Soc.*

104. **1975**, *97*, 6609-6615.

105. 74. Randić, M.; Wilkins, C. L. Graph-Theoretical Ordering of Structures as a Basis for Systematic Searches for Regularities in Molecular Data, *J. Phys. Chem.* **1979**, *83*, 1525-1540.

106. 75. Diudea, M. V. Layer Matrices in Molecular Graphs, *J. Chem. Inf. Comput. Sci.*

107. **1994**, *34*, 1064 -1071.

108. 76. Morgan, H. The Generation of a Unique Machine Description for Chemical Structures. A Technique Developed at Chemical Abstracts Service, *J. Chem. Doc.* **1965**, *5*, 107-113.

109. 77. Diudea, M. V.; Minailiuc, O. M.; Katona, G. Novel Connectivity Descriptors Based on Walk Degrees, *Croat. Chem. Acta,* **1996**, *69*, 857-871.

110. 78. Diudea, M. V.; Minailiuc, O. M.; Katona, G. SP Indices: Novel Connectivity Descriptors. *Rev. Roum. Chim.* **1997**, *42*, 239-249.

111. 79. Randić, M.; Mihalić, Z.; Nikolić, S.; Trinajstić, N. Graphical Bond Orders; Novel Structural Descriptors, *J. Chem. Inf. Comput. Sci.* **1994**, *34*, 403-409.

112. 80. Klein, D. J.; Randić, M. Resistance Distance, *J. Math. Chem.* **1993**, *12*, 81-95.

113. 81. Bonchev, D.; Balaban, A. T.; Liu, X.; Klein, D. J. Molecular Cyclicity and Centricity of Polycyclic Graphs.I. Cyclicity Based on Resistance Distance or Reciprocal Distance. *Int. J. Quantum Chem.* **1994**, *50*, 1-20.

114. 82. Hall, L. H.; Kier, L. B. Determination of Topological Equivalence in Molecular Graphs from the Topological State, *Quant. Struct.-Act. Relat.* **1990**, *9*, 115-131.

115. 83. Schultz, H. P.; Schultz, E. B.; Schultz, T. P. Topological Organic Chemistry. 2. Graph Theory, Matrix Determinants and Eigenvalues, and Topological Indices of Alkanes, *J. Chem. Inf. Comput. Sci.* **1990**, *30*, 27-29.

116. 84. Schultz, H. P.; Schultz, T. P. Topological Organic Chemistry.3. Graph Theory, Binary and Decimal Adjacency Matrices, and Topological Indices of Alkanes, *J. Chem. Inf. Comput. Sci.* **1991**, *31*, 144-147.

117. 85. Schultz, H. P.; Schultz, E. B.; Schultz, T.P. Topological Organic Chemistry. 4. Graph Theory, Matrix Permanents, and Topological Indices of Alkanes, *J. Chem. Inf. Comput. Sci.* **1992**, *32*, 69-72.

118. 86. Schultz, H. P.; Schultz, T. P. Topological Organic Chemistry. 5. Graph Theory, Matrix Hafnians and Pfaffnians, and Topological Indices of Alkanes, *J. Chem. Inf. Comput. Sci.* **1992**, *32*, 364-366.

119. 87. Schultz, H. P.; Schultz, T. P., Topological Organic Chemistry. 6. Theory and Topological Indices of Cycloalkanes, *J. Chem. Inf. Comput. Sci.* **1993**, *33*, 240-244.

120. Schultz, H. P.; Schultz, E. B.; Schultz, T.P. Topological Organic Chemistry.7. Graph Theory and Molecular Topological Indices of Unsaturated and Aromatic Hydrocarbons, *J. Chem. Inf. Comput. Sci.* **1993**, *33*, 863-867.

121. Schultz, H. P.; Schultz, E. B.; Schultz, T. P., Topological Organic Chemistry. 8.

122. Graph Theory and Topological Indices of Heteronuclear Systems, *J. Chem. Inf. Comput. Sci.* **1994,** *34*, 1151-1157.

123. Schultz, H. P.; Schultz, E. B.; Schultz, T. P., Topological Organic Chemistry.9.

124. Graph Theory and Molecular Topological Indices of Stereoisomeric Compounds, *J. Chem. Inf. Comput. Sci.* **1995**, *35*, 864-870.

125. Schultz, H. P.; Schultz, E. B.; Schultz, T. P., Topological Organic Chemistry.10.

126. Graph Theory and Topological Indices of Conformational Isomers, *Chem. Inf. Comput. Sci.* **1996,** *36*, 996-1000.

Chapter 3

POLYNOMIALS IN CHEMICAL GRAPH THEORY

3.1. INTRODUCTION

Why Polynomials?

There are two main routes by which polynomials enter into chemical graph theory.
First, in *quantum chemistry* the (approximate) solution of the Schrödinger equation

$$\hat{H}\Psi_j = E_j \cdot \Psi_j, \quad j = 1, 2, 3, \ldots, \tag{3.1}$$

is usually reduced to the finding of eigenvalues and eigenvectors of the so-called Hamiltonian matrix (which, in turn, is the Hamiltonian operator $\hat{H}$ represented within some finite vector basis). Now, if $\mathbf{H}$ is such a Hamiltonian matrix, then its eigenvalues are approximately equal to the energies E_1, E_2, E_3,..., occurring in eq 3.1. These eigenvalues are the solutions of the so-called *secular equation*

$$\det[\varepsilon\mathbf{I} - \mathbf{H}] = \tag{3.2}$$

where $\mathbf{I}$ stands for the unit matrix of a pertinent order.
The left hand side of (3.2), namely

$$\det[\varepsilon\mathbf{I} - \mathbf{H} \tag{3.3}$$

is just a polynomial in the indeterminate ε. The degree of this polynomial (N) is equal to the dimension of the vector space in which the Hamiltonian operator $\hat{H}$ is represented, and is also equal to the order of the Hamiltonian matrix $\mathbf{H}$.

In quite a few approximations encountered in quantum chemistry, the Hamiltonian matrix is somehow related to a molecular graph. The best known, and simplest, example is found in the Hückel molecular orbital theory:

$$\mathbf{H} = \alpha_{H\,MO}\cdot\mathbf{I} + \beta_{H\,MO}\cdot\mathbf{A}(G)$$

where $\mathbf{A}(G)$ is the adjacency matrix of a pertinently constructed skeleton graph (often called "Hückel graph", representing the π-electron network of a conjugated hydrocarbon, [1-3] whereas $\alpha_{H\,MO}$ and $\beta_{H\,MO}$ are parameters of the Hückel theory (not to be confused with the polynomials α and β considered in the later parts of this chapter). In this case the polynomial (3.3) is equal to

$$\beta_{HMO}{}^{N} \cdot \det\left[\frac{\varepsilon - \alpha_{H\,MO}}{\beta_{H\,MO}}\mathbf{I} - \mathbf{A}(G)\right]$$

the non-trivial part of which is

$$\varphi(G, \lambda) = \det[\lambda\mathbf{I} - \mathbf{A}(G)] \tag{3.4}$$

with the indeterminate λ standing instead of $(\varepsilon - \alpha_{H\,MO})/\beta_{H\,MO}$. The polynomial (3.4) is called the *characteristic polynomial* of the graph G. It is certainly the most popular and most extensively studied graph polynomial in chemical graph theory.

Consider, as an example, the graph $G_{3.1}$ depicted in Figure 3.1. It has eight vertices ($N = 8$, labeled by 1, 2, ..., 8) and seven edges ($m = 7$, labeled by a, b, ..., g). Then

$$\varphi(G_{3.1}, \lambda) = \det\begin{bmatrix} \lambda & 0 & -1 & 0 & 0 & 0 & 0 & 0 \\ 0 & \lambda & -1 & 0 & 0 & 0 & 0 & 0 \\ -1 & -1 & \lambda & -1 & 0 & 0 & 0 & 0 \\ 0 & 0 & -1 & \lambda & -1 & 0 & 0 & 0 \\ 0 & 0 & 0 & -1 & \lambda & -1 & 0 & 0 \\ 0 & 0 & 0 & 0 & -1 & \lambda & -1 & -1 \\ 0 & 0 & 0 & 0 & 0 & -1 & \lambda & 0 \\ 0 & 0 & 0 & 0 & 0 & -1 & 0 & \lambda \end{bmatrix}$$

Either by direct expansion of this determinant or (better) by some of the numerous known techniques for the calculation of the characteristic polynomial (see below) it is not too difficult to obtain:

$$\varphi(G_{3.1}, \lambda) = \lambda^8 - 7\lambda^6 + 13\lambda^4 - 4\lambda^2 \tag{3.5}$$

It is then an easy exercise in calculus to find the zeros of this polynomial, namely the roots of the equation $\varphi(G_{3.1}, \lambda) = 0$. These eight zeros read:

$$\lambda_1 = 2; \ \lambda_2 = \sqrt{\frac{3+\sqrt{5}}{2}} \ ; \ \lambda_3 = \sqrt{\frac{3-\sqrt{5}}{2}} \ ; \ \lambda_4 = 0; \tag{3.6}$$

$$\lambda_5 = -2; \ \lambda_6 = -\sqrt{\frac{3+\sqrt{5}}{2}} \ ; \ \lambda_7 = -\sqrt{\frac{3-\sqrt{5}}{2}} \ ; \ \lambda_8 = 0;$$

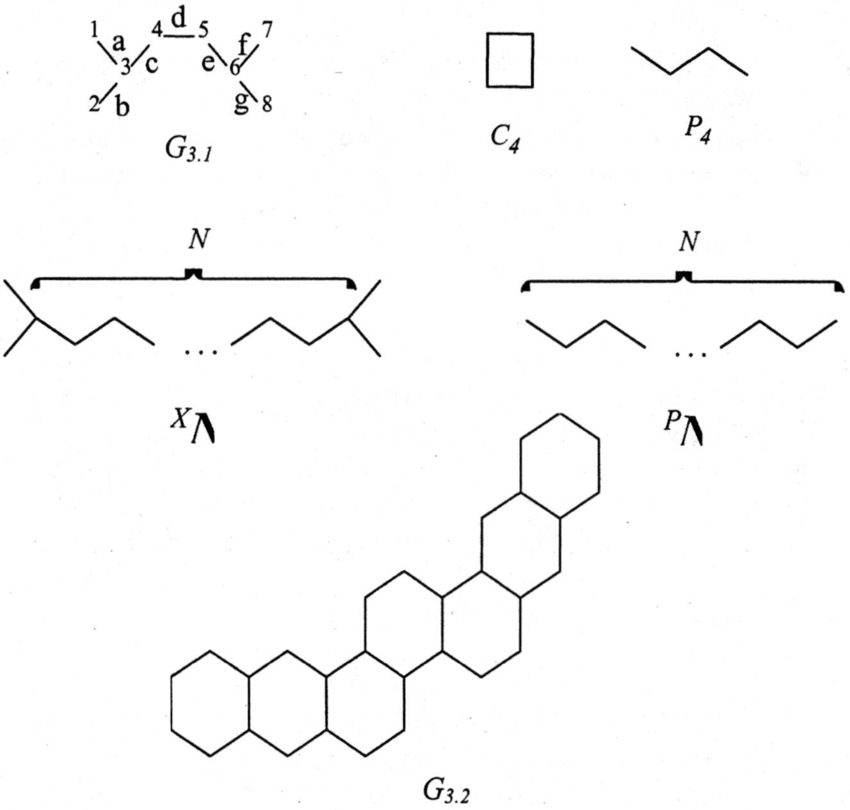

Figure 3.1. Examples illustrating the unusual connections between molecular graph, revealed by means of graph polynomials; for details see text.

Various modifications of φ have been put forward in the chemical literature, for instance the matching polynomial, [4-9] the μ-polynomial [10-12] and the β-polynomial, [13-15] defined and discussed at a later point. These could be understood as the constituents of the secular equations, eq 3.3, of some, appropriately modified, Hamiltonian operators. Instead of the determinant in eq. 3.4, some authors considered the analogous expression with the permanent [16, 17] - the permanental polynomial. Recently a more the general class of so-called immanantal polynomials attracted the attention of researchers, [18-20] of which the characteristic and the permanental polynomials are special cases.

Second, in numerous, both chemical and non-chemical, applications of graph theory one often encounters finite sequences of certain graph invariants, all associated to the same graph. Suppose $C = (C_0, C_1, C_2, ..., C_p)$ is such a sequence. Then instead of $p+1$ distinct quantities C_k, $k = 0, 1, 2, ...p$, one could introduce a single quantity - a polynomial - defined as

$$C_p \lambda^p + ... + C_2 \lambda^2 + C_1 \lambda + C_0 \equiv \sum_{k=0}^{p} C_k \lambda^k \qquad (3.7)$$

Needless to say that (3.7) is not the only possible form which a polynomial associated with the sequences C may be given.

The polynomial (3.7) contains precisely the same information as the sequence C. In some cases, however, it is easier to work with a polynomial than with a sequence. In some other cases, certain collective properties of the invariants considered, namely properties which can be deduced only by simultaneously taking into account the values of all C_k, $k = 0, 1, 2, ..., p$, are in a natural way deduced from the polynomial. To say the same in a more direct way: there are collective properties of sequences of graph invariants which hardly ever would be discovered without analyzing graph polynomials of the form (3.7).

To illustrate the above, consider so-called independent edge sets of the graph $G_{3.1}$ (see Figure 3.1). A collection of edges of a graph is said to be independent if no two edges have a vertex in common. It is reasonable to classify the independent edge sets according to the number of edges they contain. In the case of $G_{3.1}$ no four edges are independent (and therefore there are no independent edge sets with more than three edges). There are four distinct independent edge sets containing 3 edges:

$\{a, d, f\}$ $\{a, d, g\}$ $\{b, d, f\}$ $\{b, d, g\}$

and thirteen such sets containing 2 edges:

$$\{a, d\} \ \{a, e\} \ \{a, f\} \ \{a, g\}$$
$$\{b, d\} \ \{b, e\} \ \{b, f\} \ \{b, g\}$$
$$\{c, e\} \ \{c, f\} \ \{c, g\} \ \{d, f\}$$
$$\{d, g\}$$

Formally speaking, each set containing a single edge is also an independent edge set. Clearly, $G_{3.1}$ has seven such sets. The empty set may be viewed as a independent edge set (of any graph) with zero edges; this set is unique.

Denote by $m(G, k)$ the number of k-element independent edge sets of the graph G. Then $m(G_{3.1}, 0) = 1$, $m(G_{3.1}, 1) = 7$, $m(G_{3.1}, 2) = 13$, $m(G_{3.1}, 3) = 4$, $m(G_{3.1}, 5) = 0$, $m(G_{3.1}, 6) = 0$, $m(G_{3.1}, 7) = 0$, etc. The sequence thus obtained is infinite, but it is reasonable to end it at the value of k for which $m(G, k) \neq 0$, $m(G, k+1) = 0$. We thus arrive at a finite sequence (1, 7, 13, 4) which by (3.7) is transformed into the cubic polynomial

$$Q(G_{3.1}, \lambda) = 1 + 7\lambda + 13\lambda^2 + 4\lambda^3 \tag{3.8}$$

When a graphic polynomial is defined as in the above example, then it is fully obscure whether its zeros have any distinguished property. Yet, all the (three) zeros of the above polynomial are negative, real-valued numbers (which the readers could check relatively easily). The same collective property of the sequence ($m(G,k)$, $k = 0, 1, 2, ...$) holds in the case of all graphs G: the zeros of all polynomials of the form

$$Q(G) = Q(G,\lambda) = \sum_{k \geq 0} m(G,k)\lambda^k \tag{3.9}$$

are negative, real-valued numbers.

At a later point we shall see that the zeros of this graph polynomials are quite important in theoretical chemistry.

$Q(G,\lambda)$, (3.9), has been introduced by Hosoya [21] and called *Z-counting polynomial*.

More Motivations for Graph Polynomials

Some properties of the graph polynomials are trivial and obvious. For instance, such is the fact that the value of the Z-counting polynomial, eq 3.9, at $\lambda = 1$ is equal to the Hosoya topological index Z. Recall that this topological index is just defined as

$$Z(G) = \sum_{k \geq 0} m(G,k)$$

The fact that the zeros of $\varphi(G_{3.1},\lambda)$, eq 3.6, occur in pairs (x, $-x$) is a manifestation of one of the first general results of chemical graph theory ever obtained - the famous Coulson-Rushbrooke pairing theorem. [22, 23] Although far from being a trivial feature, the pairing of the numbers (3.6) should be no surprise to a reader of this book.

In many instances, however, by means of graph polynomials some quite unusual connections between (molecular) graphs can be envisaged. We illustrate this by a few examples.

The polynomial (3.5) can be factorized as:

$$\varphi(G_{3.1},\lambda) = (\lambda^4 - 4\lambda^2)(\lambda^4 - 3\lambda^2 + 1)$$

Each of these factors is a characteristic polynomial itself: ($\lambda^4 - 4\lambda^2$) is the characteristic polynomial of the 4-membered cycle, C_4, see Figure 3.1, whereas $\lambda^4 - 3\lambda^2 + 1$ is the characteristic polynomial of the path graph with 4 vertices, P_4, see Figure 3.1. As a consequence, the set of eigenvalues of $G_{3.1}$, eq 3.6, is just the union of the set of eigenvalues of C_4 and P_4. The eigenvalues of C_4 are +2, -2, 0 & 0. The eigenvalues of P_4 are $\pm\sqrt{3 \pm \sqrt{2}}$.

This observations is, in fact, a special case (for $N = 4$) of a more general result: [24-26]

$$\varphi(X_N, \lambda) = \varphi(C_4,\lambda)\varphi(P_N,\lambda) \tag{3.10}$$

From eq 3.10, we see that the two-component graph consisting of a copy of C_4 and a copy of P_N has the same characteristic polynomial as the graph X_N. Thus we encountered an infinite family of pairs of non-isomorphic graphs with coinciding characteristic polynomials. (With regard to this so-called *isospectrality* property of graphs, which is not duly discussed in this chapter, see Refs. [27-30]).

It is somewhat less obvious that the polynomials (3.5) and (3.8) are closely related. Indeed, for $i = \sqrt{-1}$, we have that

$$\lambda^4 \varphi\left(G, \frac{i}{\sqrt{\lambda}}\right)$$

is equal to the Z-counting polynomial, eq 3.8. An analogous result holds for all n-vertex acyclic graphs: [31]

$$\left(-i\sqrt{\lambda}\right)^n \varphi\left(G, \frac{i}{\sqrt{\lambda}}\right) = Q(G, \lambda) \tag{3.11}$$

A still less obvious result is that $Q(G_{3.1}, 1) = 1+7+13+4 = 25$ is equal to the number of Kekulé structures of the benzenoid hydrocarbon $G_{3.2}$, shown in Figure 3.1. This, again, is a special case of a more general finding:[32] The sextet polynomial of every unbranched catacondensed benzenoid molecule coincides with the Z-counting polynomial of a certain graph (called *Gutman tree*).[33, 34] Because the sum of the coefficients of the sextet polynomial is equal to the Kekulé structure count, [35-38] it follows that the Hosoya index of the *Gutman tree* is equal to the number of Kekulé structures of the corresponding benzenoid system; in our example, $G_{3.1}$ is the *Gutman tree* of the benzenoid hydrocarbon $G_{3.2}$. More details can be found elsewhere. [39, 40]

If we combine all the above examples, then we arrive at the fully unexpected conclusion that the number of Kekulé structures of the benzenoid hydrocarbon $G_{3.2}$ can be computed from the characteristic polynomials of the cycle C_4 and the path graph P_4, both of which are determinants of order four.

Chemical graph theory is full of such unusual connections, which are not only useful and stimulating for the underlying chemical theories, but also represent a great satisfaction to those who

work on them. Since relations of this kind are continuously being discovered until the most recent times, there is no danger that this field of research has been exhausted.

Concerning Bibliography

Before starting with the discussion on some particular polynomials of interest in chemical graph theory, a few words should be said about the published scientific works in this field. They are legion!

Producing a complete or, at least, nearly complete bibliography of papers dealing with graph polynomials would hardly be a feasible task. Such a bibliography would have to include many thousands of articles, published in journals devoted to chemistry, mathematics, physics, computer sciences, engineering, medicine, pharmacology, environmental sciences, ... The references given at the end of this chapter, although

quite numerous, are intended only to mention a few (perhaps most significant) articles, reviews and books, and to direct the interested reader towards a more
extensive literature search.

Many books are either fully or to a great extent concerned with graph polynomials, primarily with the characteristic polynomial (both ordinary and Laplacian). [1-3, 39, 41-55] Of the reviews dealing with graph polynomials we mention a few. [20, 33, 34, 40, 56-77] Many of these books and reviews contain tables of graph polynomials and/or their zeros; additional tables are found Refs. [78-83]. An almost complete list of mathematical papers concerned with the characteristic polynomial of graphs has been collected in the book [42] and was eventually updated. [46]

Details Omitted

The amount of material presented in this section had to be drastically limited (otherwise the text on graph polynomials would embrace several thick volumes). Therefore some topics, intimately related to graph polynomials are here abandoned. These are the following:

Chemical theories in which graph polynomials find applications are not outlined.

Applications of graph polynomials in various fields of chemistry, physical chemistry and physics are either not discussed at all, or are mentioned briefly, without going into any detail.

The extensively developed theory of graph eigenvalues (both regular and Laplacian) is almost completely omitted. The same applies to graph eigenvectors

Not all chemically interesting graph polynomials, but only a selection thereof, is considered. Only the most important properties of these polynomials are stated and, sometimes, illustrated by examples. In not a single case a mathematical proof of these properties is given.

Only a limited number of algorithms for the calculation of the graph polynomials is presented.

The theory of cospectral, comatching, etc. graph (namely families of graphs having equal characteristic, matching, etc. polynomials) is not elaborated, in spite of the enormous work done on this problem; some characteristic results in this field are communicated in Refs. [27-30, 84, 85].

Also not mentioned is the work on spectral moments. The kth spectral moment of a graph is the sum of the kth powers of the zeros of the characteristic polynomial. By means of the classic Newton identities, from the spectral moments one can compute the coefficients of the characteristic polynomial, and vice versa; for details see, for instance, Appendix 4 in the book. [44]

The authors believe that all these shortcomings are compensated by quoting literature sources from which the interested reader can get information on the details omitted.

3.2. THE CHARACTERISTIC POLYNOMIAL. PART 1.

The characteristic polynomial, denoted by $\varphi(G, \lambda)$ or $\varphi(G)$, is defined via eq 3.4. It is certainly the most extensively studied graph polynomial, both in mathematics and in chemical graph theory. Its theory has been reviewed on countless places (e. g. see Refs. [1, 3, 42, 43, 44, 48, 49, 51, 56, 57, 60, 68, 73, 75, 86, 87]). Its popularity among mathematical chemists comes from the fact (first observed by Günthard and Primas [88] in 1956) that the Hamiltonian matrix of the Hückel molecular orbital (HMO) theory is a simple linear function of the adjacency matrix of the corresponding molecular graph G.[1, 3, 43, 48, 56, 75, 86, 87] Consequently, each HMO π-electron energy level is a linear function of the corresponding zero of the characteristic polynomial of G.

It is less well known that Heilbronner at al. have developed a theory [89, 90] in which the zeros of the characteristic polynomial of the line graph of the hydrogen-filled molecular graph are in a linear manner related to the σ-electron energy levels of the corresponding saturated hydrocarbon. (Recall that in hydrogen-filled molecular graphs vertices represent both carbon and hydrogen atoms).

The Harary Theorem

Let G be a graph on N vertices. Then its characteristic polynomial $\varphi(G)$ is of degree N and can be written as:

$$\varphi(G, \lambda) = \sum_{k=0}^{N} a_k(G) \lambda^{N-k} \tag{3.12}$$

Hence $a_0(G)$, $a_1(G)$, $a_2(G)$, ..., $a_N(G)$ are the coefficients of the characteristic polynomial of the graph G. For all graphs, $a_0(G) = 1$.

The central result in the theory of the characteristic polynomial is the Harary theorem. It determines how $a_k(G)$, $k = 1, 2, ..., N$, depend on the structure of the graph G.

First a few historical remarks.

Many authors have tried to express the dependence of the coefficients a_k on the structure of the underlying graph. The best known among these (unsuccessful) attempts are that of Samuel [91] in 1949 and Coulson [92] in 1950. The structure-dependency of the determinant of the adjacency matrix of a graph was discovered by Frank Harary [93, 94] in 1962. From this result the coefficient-theorem follows straightforwardly; recall that det $\mathbf{A}(G) = (-1)^N a_N(G)$.

The explicit statement of the actual theorem was discovered in 1964 practically independently by Horst Sachs [95] (a mathematician), Mirko Milić [96] (an electrical engineer) and Leonard Spialter [97] (a computer chemist active in chemical documentation). Eventually several other scholars arrived at the same result (for details see p. 36 of Ref. [42]).

Until 1972 the theorem was not known to theoretical chemists. Then it was *discovered* (in the library) and formulated in a manner understandable to chemists. [86] The authors of the paper [86] were not careful enough and attributed the result solely to Sachs, naming it the *Sachs theorem*. Because of this mistake, in the subsequently published chemical literature the result was almost exclusively referred to as the *Sachs theorem*. Attempts to rectify the mistake came much later. [98]

Anyway, in what follows we speak of the *Harary theorem*.

The cycle C_N on N vertices, $N \geq 3$, is a connected graph whose all vertices are of degree two (i.e. each vertex has exactly two first neighbors). Denote by K_2 the connected graph on two vertices; this graph may be viewed as the two-vertex complete graph or the two-vertex path graph. A graph whose all components are cycles and/or K_2-graphs is called a Sachs graph. (We keep here the nowadays commonly accepted name *Sachs graph*, although *Harary graph* would, probably, be more justified.)

Consider a graph G on n vertices and let its characteristic polynomial be of the form (3.12).

Theorem 3.1 (Harary, Sachs, Milić, Spialter). Let S be a Sachs graph with $N(S)$ vertices, possessing a total of $p(S)$ components, of which $c(S)$ are cycles and $p(S) - c(S)$ are K_2-graphs. Then for $k = 1, 2, ..., N$,

$$a_k(G) = \sum_S (-1)^{p(S)} 2^{c(S)} \tag{3.13}$$

where the summation goes over all Sachs graphs S for which $N(S) = k$ and which are contained (as subgraphs) in the graph G. If there are no such Sachs graphs, then $a_k=0$.

Example 3.1. We illustrate the Harary theorem on the example of the molecular graph G_3 depicted in Figure 3.2. This graph contains (as subgraphs) two cycles, C_3 and C_5; fortunately for us, these cycles have no vertex in common which significantly simplifies the application of formula (3.13). The nine edges of $G_{3.3}$ are labeled by $a, b, c, ..., h, i$. Each edge (together with its two end-vertices) corresponds to a K_2-graph.

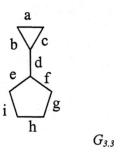

$G_{3.3}$

Figure 3.2. A molecular graph on which the application of the Harary theorem, eq 3.13, is illustrated; contrary to claims by many authors, already in this case it is not easy to perform the actual computation (see text); for molecular graphs with more vertices and cycles the computation of the coefficients of the characteristic polynomial by means of eq 3.13 becomes a hopelessly difficult task

Case k = 1. There are no Sachs graphs with one vertex. Therefore, $a_1(G_{3.3}) = 0$;

Case $k = 2$. The Sachs graphs with two vertices necessarily have one component which is a K_2-graph. In the case of $G_{3,3}$ there are nine such Sachs graphs, corresponding to the nine edges. Therefore, $a_2(G_{3,3}) = 9 \times [(-1)^1 2^0] = -9$.

Case $k = 3$. The Sachs graphs with three vertices necessarily have one component which is a triangle (C_3). The graph $G_{3,3}$ contains one such Sachs graph, and therefore $a_3(G_{3,3}) = [(-1)^1 2^1] = -2$

Case $k = 4$. The Sachs graphs with four vertices are either composed of a four-membered cycle or of two K_2-graphs. Because $G_{3,3}$ possesses no four-membered cycle, its four-vertex Sachs graphs are those corresponding to pairs of independent edges.

There are 24 such pairs:

$$a,d \quad a,e \quad a,f \quad a,g \quad a,h \quad a,i$$
$$b,e \quad b,f \quad b,g \quad b,h \quad b,i \quad c,e$$
$$c,f \quad c,g \quad c,h \quad c,i \quad d,g \quad d,h$$
$$d,i \quad e,g \quad e,h \quad f,h \quad f,i \quad g,i$$

Therefore, $a_4(G_{3,3}) = 24 \times [(-1)^2 2^0] = 24$

Case $k = 5$. The Sachs graphs with five vertices are either composed of a five-membered cycle or of a two-component system consisting of a triangle and a K_2-graph. G_3 possesses both types of Sachs graph: one C_5 and five $C_3 + K_2$:

$C_3,e \quad C_3,f \quad C_3,g \quad C_3,h \quad C_3,i$

Consequently, $a_5(G_{3,3}) = 1 \times [(-1)^1 2^1] + 5 \times [(-1)^2 2^1] = 8$

Case $k = 6$. Here the real complications begin. The Sachs graphs with six vertices may be composed of:

a) a six-membered cycle, or

b) a four-membered cycle and a K_2-graph, or

c) two (disjoint) three-membered cycles, or

d) three K_2-graphs.

In $G_{3,3}$ only the latter types of 6-vertex Sachs graphs are contained, pertaining to the following selections of three independent edges:

$$a,d,g \quad a,d,h \quad a,d,i \quad a,e,g \quad a,e,h$$
$$a,f,h \quad a,f,i \quad a,g,i \quad b,e,g \quad b,e,h$$
$$b,f,h \quad b,f,i \quad b,g,i \quad c,e,g \quad c,e,h$$
$$c,f,h \quad c,f,i \quad c,g,i \quad d,g,i$$

Therefore, $a_6(G_{3,3}) = 19 \times [(-1)^3 2^0] = -19$.

Case $k = 7$. The seven-vertex Sachs graphs may be composed of

a) a seven-membered cycle, or

b) a five-membered cycle and a K_2-graph, or

c) a three-membered cycle and two K_2-graphs.

The latter two types are contained in G_3, namely:

$$C_5,a \quad C_5,b \quad C_5,c$$
$$C_3,e,g \quad C_3,e,h \quad C_3,f,h \quad C_3,f,i \quad C_3,g,i$$

resulting in $a_7(G_{3,3}) = 3 \times [(-1)^2 2^1] + 5 \times [(-1)^3 2^1] = -4$.

Case $k = 8$. The eight-vertex Sachs graphs may be composed of

a) an eight-membered cycle, or

b) a six-membered cycle and a K_2-graph, or

c) a four-membered cycle and two K_2-graphs, or

d) two four-membered cycles, or

e) a five-membered cycle and a three-membered cycle, or

f) two three-membered cycles and a K_2-graph, or

g) four K_2-graphs.

In our example we encounter only with the Sachs graphs of type e) and g), one of each type: $C_5 + C_3$ and a, d, i, g. This implies $a_8(G_{3,3}) = [(-1)^2 2^2] + [(-1)^4 2^0] = 5$. Thus all coefficients of $\varphi(G_{3,3})$ have been calculated and we finally obtain:

$$\varphi(G_{3,3}, \lambda) = \lambda^8 - 9\lambda^6 - 2\lambda^5 + 24\lambda^4 + 8\lambda^3 - 19\lambda^2 - 4\lambda + 5$$

Another way to express the Harary theorem is the following.

Theorem 3.1a. Let S be a Sachs graph with $N(S)$ vertices, possessing a total of $p(S)$ components, of which $c(S)$ are cycles and $p(S)-c(S)$ are K_2-graphs. Then

$$\varphi(G,\lambda) = \sum_S (-1)^{p(S)} 2^{c(S)} \lambda^{N-N(S)} \qquad (3.14)$$

where the summation goes over all Sachs graphs S contained (as subgraphs) in the graph G. In formula (3.14) the summation includes also the empty Sachs graph (a fictitious *graph* with $N(S) = p(S) = c(S) = 0$) which is assumed to be the subgraph of any graph.

The above example is intended not only to make the reader familiar with the usage of the formula (3.13), but also to illustrate how difficult is the calculation of φ by means of the Harary theorem. It should be said clearly and plainly: Except for a few very small molecular graphs, *the Harary theorem is not suitable for the calculation of the coefficients of the characteristic polynomial.*

On the other hand, the Harary theorem represents a powerful tool for deducing general properties of the characteristic polynomial, in particular on its dependence on graph (molecular) structure. Here are a few simple results of this kind.

By careful reading the above example we immediately see that not only for G_3, but for all graphs G,

$a_1(G) = 0$;

$a_2(G) = $ - number of triangles of G;

$a_3(G) = $ - 2 × the number of triangles of G;

$a_3(G) = a_5(G) = a_7(G) = \ldots = 0$ if and only if the graph G possesses no odd membered cycles; recall that such are the molecular graphs of the so-called alternant hydrocarbons.

If all odd coefficients of $\varphi(G)$ are zero (which happens in the case of molecular graphs of alternant hydrocarbons) then $\varphi(G,\xi) = 0$ implies $\varphi(G,-\xi) = 0$ and therefore the zeros of such characteristic polynomials occur in pairs $(\xi, -\xi)$.[22]

Denote, as before, the number of k-element independent edge sets of a graph G by $m(G,k)$. As before, $m(G,0) = 1$ for all graphs. A far-reaching consequence of the Harary theorem is the following:

Theorem 3.2. If the graph G is acyclic then all the odd coefficients of $\pi(G)$ are equal to zero, $a_0(G) = 1$, whereas for $k = 1, 2, \ldots, [N/2]$, $a_{2k}(G) = (-1)^k m(G,k)$. In other words:

$$\varphi(G,\lambda) = \sum_{k=0}^{[n/2]} (-1)^k m(G,k)\lambda^{n-2k} \tag{3.15}$$

Formula (3.15) was known already to Sachs. [95] Hosoya [31] was the first who extensively used it. Formula (3.15) is the motivation for the introduction of another important graph polynomial - the *matching polynomial*.

3.3. THE MATCHING POLYNOMIAL

The right-hand side of eq 3.15 is equal to the characteristic polynomial if and only if the graph G is acyclic. On the other hand, the right-hand side of eq 3.15 is a well defined polynomial for any graph. Thus we define a new graph polynomial as:

$$\alpha(G) = \varphi(G,\lambda) = \sum_{k=0}^{[N/2]} (-1)^k m(G,k)\lambda^{n-2k} \tag{3.16}$$

and call it the *matching polynomial* of the graph G.

Immediately from this definition follows:

Theorem 3.3. The matching polynomial of a graph G coincides with the characteristic polynomial of G if and only if G is acyclic.

In view of eq 3.4, φ is the characteristic polynomial of a symmetric matrix whose entries are real-valued numbers. As well known in linear algebra, all zeros of such a polynomial are necessarily real-valued numbers. From Theorem 3.3 we then see that all zeros of the matching polynomial of an acyclic graph are real-valued numbers.

However, this latter property is not restricted to acyclic graphs. We namely have:

Theorem 3.4. All the zeros of the matching polynomials of all graphs are real-valued numbers.

The history of the polynomial α is quite perplexed. It has been independently conceived by quite a few authors, mathematicians, physicists and chemists, in many cases in connection with Theorem 3.4. Already this detail indicates that this polynomial found numerous applications (which, however, will not be outlined in this chapter). Around 1970 a theoretical model has been developed in statistical physics, [99-102] in which the partition function was represented by a polynomial which was equivalent to what above was defined as α. (Of course, the terminology used by physicists was quite different than ours). In order to be able to describe phase transitions within this model, it was necessary that α has at least one complex-valued zero. The authors of Refs. [99-102] proved that this never is the case (i.e., that Theorem 3.4 holds), which for their theoretical model was a disappointing result. Heilmann and Lieb [100] offer not less than three different proofs of Theorem 3.4. Anyway, after proving Theorem 3.4 the model was abandoned. Nevertheless, the research the physicists made on α, especially the results by Heilmann and Lieb[100] were later recognized as very important for the theory of the matching polynomial.

Around the same time Hosoya[21] introduced his topological index and the Z-counting polynomial, eq 3.9. This polynomial is essentially the same as α, eq 3.16. A formal transformation of $Q(G)$ into $\alpha(G)$ and vice versa is straightforward (cf. eq 3.11):

$$Q(G,\lambda) = \left(-i\sqrt{\lambda}\right)^n \alpha\left(G, \frac{i}{\sqrt{\lambda}}\right)$$

$$\alpha(G,\lambda) = \lambda^n Q\left(G, -\frac{1}{\lambda^2}\right)$$

Few years later Nijenhuis [103] demonstrated that the combinatorial object called rook polynomial has the distinguished property of having real-valued zeros. Only much later this result was incorporated into the theory of matching polynomials [9, 104-106] when it was realized that every rook polynomial is the matching polynomial of some graph.

Independently of all these developments, Edward Farrell [7] (a mathematician) defined a graph polynomial essentially identical[107] to α, and established its basic properties (but not the reality of its zeros); he was first to use the name matching polynomial. Farrell's paper [7] appeared in 1979, but was written much earlier, certainly before 1977.

In 1977 two independent but equivalent approaches were put forward, by means of which the resonance energy of conjugated molecules could be calculated in a new and very convenient manner.[4, 5] For this one has to find the zeros of $\varphi(G)$ and $\alpha(G)$, with G being the pertinent molecular graph. For the success of the method it is essential that all zeros of both $\varphi(G)$ and $\alpha(G)$ be real-valued. (Hence, curiously: what was bad for the theory of phase transitions, is good for the theory of aromaticity.) Both Aihara[4] and Gutman et al.[5] were influenced by earlier work by Hosoya.[21, 31] Aihara[4] named α the reference polynomial whereas Gutman et al.[5, 6, 108] called it the acyclic polynomial. Eventually, a general agreement was reached to call α, eq 3.16, the matching polynomial of the graph G.

Without knowing the earlier results of Heilmann and Lieb,[99, 100] Kunz[101, 102] and Nijenhuis, [103] Chris Godsil (a mathematician) and one of the present authors proved Theorem 4 anew.[9, 104, 109] The same authors demonstrated[110] that Theorem 3.4 holds also if G is the (weighted) graph representing heteroconjugated π-electron systems. In 1981 Godsil[111] arrived at the following powerful result, from which Theorem 3.4 follows as an easy consequence.

Theorem 3.5. For any graph G there exists an acyclic graph G^*, such that $\alpha(G,\lambda)$ is a divisor of $\varphi(G^*,\lambda)$.

If the graph G in Theorem 3.5 is connected, then G^* is called *the Godsil tree* of G.

The matching polynomial obeys a simple recurrence relation which makes its calculation relatively easy:

Theorem 3.6. Let G be a graph and e its edge connecting the vertices x and y. Then,

$$\alpha(G, \lambda) = \alpha(G\text{-}e, \lambda) - \alpha(G\text{-}x\text{-}y, \lambda) \tag{3.17}$$

If x is a pendent vertex (i.e. y is its only neighbor), then

$$\alpha(G, \lambda) = \lambda\alpha(G\text{-}x, \lambda) - \alpha(G\text{-}x\text{-}y, \lambda) \tag{3.18}$$

For calculations based on Theorem 3.6, eq 3.19 is also frequently needed. If G consists of (disconnected) components G' and G'', then

$$\alpha(G, \lambda) = \alpha(G', \lambda)\, \alpha(G'', \lambda) \tag{3.19}$$

Example 3.2. We illustrate the application of the recurrence relations (3.17)-(3.19) on the example of $G_{3.3}$, Figure 3.2. First, however, we need some preparation.

We compute the matching polynomials of the path graphs P_N, see Figure 3.1. Choosing x to be a terminal vertex of the path P_N we get from (3.18):

$\alpha(P_N, \lambda) = \lambda\alpha(P_{N-1}, \lambda) - \alpha(P_{N-2}, \lambda)$

Because $(P_0, \lambda) \equiv 1$ and $(P_1, \lambda) \equiv \lambda$ we obtain for $N = 2$:

$\alpha(P_2, \lambda) = \lambda[\lambda] - [1] = \lambda^2 - 1$

then for $N = 3$:

$\alpha(P_3, \lambda) = \lambda[\lambda^2 - 1] - [\lambda] = \lambda^3 - 2\lambda$

then for $N = 4$:

$\alpha(P_4, \lambda) = \lambda[\lambda^3 - 2\lambda] - [\lambda^2 - 1] = \lambda^4 - 3\lambda^2 + 1$

then for $N = 5$:

$\alpha(P_5, \lambda) = \lambda[\lambda^4 - 3\lambda^2 + 1] - [\lambda^3 - 2\lambda] = \lambda^5 - 4\lambda^2 + 3\lambda$

etc.

Choosing any edge of a cycle C_N and applying (3.17) we get:

$\alpha(C_N, \lambda) = \alpha(P_N, \lambda) - \alpha(P_{N-2}, \lambda)$

which for the three- and five-membered cycles gives:

$\alpha(C_3, \lambda) = \alpha(P_3, \lambda) - \alpha(P_1, \lambda) = (\lambda^3 - 2\lambda) - (\lambda) = \lambda^3 - 3\lambda$

$\alpha(C_5, \lambda) = \alpha(P_5, \lambda) - \alpha(P_3, \lambda) = (\lambda^5 - 4\lambda^2 + 3\lambda) - (\lambda^3 - 2\lambda) = \lambda^5 - 5\lambda^3 + 5\lambda$

We are now ready to compute $\alpha(G_{3.3})$. For this choose the edge d (whose end vertices are x and y) and apply (3.17):

$$\alpha(G_{3.3}, \lambda) = \alpha(G_{3.3}\text{-}d, \lambda) - \alpha(G_{3.3}\text{-}x\text{-}y, \lambda) \tag{3.20}$$

Now, $G_3 - d$ is a disconnected graph composed of C_3 and C_5. Therefore by eq 3.19:

$\alpha(G_{3.3}\text{-}d, \lambda) = \alpha(C_3, \lambda) \cdot \alpha(C_5, \lambda) = (\lambda^3 - 3\lambda)(\lambda^5 - 5\lambda^3 + 5\lambda) = \lambda^8 - 8\lambda^6 + 20\lambda^4 - 15\lambda^2$

Similarly, $G_{3.3} - x - y$ is disconnected, composed of P_2 and P_4. Therefore, by (3.19):

$\alpha(G_{3.3}\text{-}x\text{-}y, \lambda) = \alpha(P_2, \lambda) \cdot \alpha(P_4, \lambda) = (\lambda^2 - \lambda)(\lambda^4 - 3\lambda^1 + 1) = \lambda^6 - 4\lambda^4 + 4\lambda^2 - 1$

Substituting these expressions back into (3.20) we readily obtain:

$\alpha(G_{3,3}, \lambda) = (\lambda^8 - 8\lambda^6 + 20\lambda^4 - 15\lambda^2) - (\lambda^6 - 4\lambda^4 + 4\lambda^2 - 1) = \lambda^8 - 9\lambda^6 + 19\lambda^2 + 1$

The recurrence relations (3.17) - (3.19) can be expressed in terms of the Z-counting polynomials:

Theorem 3.6a. Using the same notation as in eqs 3.17 - 3.19, the Z-counting polynomial, defined via eq 3.9, satisfies:

$$Q(G, \lambda) = Q(G-e, \lambda) + \lambda Q(G - x - y, \lambda) \qquad (3.21)$$

$$Q(G, \lambda) = Q(G-x, \lambda) - \lambda Q(G - x - y, \lambda) \qquad (3.22)$$

$$Q(G, \lambda) = Q(G', \lambda) \cdot Q(G'', \lambda) \qquad (3.23)$$

The matching polynomials, their coefficients and (in some cases) their zeros were determined for numerous classes of graphs.[6, 84, 108, 112-143] Several computer-aided computation algorithms for the calculation of α were put forward.[144-152]

The fact that the matching polynomial has real zeros and is closely related to the characteristic polynomial of the underlying graph G (see below), motivated many authors to seek for a graph-like object, denote it by G_{hyp}, which would have the property $\varphi(G_{hyp}, \lambda) \equiv \alpha(G, \lambda)$[153-162] This search was successful in many cases - for instance, for unicyclic and bicyclic graphs. G_{hyp} is usually constructed from G so that some edges of G are weighted by complex-valued (or even quaternion valued!)[159] numbers.

The matching polynomial is intimately connected to the characteristic polynomial and has many properties analogous to the latter. Some of these relations are outlined in the subsequent section. More properties of the matching polynomials can be found in Chapter 4 of the book[46] and elsewhere.[163-169] As a curiosity we mention that several important orthogonal polynomials are matching polynomials of some pertinently chosen graphs.[170-172] For instance, the matching polynomial of the n-vertex complete graph is equal to the Hermite polynomial.

3.4. THE CHARACTERISTIC POLYNOMIAL. PART 2.

In the case of acyclic graphs the relation between the characteristic and the matching polynomials is straightforward (see Theorem 3.3). If a graph G contains cycles, then the relation between $\varphi(G)$ and $\alpha(G)$ is somewhat more complicated.

Let G be a graph and $C^1, C^2, ..., C^r$ be the cycles contained (as subgraphs) in it, see Figure 3.3. The subgraph $G - C^i$ is obtained by deleting from G all vertices belonging to C^i (and, of course, all edges incident to these vertices). If the cycles C^i and C^j are disjoint (i.e., have no vertices in common), then the subgraph $G - C^i - C^j$ is defined as $(G - C^i) - C^j$ or, what is the same, as $(G - C^j) - C^i$. If C^i and C^j have joint vertices, then without defining $G - C^i - C^j$, in the below formulas we set $\alpha(G - C^i - C^j, \lambda) \equiv 0$ and $\varphi(G - C^i - C^j, \lambda) \equiv 0$. The case of the subgraphs $G - C^i - C^j - C^k$, $G - C^i - C^j - C^k - C^h$, etc. is treated analogously. Some of the subgraphs $G - C^i$, $G - C^i - C^j$, etc. may be empty, i.e., all vertices of G need to be deleted. If H is the empty graph then it is both convenient and consistent to set $\alpha(H, \lambda) \equiv 1$ and $\varphi(H, \lambda) \equiv 1$.

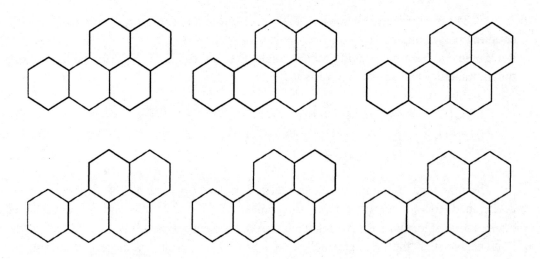

Figure 3.3. Some cycles of the molecular graph of benzo[a]pyrene, having a total of 21 cycles; the main practical difficulty in the calculation of the characteristic polynomials of polycyclic graphs lies in the enormous number of cycles that need to be taken into account

Theorem 3.7. With the above specified notation and conventions,

$$\varphi(G, \lambda) = \alpha(G, \lambda) - 2\sum_i \alpha\,(G - C^i, \lambda) + 4\sum_{i<j} \alpha\,(G - C^i - C^j, \lambda) -$$

$$8\sum_{i<j<k} \alpha\,(G - C^i - C^j - C^k, \lambda) + \dots \tag{3.24}$$

$$\alpha(G, \lambda) = \varphi(G, \lambda) + 2\sum_i \varphi\,(G - C^i, \lambda) + 4\sum_{i<j} \varphi\,(G - C^i - C^j, \lambda) +$$

$$8\sum_{i<j<k} \varphi\,(G - C^i - C^j - C^k, \lambda) + \dots \tag{3.25}$$

where the summations go over all cycles, pairs of cycles, triplets of cycles, etc., contained in G.

Note that on the right-hand sides of (3.24) and (3.25) there are 2^r summands, some of which may be equal to zero. Indeed, the actual application of formulas (3.24) and (3.25) is much simplified by the fact that in the second, third, etc. summations only pairs, triplets, etc. of mutually disjoint cycles need to be considered.

Formula (3.24) seems to be discovered by Hosoya;[31] it was later extensively applied within the theory of cyclic conjugation,[11] where also formula (3.25) was reported for the first time. A more compact way of writing (3.24) and (3.25) is

$$\varphi(G, \lambda) = \sum_R (-2)^{p(R)} \alpha(G - R, \lambda); \quad \alpha(G, \lambda) = \sum_R (+2)^{p(R)} \varphi(G - R, \lambda)$$

where the summations go over all regular graphs R of degree two, contained as subgraphs in G (including the empty graph): $p(R)$ is the number of component of R. Recall that in a regular graph of degree δ all vertices have exactly δ first neighbors. Two special cases of Theorem 3.7 deserve to be mentioned:

Corollary 3.7.1. If G is a unicyclic graph ($r = 1$) and C is its unique cycle, then

$$\varphi(G,\lambda) = \alpha(G,\lambda) - 2\alpha(G-C,\lambda); \quad \alpha(G,\lambda) = \varphi(G,\lambda) + 2\varphi(G-C,\lambda)$$

Corollary 3.7.2. If C_N is the N-vertex cycle, then

$$\varphi(C_N,\lambda) = \alpha(C_N,\lambda) - 2$$

Example 3.3. Calculate (again) the characteristic polynomial of the graph $G_{3.3}$ from Figure 3.2, this time employing eq 3.24. The graph $G_{3.3}$ possesses two cycles, C_3 and C_5, which are disjoint. Hence,

$$\varphi(G_{3.3},\lambda) = \alpha(G_{3.3},\lambda) - 2[\alpha(G_{3.3}-C_3,\lambda) + \alpha(G_{3.3}-C_5,\lambda)] + 4[\alpha(G_{3.3}-C_3-C_5,\lambda)]$$

The polynomial $\alpha(G_{3.3})$ has been calculated in Example 3.2.

The subgraph $G_{3.3} - C_3$ is just the cycle C_5. Similarly, $G_{3.3} - C_5 = C_3$. The matching polynomials of C_3 and C_5 have also been computed in Example 3.2. Then, in view of Corollary 3.7.2,

$$\varphi(C_3, \lambda) = \alpha(C_3, \lambda) - 2 = \lambda^3 - 3\lambda - 2$$
$$\varphi(C_5, \lambda) = \alpha(C_5, \lambda) - 2 = \lambda^5 - 5\lambda^3 + 5\lambda - 2$$

The subgraph $G_{3.3} - C_3 - C_5$ has no vertices, and therefore $\alpha(G_{3.3} - C_3 - C_5, \lambda) \equiv 1$. Bearing the above in mind we have

$$\varphi(C_3, \lambda) = [\lambda^8 - 9\lambda^6 + 24\lambda^4 - 19\lambda^2 + 1] - [(\lambda^5 - \lambda^3 + 5\lambda - 2) + (\lambda^3 - 3\lambda - 2)] + 4[1] =$$
$$= \lambda^8 - 9\lambda^6 - 2\lambda^5 + 24\lambda^4 + 8\lambda^3 - 19\lambda^2 - 4\lambda + 5$$

which, of course, is same as what we obtained in Example 3.1.

By means of Theorem 3.7 the characteristic polynomial is expressed in terms of matching polynomials, and vice versa. It is sometimes advantageous to express the characteristic polynomial of a graph in terms of characteristic polynomials of subgraphs. (The analogous result for the matching polynomial is Theorem 3.6). Of the several recurrence relations of this kind[173-191] we mention here a simple and, probably, most handy one.

Theorem 3.8. Let G be a graph and e its edge connecting the vertices x and y. Then,

$$\varphi(G, \lambda) = \varphi(G - e, \lambda) - \varphi(G - x - y, \lambda) - 2\sum_c \varphi(G-C, \lambda) \qquad (3.26)$$

with the summation going over all cycles C containing the edge e.

Formula (3.26) is often attributed to Schwenk[175] although it can be found already in a paper by Heilbronner.[173] Anyway, the following corollary of Theorem 3.8 is known as the Heilbronner formula.

Corollary 3.8.1. If the edge e does not belong to any cycle (in which case it is called *bridge*), then

$$\varphi(G, \lambda) = \varphi(G - e, \lambda) - \varphi(G - x - y, \lambda) \tag{3.27}$$

In particular, formula (3.27) holds for any edge of any acyclic graph. If x is a pendent vertex (i.e. y is its only neighbor), then

$$\varphi(G, \lambda) = \lambda\varphi(G - x, \lambda) - \varphi(G - x - y, \lambda) \tag{3.28}$$

Eqs 3.26 - 3.28 should be compared with (3.17) and (3.18). When applying these recurrence relations also the identity

$$\varphi(G, \lambda) = \varphi(G', \lambda)\varphi(G'', \lambda)$$

may be of great use, where the notation is the same as in eq 3.19.

Among the plethora of other known results for the characteristic polynomial we state here only two, which reveal further deep lying analogies between the characteristic and the matching polynomials.

Let G be a graph and $v_1, v_2, ..., v_N$ its vertices. Then

$$\frac{\partial\varphi(G,\lambda)}{\partial\lambda} = \sum_{i=1}^{N} \varphi(G - v_i, \lambda); \quad \frac{\partial\alpha(G,\lambda)}{\partial\lambda} = \sum_{i=1}^{N} \alpha(G - v_i, \lambda); \tag{3.29}$$

The φ-part of this identity is due to Clarke, [192] the α-part was first reported by Gutman and Hosoya.[108] For applications see Refs. [193-195].

The left hand sides of the following peculiar identities are called *graph propagators*.[196] Let x and y be two (not necessarily adjacent) vertices of the graph G. Let P be a path connecting x and y. Then,

$$\varphi(G - x, \lambda)\varphi(G - y, \lambda) - \varphi(G, \lambda)\varphi(G - x - y, \lambda) = [\sum_{P} \varphi(G - P, \lambda)]^2 \tag{3.30}$$

$$\alpha(G - x, \lambda)\alpha(G - y, \lambda) - \alpha(G, \lambda)\alpha(G - x - y, \lambda) = \sum_{P} [\alpha(G - P, \lambda)]^2 \tag{3.31}$$

with the summations going over all paths of G, whose end vertices are x and y. Evidently, for all real values of the indeterminate λ, both propagators (3.30) and (3.31) are positive. Formulas (3.30) and (3.31) were reported in Refs.[197] and [100], respectively. Formula (3.30) is just a graph-theoretical version of an old result in linear algebra.

The characteristic polynomial can be computed also in several other ways. Many (perhaps too many) algorithms for its calculation have been put forward.[198-227] When the graph possesses some symmetry, then special calculation techniques could be developed.[228-245] (Recall that here we do not quote papers devoted to the finding of graph eigenvalues using symmetry arguments). Discussing the calculation of the characteristic polynomials of graphs with weighted edges and vertices[246-250] goes beyond the ambit of this book. Needless to say that characteristic polynomials of a great variety of particular graphs and particular classes of graphs have been determined. Such details are, first of all. to be found in the seminal book[42] and elsewhere.[251-276]

Immanantal Polynomials.

Let $\mathbf{M} = \|\mathbf{M}_{ij}\|$ be a square matrix of order N. Let χ be one of the irreducible characters of the symmetric permutation group S_N. The *immanent* of the matrix $\mathbf{M}$, corresponding to the character χ of S_N, is defined as

$$d_\chi(M) = \sum_{g \in S_N} \chi(g)\mathbf{M}_{1,g(1)}\mathbf{M}_{2,g(2)} \ldots \mathbf{M}_{N,g(N)}$$

where g stands for an element of S_N which transforms the permutation $(1, 2, \ldots, N)$ into $(g(1), g(2), \ldots, g(N))$.

Note that if χ is the alternating character, then d_χ is the familiar *determinant*; if $\chi(g) = 1$ for all g, then d_χ is the *permanent*.

The *immanantal polynomial* of the matrix $\mathbf{M}$ is $d_\chi(\lambda\mathbf{I} - \mathbf{M})$. If $\mathbf{M}$ is the adjacency matrix of a graph, then one speaks of the *immanantal polynomial of a graph*. The characteristic and the permanental polynomials are special cases of immanantal polynomials. For more details see elsewhere.[18-20]

3.5. A UNIFYING APPROACH: THE μ-POLYNOMIAL

The fundamental difference between the characteristic and the matching polynomials is in the effect of cycles. One may view $\alpha(G)$ and the characteristic polynomial of G in which all cyclic contributions (originating from the cyclic Sachs graphs, $c(S)>0$, cf. Theorem 3.1) have been completely neglected. One may wonder what would happen by neglecting the contributions of only some cycles of G, or by only partially neglecting these contributions. Such deliberations resulted in the concept of the μ-polynomial.

Let, as before, the graph G contain (as subgraphs) the cycles $C^1, C^2, \ldots, C^r$, see Figure 3.3. For $i = 1, 2, \ldots, r$, associate a variable $t(C^i)$ to the cycle C^i. This variable is viewed as the weight of the cycle C^i: if $t(C^i) = 1$ then the effect of this cycle is fully taken into account, if $t(C^i) = 0$ then the effect of this cycle is fully neglected. Denote, for brevity, the r-tuple $[t(C^1), t(C^2), \ldots, t(C^r)]$ by $\vec{t}$, which may be viewed as an r-dimensional vector. Then the μ-polynomial of the graph G is defined as follows.[11]

If the graph G is acyclic, then

$$\mu(G, \vec{t}) = \mu(G, \vec{t}, \lambda) \equiv \alpha(G, \lambda) \equiv \varphi(G, \lambda) \tag{3.32}$$

If the graph G possesses cycles, then

$$\mu(G,\vec{t}\,) = \mu(G,\vec{t}\,,\lambda) = \alpha(G,\lambda) - 2\sum_i t\,(C^i)\alpha(G - C^i,\lambda) + 4\sum_{i<j} t\,(C^i)\,t(C^j)\alpha(G - C^i - C^j,\lambda) - 8$$

$$\sum_{i<j<k} t\,(C^i)t(C^j)\,t(C^k)\alpha(G - C^i - C^j - C^k,\lambda) + ...$$

$$(3.33)$$

The symbol $\vec{t}$ in eq 3.32 is fictitious; the purpose of eq 3.32 is to define the μ-polynomial-concept for all graphs.

Formula (3.33) should be compared with (3.24). The idea behind it is that by continuously changing the parameters t (usually between 1 and 0) we can continuously change the effect of the respective cycles on the polynomial itself (which we will discuss below) and on various π-electron characteristics of conjugated molecules (which are calculated from the polynomials, but which we are not discussing in this book; for details see Ref.[11].

This graph polynomial was conceived while the authors of Ref.[11] worked together in Mülheim, Germany. The suggestion to name it *Mülheim polynomial* was not accepted by the mathematico-chemical community; what only reminds this attempt is the symbol of the polynomial. The following property of the μ-polynomial follows directly from its definition and/or from the analogous properties of both the characteristic and the matching polynomial.

Denote by $\vec{1}$ and $\vec{0}$ the vectors $(1, 1, ..., 1)$ and $(0, 0, ..., 0)$, respectively.

Theorem 3.9. Let G be a graph possessing at least one cycle. Then, using the same notation as in eqs 3.29 and 3.26,

$$\mu(G,\vec{1},\lambda) = \varphi(G,\lambda)\,;\, \mu(G,\vec{0},\lambda) = \alpha(G,\lambda)$$

$$\frac{\partial\mu(G,\vec{t},\lambda)}{\partial\lambda} = \sum_{i=1}^{n}\mu(G - v_i,\vec{t},\lambda)$$

$$\mu(G,\vec{t},\lambda) = \mu(G - e,\vec{t},\lambda) - \mu(G-x-y,\vec{t},\lambda) - 2\sum_C t\,(C)\mu(G - C,\vec{t},\lambda)$$

Further recurrence relations are found elsewhere.[277-280] The dependence of the μ-polynomial on a particular cycle C (or more precisely, on its weight $t(C)$) is given by

$$\frac{\partial\mu(G,\vec{t},\lambda)}{\partial t(C)} = \mu(G - C,\vec{t},\lambda)$$

which is a relation of crucial importance in the theory of cyclic conjugation.[11]

The μ-polynomial not only includes as special cases the characteristic and the matching polynomials, but also many other graph polynomials. In particular, if the cycle-weights are chosen so that all cycles of the same size have equal weights, then we arrive at the circuit polynomials, invented and extensively studied by Farrell.[10, 12, 61, 85, 281-290] Some other cycle-related graph polynomials have sporadically occurred in the chemical literature.[291-294]

Another important special case is the β-polynomial (sometimes called "circuit characteristic polynomial").[13-15, 161.295-299] Let G be a graph and C one of its cycles. Choose the vector $\vec{t}$ so that $t(C) = 1$ and $t(C') = 0$ for all other cycles C' (if any). Then

$$\beta(G, C) = \beta(G, C, \lambda) = \mu(G, \vec{t}, \lambda)$$

or, what is the same,

$$\beta(G, C, \lambda) = \alpha(G, \lambda) - 2\alpha(G - C, \lambda) \tag{3.34}$$

Formula (3.34) should be compared with Corollary 3.7.1. Indeed, if G is unicyclic, then its β-polynomial is the same as the characteristic polynomial.

The β-Polynomial Hypothesis

The β-polynomial has been defined so that it contains the effect of just one individual cycle of a polycyclic (molecular) graph, namely the effect of the cycle C. This feature is the basis of the application of the β-polynomials in the theory of cyclic conjugation, for the calculation of the effect of an individual cycle on various π-electron properties of a polycyclic conjugated molecule, especially on its total π-electron energy.[13]

With regard to this application, it is necessary that all the zeros of $\beta(G, C)$ be real-valued numbers, preferably for all graphs G and all cycles C contained in them. Numerical calculations (e. g. in Ref.[13]) showed that this is the case in many chemically relevant examples. Further studies revealed that the zeros of $\beta(G, C)$ are real for many types of graphs,[14, 15, 161, 295-299] among which are all unicyclic graphs (which is trivial), all bicyclic graphs, all graphs with eight and fewer vertices, the complete graphs, etc. In spite of all these efforts, and in spite of a reward offered,[297] the following hypothesis remains unsolved; it could be considered as one of the most challenging problems in the theory of graph polynomials (of interest in chemistry).

Conjecture. If G is any cycle-containing graph and C is any of its cycles, then all the zeros of $\beta(G, C, \lambda) = \alpha(G) - 2\alpha(G - C)$ are real-valued numbers.

This conjecture may be false. If so, then finding a single counterexample (a particular graph G and a particular cycle C in it), for which at least one zero of $\beta(G, C, \lambda)$ is complex-valued, would suffice.

* * *

We see that the characteristic and the matching polynomials can be viewed as two limit cases of the μ-polynomial. Because they both have real zeros, the natural question is what can be said about the reality of the zeros of $\mu(G)$. In the general case, some zeros of some μ-polynomials may be complex-valued numbers.[11, 300] There, however, exists an interesting result:[11]

Theorem 3.10. Let G be a graph and C^1, C^2, ..., C^r be its cycles, $r \geq 2$. If any two cycles of G are disjoint (i. e., have no vertex in common), then all the zeros of $\mu(G, \vec{t}, \lambda)$ are real-valued numbers, provided $-1 \leq t(C^i) \leq +1$ holds for all $i = 1, 2, ..., r$.

If the conditions $-1 \leq t(C^i) \leq +1$ are not obeyed for all cycles C^i, then complex-valued zeros may occur. If $r = 1$ then all zeros of $\mu(G, \vec{t}, \lambda)$ are real-valued, irrespective of the value of $t(C^I)$.[301]

An intriguing unsolved problem in the theory of the μ-polynomial is the μ-analog of eqs 3.30 and 3.31. In other words: what can be said about the propagator

$$\mu(G\text{-}x,\vec{t}, \lambda)\mu(G\text{-}y,\vec{t}, \lambda) - \mu(G,\vec{t}, \lambda)\mu(G\text{-}x\text{-}y,\vec{t}, \lambda) \qquad (3.35)$$

and can it be expressed in terms of paths connecting the vertices x and y? Under which conditions is this propagator positive-valued? Notice that for $\vec{t} = \vec{1}$ and $\vec{t} = \vec{0}$ the μ-propagator (3.35) reduces to the φ- and α-propagators, (3.30) and (3.31), respectively.

3.6. THE LAPLACIAN POLYNOMIAL

The Laplacian matrix is a very important object in the analysis of electrical networks (and is, among other things connected with the classical Kirchhoff laws). Its role in chemical graph theory is much more modest.[76, 302] Therefore, the fundamentals of the theory of the Laplacian polynomial, outlined in this section, should be understood primarily as a possibility (and an invitation) for future chemical applications. Some chemical connections of the Laplacian matrix, especially those related to the Wiener index and other distance-related structure-descriptors, are mentioned elsewhere in this book.

Let, as before, G be a (molecular) graph, v_1, v_2, ..., v_N its vertices and $\mathbf{A} = \mathbf{A}(G)$ its adjacency matrix. The degree d_i of the vertex v_i is the number of the first neighbors of this vertex (see Chap. 1).

Let $\mathbf{DEG} = \mathbf{DEG}(G)$ be the square matrix of order N whose i-th diagonal element is d_i and whose all off-diagonal elements are zero. Then the *Laplacian matrix* of the graph G is defined as

$$\mathbf{La} = \mathbf{La}(G) = \mathbf{DEG}(G) - \mathbf{A}(G)$$

The *Laplacian characteristic polynomial* of the graph G is just the characteristic polynomial of the Laplacian matrix:

$$\Psi(G) = \Psi(G, \lambda) = det[\lambda\mathbf{I} - \mathbf{La}(G)] \qquad (3.36)$$

and we write it in the form

$$\Psi(G, \lambda) = \sum_{k=0}^{N} c_k(G)\lambda^{N-k} \tag{3.37}$$

which should be compared with eq 3.12. The mathematical theory of Laplacian polynomials and of their zeros - the so-called Laplacian graph spectra - is nowadays well elaborated; for details see the reviews.[20, 69, 72, 76]

The Kel'mans Theorem

The fundamental result, relating the coefficients of $\Psi(G)$ with the structure of the graph G, is the Kel'mans theorem.[42, 303] (This theorem was first communicated by Kel'mans in 1967 in a booklet entitled *Cybernetics in the Service of Communism* published in Moscow and Leningrad, in Russian language.) To formulate it we need a few definitions.

Consider a graph G on N vertices. Any N-vertex subgraph H of G is a said to be a *spanning subgraph*; hence H is obtained from G by deleting some of its edges, but none of its vertices. If H is acyclic, we say that H is a *spanning forest* of G; if H is acyclic and connected, then H is a *spanning tree* of G.

Let F be a spanning forest of a graph G. Let $T_1, T_2, ..., T_p$ be the components of F, with $N(T_1), N(T_2), ..., N(T_p)$ vertices, respectively,

$N(T_1) + N(T_2) + ... + N(T_p) = N(F) = N$

Then the product $N(T_1) N(T_2) ... N(T_p)$ will be denoted by $\gamma(F)$.

For an example see Figure 3.4.

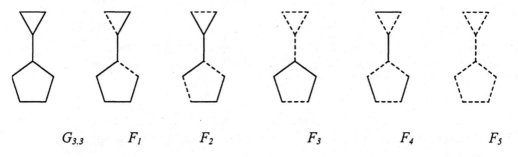

| $G_{3.3}$ | F_1 | F_2 | F_3 | F_4 | F_5 |

Figure 3.4. The molecular graph $G_{3.3}$ and some of its spanning forests;
$\gamma(F_1) = 8$, $\gamma(F_2) = 5\cdot3 = 15$, $\gamma(F_3) = 1\cdot1\cdot1\cdot5 = 5$, $\gamma(F_4) = 2\cdot2\cdot2\cdot2 = 16$,
$\gamma(F_5)=1\cdot1\cdot1\cdot1\cdot1\cdot1\cdot1\cdot1 = 1$; of these spanning forests only F_1 is a spanning tree

Theorem 3.11. Let G be a graph on N vertices and Laplacian characteristic polynomial $\Psi(G, \lambda)$, given by eq 3.37. Then

$$c_k(G) = (-1)^k \sum_{F} \gamma(F)$$

with summation going over all spanning forests of G which have N - k components ($p = N - k$).

The practical application of the Kel'mans theorem is rather tedious. This is seen from the below example, where we compute $\Psi(G)$ of a very small graph.

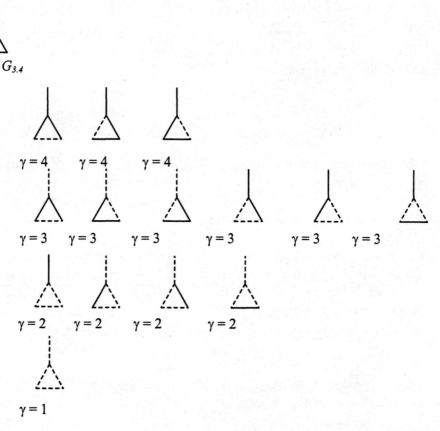

Figure 3.5. A small graph $G_{3.4}$ and all its spanning forests; with the increasing number of vertices and edges, the number of spanning forests becomes enormously large, thus making the calculation of the coefficients of the Laplacian polynomial by means of the Kel'mans theorem unfeasible

Example 3.4. The spanning forests of the 4-vertex graph $G_{3.4}$ are depicted in Figure 3.5, together with the respective *gamma*-values. Then by direct application of the Kel'mans theorem we have:

$$c_0(G_{3.4}) = (-1)^0[1] = 1$$
$$c_1(G_{3.4}) = (-1)^1[2+2+2+2] = -8$$
$$c_2(G_{3.4}) = (-1)^2[3+3+3+3+3+4] = 19$$
$$c_3(G_{3.4}) = (-1)^3[4+4+4] = -12$$

Because there cannot be spanning forests with 4 - 4 = 0 components, it follows that $c_4(G_{3.4}) = 0$. We thus have:

$$\Psi(G_{3.4},\ \lambda) = \lambda^4 - 8\lambda^3 + 19\lambda^2 - 12\lambda$$

After carefully working out the above example we easily envisage a few general results, holding for any graph G with N vertices and m edges:

$c_0(G) = 1$; $c_1(G) = -2m$; $c_N(G) = 0$; $c_{N-1}(G) = (-1)^{N-1} N \times$ number of spanning trees.

Because only connected graphs have spanning trees, we see that $c_{N-1}(G) \neq 0$ if and only if the graph G is connected. Further, $c_{N-1}(G) = \pm N$ if and only if the graph G is connected and acyclic, i. e. a tree.   A less elementary result of the same kind is:

Corollary 3.11.1. If G is a tree, then

$$-N\frac{c_{N-2}(G)}{c_{N-1}(G)} = W(G)$$

where $W(G)$ is the Wiener number of G, a distance-related topological index, discussed elsewhere in this book.

* * *

A graph is said to be regular of degree d_i if all its vertex degrees are equal to deg_i, that is $d_1 = d_2 = ... = d_N = d_i$. For such graphs, $\mathbf{La} = d_i\mathbf{I} - \mathbf{A}$ and from eqs 3.4 and 3.36 we straightforwardly obtain

$$\Psi(G, \lambda) = (-1)^N \varphi(G, d_i - \lambda)$$

a relation which holds for N-vertex regular graphs of degree d_i. For instance, the cycle C_N is a regular graph of degree 2 and therefore

$$\Psi(C_N, \lambda) = (-1)^N \varphi(C_N, 2-\lambda)$$

For non-regular graphs the relation between the regular and the Laplacian characteristic polynomials is somewhat less simple:[304]

Theorem 3.12. Let G be a graph on N vertices, $v_1, v_2, ..., v_N$ and let d_i be the degree of the vertex v_i, $i = 1, 2, ..., N$. Then

$$\Psi(G, \lambda) = (-1)^N [\varphi(G, -\lambda) + \sum_i d_i \, \varphi(G - v_i, -\lambda) + \sum_{i<j} d_i d_j \, \varphi(G - v_i - v_j, -\lambda) +$$

$$\sum_{i<j<k} d_i d_j d_k \, \varphi(G - v_i - v_j - v_k, -\lambda) + ...]$$

with the summations going over all vertices, pairs of vertices, triplets of vertices, etc. of G. As before, $\varphi(G - v_1 - v_2 - ... - v_N) \equiv 1$.

For bipartite graphs (that are graphs without odd-membered cycles) the following relation was recently reported.[305]

Theorem 3.13. Let G be a connected bipartite graph with N vertices and m edges. Let $L(G)$ be the line graph of G. Then,

$$\lambda^m \Psi(G, \lambda) = \lambda^N \varphi(L(G), \lambda - 2)$$

3.7. MOVING IN ANOTHER DIRECTION: THE INDEPENDENCE POLYNOMIAL

The above outlined (mutually closely related) Z-counting and matching polynomials are defined via the quantities $m(G, k)$, cf. eqs 3.9 and 3.16. To repeat: $m(G,$

k) is the number of k-element independent edge sets of the graph G. In other words: $m(G, k)$ is the number of ways in which k mutually independent edges are selected in G.

One may ask if instead of selecting independent edges we could design graph polynomials by selecting some other structural features of the graph. The natural choice would be the vertices of the graph.

In close analogy to the numbers $m(G, k)$ and the polynomial $Q(G)$ we now introduce the numbers $N(G, k)$ and the polynomial $\omega(G)$, named the independence polynomial. Two vertices of a graph G are said to be independent if they are not adjacent. A set of vertices of G is said to be independent if all of its elements are mutually independent. The number of distinct k-element independent vertex sets of G is denoted by $N(G, k)$. In addition, $N(G, 0) = 1$ and $N(G, 1) =$ number of vertices of G. The *independence polynomial* is then defined in full analogy with eq 3.9:

$$\omega(G) = \omega(G, \lambda) = \sum_{k \geq 0} N(G, k)\lambda^k \qquad (3.38)$$

Example 3.5. By direct application of (3.38) we compute the independence polynomial of the graph $G_{3.5}$, depicted in Figure 3.6.

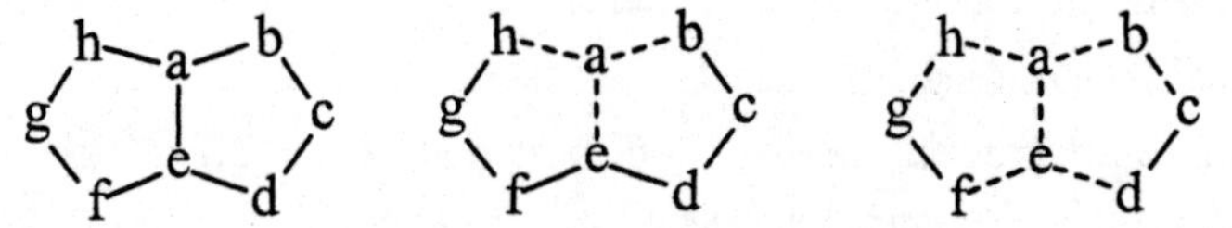

$$G3.5 \qquad\qquad G3.5-a \qquad\qquad G3.5-Na$$

Figure 3.6. A molecular graph and its two subgraphs, used to illustrate the calculation of the independence polynomial

Case $k = 1$. $G_{3.5}$ has eight vertices and therefore $N(G_{3.5}, 1) = 8$.
Case $k = 2$. The following pairs of vertices of G_5 are not adjacent:

a, c	a, d	a, f	a, g	b, d
b, e	b, f	b, g	b, h	c, e
c, f	c, g	c, h	d, f	d, g
d, h	e, g	e, h	f, h	

Hence, $N(G_{3.5}, 2) = 19$.
Case $k = 3$. The following triplets of vertices of $G_{3.5}$ are not mutually adjacent:

a, c, f	a, c, g	a, d, f	a, d, g	b, d, f
b, d, g	b, d, h	b, e, g	b, e, h	b, f, h
c, e, g	c, e, h	c, f, h	d, f, h	

Hence, $N(G_{3.5}, 3) = 14$.
Case k 4. There is a unique set of four independent vertices, namely: $\{b, d, f, h\}$. Therefore $N(G_{3.5}, 4) = 1$ and, in addition, $N(G_{3.5}, k) = 0$ for $k > 4$. Bearing in mind that by definition $N(G_{3.5}, 0) = 1$, we arrive at
$$\omega(G_5, \lambda) = 1 + 8\lambda + 19\lambda^2 + 14\lambda^3 + \lambda^4$$

* * *

The fundamental recursion relations for $\omega(G)$ are given by the following:

Theorem 3.14 Let x be a vertex of the graph G. The set consisting of x and its first neighbors is denoted by N_x. Then the independence polynomial, defined via eq 3.38, satisfies:

$$\omega(G, \lambda) = \omega(G - x, \lambda) + \lambda\omega(G - N_x, \lambda) \tag{3.39}$$

If x is a pendent vertex, its only first neighbor being y, then $N_x = \{x, y\}$ and, as a special case of (3.39),

$$\omega(G, \lambda) = \omega(G - x, \lambda) + \lambda\omega(G - x - y, \lambda) \tag{3.40}$$

Further, if G is disconnected and G' and G'' are its components, then

$$\omega(G, \lambda) = \omega(G', \lambda) \cdot \omega(G'', \lambda) \tag{3.41}$$

Relations (3.39) - (3.41) should be compared with (3.21) - (3.23).

Example 3.6. We calculate once again $\omega(G_5)$, this time by using Theorem 3.14. For this we need some preparation. We first calculate the independence polynomials of the first few path graphs. For P_1 and P_2 by direct calculation we readily get: $\omega(P_1) = 1 + \lambda$, $\omega(P_2) = 1 + 2\lambda$. Now, applying (3.40) to a pendent vertex x of P_N and bearing in mind that $P_N - x = P_{N-1}$ and $P_N - x - y = P_{N-2}$ we get

$$\omega(P_N, \lambda) = \omega(P_{N-1}, \lambda) + \lambda\omega(P_{N-2}, \lambda)$$

which successively yields:

$$\omega(P_3, \lambda) = [1 + 2\lambda] + \lambda [1 + \lambda] = 1 + 3\lambda + \lambda^2$$
$$\omega(P_4, \lambda) = [1 + 3\lambda + \lambda^2] + \lambda[1 + 2\lambda] = 1 + 4\lambda + 3\lambda^2$$
$$\omega(P_5, \lambda) = [1 + 4\lambda + 3\lambda^2] + \lambda[1 + 3\lambda + \lambda^2] = 1 + 5\lambda + 6\lambda^2 + \lambda^3$$
etc
$$\omega(P_8, \lambda) = 1 + 7\lambda + 15\lambda^2 + 10\lambda^3 + \lambda$$
etc

The above polynomials should be compared with $\alpha(P_N)$, obtained in Example 3.2. Choose vertex a in $G_{3.5}$, see Figure 3.6. This vertex has three first neighbors: b, e, h and therefore $N_a = \{a, b, e, h\}$. The subgraphs $G_{3.5} - a$ and $G_{3.5} - N_a$, also shown in Figure 3.6, are in fact P_8 and $P_2 + P_2$, respectively. Therefore,

$$\omega(G_{3.5}) = \omega(G_{3.5} - a, \lambda) + \lambda\omega(G_{3.5} - N_a, \lambda) = \omega(P_8, \lambda) +$$

$\lambda\omega(P_2 + P_2, \lambda)$

$$= \omega(P_8, \lambda) + \lambda\omega(P_2, \lambda)^2 = [1 + 7\lambda + 15\lambda^2 + 10\lambda^3 + \lambda] +$$

$\lambda[1 + 2\lambda]^2$

$$= 1 + 8\lambda + 19\lambda^2 + 14\lambda^3 + \lambda^4$$

same as before, but slightly easier to calculate.

* * *

For further details on the theory of independence polynomials see Refs. [134, 168, 169, 307-309]; many properties of ω- and Q-polynomials are fully analogous, which is frequently used in their research. Other related graph polynomials are the *king, color* and *star polynomials*. [310-317]

If instead of sets of independent vertices one considers sets of mutually adjacent vertices, then one arrives at the so.called *clique polynomial*. [318-320]

With regard to the chemical applications of the independence numbers $N(G, k)$ and the associated graph polynomial $\omega(G)$ one should, first of all, mention the topological theory of Merrifield and Simmons; its details (which go far beyond the ambit of this chapter, can be found in the book.[321] The sum of the numbers $m(G, k)$, i. e. , the value of the independence polynomial for $\lambda = 1$, is known under the name *Merrifield-Simmons index*.

However, as explained below, the independence polynomials have much wider chemical (and other) applications.

Beyond the Independence Polynomial?

Once we learned how to construct the independence polynomial as an analogy of the matching polynomial, we may be inclined to introduce further such counting polynomials. We would simply have to decide which structural features in a (molecular) graph to count - and we are done. This, indeed, has happened: the mathematical and chemical literature is quite reach in such attempts, [310-320] of which we point out the chemically significant sextet polynomial in the theory of benzenoid hydrocarbons [32-36, 40, 79, 322-333] and its generalizations. [37, 38, 334]

However, there is a reason to stop at the independence polynomial. The following theorem by Gutman and Harary [306] shows that ω may be viewed as the ultimate counting polynomial of its kind. Let G be a (labeled) graph and $S = \{S_1, S_2, ..., S_N\}$ a set of some of its (labeled) subgraphs (or structural features). Suppose that for any two of these subgraphs, say S_i and S_j, we can decide whether they obey a condition S_i , S_j (which one may interpret as S_i and S_j being mutually *independent*). For $k = 2, 3, ...$, let $o(G, k)$ be the number of k-element subsets of S, in which all elements pairwise obey the relation   (ie. all elements are mutually *independent*). It is consistent (yet not necessary) to set $o(G, 0) = 1$ and $o(G, 1) = N$. Note that the numbers $o(G, k)$ have been chosen in a fairly arbitrary manner.

Theorem 3.15. The polynomial

$$\sum_{k \geq 0} o(G, k)\lambda^k$$

is the independence polynomial of some graph.

    Corollary 3.15. 1. The Z-counting polynomial is an independence polynomial. Let $L(G)$ denote the line graph of the graph G. Then, for any graph G, the Z-counting polynomial of G coincides with $\omega(L(G))$.

    Corollary 3.15. 2. The clique polynomial is an independence polynomial. Let $\overline{G}$ denote the complement of the graph G. Then, for any graph G, the clique polynomial of G coincides with $\omega(\overline{G})$.   At this point we are not going to define the sextet polynomial, [35] and Clar graph, [323] playing an important role in the Clar aromatic sextet theory of a benzenoid hydrocarbon; the interested readers should consult the book [39] the original paper by Hosoya and Yamaguchi, [35] the review [40] or some of the numerous papers on this topic. [32-34, 36, 79, 322-333] Anyway, we have

Corollary 15. 3. The sextet polynomial is an independence polynomial. Let $C(B)$ denote the Clar graph of the benzenoid hydrocarbon B. Then, for any benzenoid system B, the sextet polynomial of B coincides with $\omega(C(B))$.

3.8. MORE

Before ending this chapter we wish to mention a few more graph polynomials that are encountered in chemical graph theory. Among them is the Wheland polynomial, the coefficients of which count resonance structures of various degrees of excitation. [335-337] A related polynomial was considered by John. [338]

The Hosoya polynomial is defined as [67]

$$H(G, \lambda) = \sum_{k \geq 0} d\,(G,\,k)\lambda^k$$

where $d(G, k)$ is the number of vertex pairs of the graph G, the distance of which is k. Then $d(G, 1)$ is equal to the number of edges of G, whereas it is consistent to choose $d(G, 0) =$ number of vertices of G.

The Hosoya polynomial is defined only for connected graphs.

Hosoya, who invented $H(G, \lambda)$ named it the *Wiener polynomial* because of its remarkable property: [67]

$$\frac{dH(G,\lambda)}{d\lambda}\bigg|_{\lambda=1} = W(G)$$

where $W(G)$ is the Wiener topological index (see Chap. 4, this book). Eventually, the more appropriate name *Hosoya polynomial* has been accepted. Further results of the theory of this distance-based graph polynomial are found elsewhere. [339-344]

* * *

Is this the end?

No. But we must stop somewhere.

REFERENCES

1. Graovac, A.; Gutman, I.; Trinajstić, N. *Topological Approach to the Chemistry of Conjugated Molecules*; Springer-Verlag: Berlin, 1977.

2. Tang, A.; Kiang, Y.; Di, S.; Yen, G. *Graph Theory and Molecular Orbitals* (in Chinese); Science Press: Beijing, 1980.

3. Tang, A.; Kiang, Y.; Yan, G.; Tai, S. *Graph Theoretical Molecular Orbitals*; Science Press: Beijing, 1986.

4. Aihara, J. A New Definition of Dewar-Type Resonance Energies. *J. Am. Chem. Soc.* **1976**, 98, 2750-2758.

5. Gutman, I.; Milun, M.; Trinajstić, N., Graph Theory and Molecular Orbitals. 19. Nonparametric Resonance Energies of Arbitrary Conjugated Systems. *J. Am. Chem. Soc.* **1977**. *99*. 1692-1704.

6.   Gutman, I. The Acyclic Polynomial of a Graph. *Publ. Inst. Math. (Beograd)*, **1977**, *22*, 63-69.

7.   Farrell, E. J. An Introduction to Matching Polynomials. *J. Comb. Theory B* **1979**, *27*, 75-86.

8.   Gutman, I. The Matching Polynomial. *Commun. Math. Chem. (MATCH)* **1979**, *6*, 75-91.

9.   Godsil, C. D.; Gutman, I. On the Theory of the Matching Polynomial. *J. Graph Theory* **1981**, *5*, 137-144.

10.   Farrell, E. J. On a Class of Polynomials Obtained from the Circuits in a Graph and Its Application to Characteristic Polynomials of Graphs. *Discr. Math.* **1979**, *25*, 121-133.

11.   Gutman, I.; Polansky, O. E. Cyclic Conjugation and the Hückel Molecular Orbital Model. *Theor. Chim. Acta* **1981**, *60*, 203-226.

12.   Farrell, E. J.; Gutman, I. A Note on the Circuit Polynomial and Its Relation to the μ-Polynomial. *Commun. Math. Chem. (MATCH)* **1985**, *18*, 55-61.

13.   Aihara, J. Resonance Energies of Benzenoid Hydrocarbons. *J. Am. Chem. Soc.* **1977**, *99*, 2048-2053.

14.   Gutman, I. A Real Graph Polynomial?. *Graph Theory Notes New York* **1992**, *22*, 33-37.

15.   Li, X.; Zhao, B.; Gutman, I. More Examples for Supporting the Validity of a Conjecture on β-Polynomial. *J. Serb.. Chem. Soc.* **1995**, *60*, 1095-1101.

16.   Kasum, D.; Trinajstić, N.; Gutman, I. Chemical Graph Theory. III. On the Permanental Polynomial. *Croat. Chem. Acta* **1981**, *54*, 321-328.

17.   Rosenfeld, V. R.; Gutman, I. A Novel Approach to Graph Polynomials. *Commun. Math. Chem. (MATCH)* **1989**, *24*, 191-199.

18.   Balasubramanian, K. Immanant Polynomials of Graphs. *Theor. Chim. Acta* **1993**, *85*, 379-390.

19.   Botti, P.; Merris, R. Almost all Trees Share a Complete Set of Immanantal Polynomials. *J. Graph Theory* **1993**, *17*, 467-476.

20.   Merris, R. A Survey of Graph Laplacians. *Lin. Multilin. Algebra* **1995**, *39*, 19-31.

21.   Hosoya, H. Topological Index. A Newly Proposed Quantity Characterizing the Topological Nature of Structural Isomers of Saturated Hydrocarbons. *Bull. Chem. Soc. Japan* **1971**, *44*, 2332-2339.

22.   Coulson, C. A.; Rushbrooke, G. S. Note on the method of molecular orbitals. *Proc. Cambridge Phil. Soc.* **1940**, *36*, 193-200.

23.   Mallion, R. B.; Rouvray, D. H. The Golden Jubilee of the Coulson-Rushbrooke Pairing Theorem. *J. Math. Chem.* **1990**, *5*, 1-21.

24.   Cvetković, D. M.; Gutman, I. On Spectral Structure of Graphs Having the Maximal Eigenvalue not Greater than Two. *Publ. Inst. Math. (Beograd)* **1975**, *18*, 39-45.

25.   Gutman, I. A Graph-Theoretical Study of Conjugated Systems Containing a Linear Polyene Fragment. *Croat. Chem. Acta* **1976**, *48*, 97-108.

26.   Gutman, I. A Graph Theoretical Study on Linear Polyenes and Their Derivatives. *Acta Chim. Hung.* **1979**, *99*, 145-153.

27. Balaban, A. T.; Harary, F. The Characteristic Polynomial does not Uniquely Determine the Topology of a Molecule.

28. *J. Chem. Docum.* **1971**, *11*, 258-259.

29. Schwenk, A. J. Almost all Trees are Cospectral. In: Harary, F., Ed.; *New Directions in the Theory of Graphs*; Academic Press: New York, 1973; pp 275-307.

30. Heilbronner, E., Some Comments on Cospectral Graphs, *Commun. Math. Chem. (MATCH)* **1979**, *5*, 105-133.

31. Heilbronner, E.; Jones, T. B. Spectral Differences between 'Isospectral' Molecules, *J. Am. Chem. Soc.* **1978**, *100*, 6506-6507.

32. Hosoya, H. Graphical Enumeration of the Coefficients of the Secular Polynomials of the Hückel Molecular Orbitals. *Theor. Chim. Acta* **1972**, *25*, 215-222.

33. Gutman, I. Topological Properties of Benzenoid Systems. An Identity for the Sextet Polynomial. *Theor. Chim. Acta* **1977**, *45*, 309-315.

34. El-Basil, S. Gutman Trees. *J. Chem. Soc. Faraday Trans.* 2 **1986**, *82*, 299-316.

35. El-Basil, S. Caterpillar (Gutman) Trees in Chemical Graph Theory. *Topics Curr. Chem.* **1990**, *153*, 273-292.

36. Hosoya, H.; Yamaguchi, T. Sextet Polynomial. A New Enumeration and Proof Technique for the Resonance Theory Applied to the Aromatic Hydrocarbons. *Tetrahedron Lett.* **1975**, 4659-4662.

37. Gutman, I.; Hosoya, H.; Yamaguchi, T. Motoyama, A.; Kuboi, N. Topological Properties of Benzenoid Systems. III. Recursion Relation for the Sextet Polynomial. *Bull. Soc. Chim. Beograd* **1977**, *42*, 503-510.

38. Gutman, I. Some Combinatorial Consequences of Clar's Resonant Sextet Theory. *Commun. Math. Chem. (MATCH)* **1981**, *11*, 127-143.

39. Gutman, I. The R-Polynomial. A New Combinatorial Technique in Resonance Theory. *Bull. Soc. Chim. Beograd* **1981**, *46*, 17-22.

40. Gutman, I.; Cyvin, S. J. *Introduction to the Theory of Benzenoid Hydrocarbons*; Springer-Verlag: Berlin, 1989.

41. Hosoya, H. Clar's Aromatic Sextet and Sextet Polynomial. *Topics Curr. Chem.* **1990**, *153*, 255-272.

42. Balaban, A. T., Ed.; *Chemical Applications of Graph Theory*; Academic Press: London, 1976.

43. Cvetković, D.; Doob, M.; Sachs, H. *Spectra of Graphs - Theory and Application*; Academic Press: New York, 1980; 2nd revised ed.: Barth: Heidelberg, 1995.

44. Trinajstić, N. *Chemical Graph Theory*; CRC Press: Boca Raton, 1983; 2nd revised ed. 1992.

45. Gutman, I.; Polansky, O. E. Mathematical Concepts in Organic Chemistry; Springer-Verlag: Berlin, 1986.

46. Polansky, O. E.; Mark, G.; Zander, M. *Der Topologische Effekt an Molekülorbitalen (TEMO) - Grundlagen und Nachweis*; MPI Strahlenchemie: Mülheim, 1987.

47. Cvetković, D. M.; Doob, M.; Gutman, I.; Torgašev, A. *Recent Results in the Theory of Graph Spectra*; North-Holland: Amsterdam, 1988.

48. Brualdi, R. A.; Ryser, H. J. *Combinatorial Matrix Theory*; Cambridge Univ. Press: Cambridge, 1991.

49. Bonchev, D.; Rouvray D. H., Eds.; *Chemical Graph Theory - Introduction and Fundamentals*; Gordon & Breach: New York, 1991.

50. Dias, J. R. Molecular Orbital Calculations Using Chemical Graph Theory; Springer-Verlag: Berlin, 1993.

51. Godsil, C. D. *Algebraic Combinatorics*; Chapman & Hall: New York, 1993.

52. Diudea, M. V.; Ivanciuc, O. *Molecular Topology* (in Rumanian); Comprex: Cluj, 1995.

53. Graham, R.; Grötschel, M.; Lovász, L., Eds.; *Handbook of Combinatorics*; Elsevier: Amsterdam, 1995.

54. van Dam, E. *Graphs with Few Eigenvalues. An Interplay between Combinatorics and Algebra*; Center for Economic Research, Tilburg Univ.: Tilburg, 1996.

55. Cvetković, D.; Rowlinson, P.; Simić, S. *Eigenspaces of Graphs*; Cambridge Univ. Press: Cambridge, 1997.

56. Chung, F. R. K. *Spectral Graph Theory*; Am. Math. Soc.: Providence, 1997.

57. Gutman, I.; Trinajstić, N. Graph Theory and Molecular Orbitals. *Topics Curr. Chem.* **1973**, *42*, 49-93.

58. Gutman, I.; Trinajstić, N. Graph Spectral Theory of Conjugated Molecules. *Croat. Chem. Acta* **1975**, *47*, 507-533.

59. Trinajstić, N.; Živković, T. Graph Theory in Theoretical Chemistry (in Hungarian). *Kem. Közl.* **1975**, *44*, 460-465.

60. Trinajstić, N. New Developments in Hückel Theory. *Int. J. Quantum Chem. Quantum Chem. Symp.* **1977**, *11*, 469-472.

61. Trinajstić, N. Computing the Characteristic Polynomial of a Conjugated System Using the Sachs Theorem. *Croat. Chem. Acta* **1977**, *49*, 539-633.

62. Farrell, E. J. A Survey of the Unifying Effects of F-Polynomials in Combinatorics and Graph Theory. In: Cvetković, D.; Gutman, I.; Pisanski, T.; Tošić, R., Eds.; *Graph Theory*. Proceedings of the Fourth Yugoslav Seminar on Graph Theory; Univ. Novi Sad: Novi Sad, **1984**; pp 137-150.

63. Balasubramanian, K. Applications of Combinatorics and Graph Theory to Spectroscopy and quantum Chemistry. *Chem. Rev.* **1985**, *85*, 599-618.

64. Balasubramanian, K. Applications of Edge-Colorings of Graphs and Characteristic Polynomials to Spectroscopy and Quantum Chemistry. *J. Mol. Struct.* (*Theochem*) **1989**, *185*, 229-248.

65. Dias, J. R. Facile Calculations of the Characteristic Polynomial and π-Energy Levels of Molecules Using Chemical Graph Theory. *J. Chem. Educ.* **1987**, *64*, 213-216.

66. Gutman, I. Graphs and Graph Polynomials of Interest in Chemistry. In: Tinhofer, G.; Schmidt, G., Eds.; *Graph-Theoretic Concepts in Computer Science*; Springer-Verlag: Berlin, 1987; pp 177-187.

67. Gutman, I. Polynomials in Graph Theory (in Bulgarian). In: Tyutyulkov, N.; Bonchev, D., Eds.; *Graph Theory and Applications in Chemistry*; Nauka i Izkustvo: Sofia, 1987; pp 53-85.

68. Hosoya, H. On some Counting Polynomials in Chemistry. *Discr. Appl. Math.* **1988**, *19*, 239-257.

69. Trinajstić, N. The Characteristic Polynomial of a Chemical Graph. *J. Math. Chem.* **1988**, *2*, 197-215.

70. Grone, R.; Merris, R.; Sunder, V. S. The Laplacian Spectrum of a Graph. *SIAM J. Matrix Anal. Appl.* **1990**, *11*, 218-238.

71. Hosoya, H. Counting Polynomials in Valence Bond Theory. In: Klein, D. J.; Trinajstić, N., Eds.; *Valence Bond Theory and Chemical Structure*; Elsevier: Amsterdam, 1990; pp 53-80.

72. John, P.; Sachs, H. Calculating the Numbers of Perfect Matchings and of Spanning Trees, Pauling's Orders, the Characteristic Polynomial, and the Eigenvectors of a Benzenoid System. *Topics Curr. Chem.* **1990**, *153*, 145-179.

73. Mohar, B. The Laplacian Spectrum of Graphs. In: Alavi, Y.; Chartrand, G.; Ollermann, O. R.; Schwenk, A. J., Eds.; *Graph Theory, Combinatorics, and Applications*; Wiley: New York, 1991; pp 871-898.

74. Gutman, I. Polynomials in Graph Theory. In: Bonchev, D.; Rouvray, D. H., Eds.; *Chemical Graph Theory: Introduction and Fundamentals*; Gordon & Breach: New York, 1991; pp 133-176.

75. Mihalić, Z.; Veljan, D.; Amić, D.; Nikolić, S.; Plavšić, D.; Trinajstić, N. The Distance Matrix in Chemistry. *J. Math. Chem.* **1992**, *11*, 223-258.

76. Dias, J. R. An Example of Molecular Orbital Calculation Using the Sachs Graph Method. *J. Chem. Educ.* **1992**, *69*, 695-700.

77. Merris, R. Laplacian Matrices of Graphs: A Survey. *Lin. Algebra Appl.* **1994**, 197-198, 143-176.

78. Trinajstić, N.; Babić, D.; Nikolić, S.; Plavšić, D.; Amić, D.; Mihalić, Z. The Laplacian Matrix in Chemistry. *J. Chem. Inf. Comput. Sci.* **1994**, *34*, 368-376.

79. Motoyama, A.; Hosoya, H. Tables of King and Domino Polynomials for Polyomino Graphs. *Nat. Sci. Rep. Ochanomizu Univ.* **1976**, *27*, 107-123.

80. Ohkami, N.; Hosoya, H. Tables of the Topological Characteristics of Polyhexes (Condensed Aromatic Hydrocarbons). II. Sextet Polynomials. *Nat. Sci. Rep. Ochanomizu Univ.* **1984**, *35*, 71-83.

81. Gutman, I.; Petrović, S.; Mohar, B. Topological Properties of Benzenoid Systems. XX. Matching Polynomials and Topological Resonance Energies of Cata-Condensed Benzenoid Hydrocarbons, *Coll. Sci. Papers Fac. Sci. Kragujevac* **1982**, *3*, 43-90.

82. Gutman, I.; Petrović, S.; Mohar, B. Topological Properties of Benzenoid Systems. XXa. Matching Polynomials and Topological Resonance Energies of Peri-Condensed Benzenoid Hydrocarbons. *Coll. Sci. Papers Fac. Sci. Kragujevac* **1983**, *4*, 189-225.

83. Cvetković, D.; Petrić, M. A Table of Connected Graphs on Six Vertices. *Discrete Math.* **1984**, *50*, 37-49.

84. Stevanović, D.; Gutman, I. Hosoya Polynomials of Trees with up to 10 Vertices. *Coll. Sci. Papers Fac. Sci. Kragujevac* **1999**, *21*, 111-119.

85. Farrell, E. J.; Wahid, S. A. Some General Classes of Comatching Graphs. *Int. J. Math. Math. Sci.* **1987**, *10*, 519-524.

86. Farrell, E. J.; Grell, J. C. Some Further Constructions of Cocircuit and Cospectral Graphs. *Caribb. J. Math.* **1986**, *4*, 17-28.

87. Graovac, A.; Gutman, I.; Trinajstić, N.; Živković, T. Graph Theory and Molecular Orbitals. Application of Sachs Theorem. *Theor. Chim. Acta* **1972**, *26*, 67-78.

88. Tang, A.; Kiang, Y. Graph Theory of Molecular Orbitals. *Sci. Sinica* **1976**, *19*, 207-226.

89. Günthard, H. H.; Primas, H. Zusammenhang von Graphen Theorie und MO-Theorie von Molekulen mit Systemen konjugierter Bindungen. *Helv. Chim. Acta* **1956**, *39*, 1645-1653.

90. Bieri, G.; Dill, J. D.; Heilbronner, E.; Schmelzer, A. Application of the Equivalent Bond Orbital Model to the C_{2s}-Ionization Energies of Saturated Hydrocarbons. *Helv. Chim. Acta* **1977**, *60*, 2234-2247.

91. Heilbronner, E. A Simple Equivalent Bond Orbital Model for the Rationalization of the C_{2s}-Photoelectron Spectra of the Higher n-Alkanes, in Particular of Polyethylene. *Helv. Chim. Acta* **1977**, *60*, 2248-2257.

92. Samuel, I. Résolution d'un Déterminant Seculaire par la Méthode des Polygones. *Compt. Rend. Acad. Sci. Paris* **1949**, *229*, 1236-1237.

93. Coulson, C. A. Notes on the Secular Determinant in Molecular Orbitals Theory. *Proc. Cambridge Phil. Soc.* **1950**, *46*, 202-205.

94. Harary, F. A Graph Theoretic Method for the Complete Reduction of a Matrix with a View Toward Finding its Eigenvalues. *J. Math. Phys.* **1959**, *38*, 104-111.

95. Harary, F. The Determinant of the Adjacency Matrix of a Graph. *SIAM Rev.* **1962**, *4*, 202-210.

96. Sachs, H. Beziehungen zwischen den in einem Graphen entaltenen Kreisen und seinem Charakteristischen Polynom. *Publ. Math. (Debrecen)* **1964**, *11*, 119-134.

97. Milić, M. Flow-Graph Evaluation of the Characteristic Polynomial of a Matrix. *IEEE Trans. Circuit Theory* **1964**, *11*, 423-424.

98. Spialter, L. The Atom Connectivity Matrix Characteristic Polynomial (ACM CP) and Its Physico-Geometric (Topological) Significance. *J. Chem. Docum.* **1964**, *4*, 269-274.

99. Gutman, I. Rectifying a Misbelief: Frank Harary's Role in the Discovery of the Coefficient-Theorem in Chemical Graph Theory. *J. Math. Chem.* **1994**, *16*, 73-78.

100. Heilmann, O. J.; Lieb, E. H. Monomers and Dimers. *Phys. Rev. Lett.* **1970**, *24*, 1412-1414.

101. Heilmann, O. J.; Lieb E. H. Theory of Monomer-Dimer Systems. *Commun. Math. Phys.* **1972**, *25*, 190-232.

102. Kunz, H. Location of the Zeros of the Partition Function for Some Classical Lattice Systems. *Phys. Lett* **1970**, *32A*, 311-312.

103. Gruber, C.; Kunz, H. General Properties of Polymer Systems. *Commun. Math. Phys.* **1971**, *22*, 133-161.

104. Nijenhuis, A. On Permanents and the Zeros of Rook Polynomials. *J. Comb. Theory A* **1976**. *21*. 240-244.

105. Godsil, C. D. Gutman, I. On the Matching Polynomial of a Graph. In: Lovász, L.; Sós, V. T., Eds.; *Algebraic Methods in Graph Theory*; North-Holland: Amsterdam, 1981; pp 241-249.

106. Farrell, E. J. On the Matching Polynomials and Its Relation to the Rook Polynomial. *J. Franklin Inst.***1988**, *325*, 527-536.

107. Farrell, E. J.; Whitehead, E. G. Matching, Rook and Chromatic Polynomials and Chromatically Vector Equivalent Graphs. *J. Comb. Math. Comb. Comput.* **1991**, *9*, 107-118.

108. Farrell, E. J. The Matching Polynomial and Its Relation to the Acyclic Polynomial of a Graph. *Ars Comb.* **1980**, *9*, 221-228.

109. Gutman, I.; Hosoya, H. On the Calculation of the Acyclic Polynomial. *Theor. Chim. Acta* **1978**, *48*, 279-286.

110. Godsil, C. D.; Gutman, I. Topological Resonance Energy is Real. *Z. Naturforsch.* **1979**, *34a*, 776-777.

111. Godsil, C. D.; Gutman, I. Some Remarks on the Matching Polynomial and Its Zeros. *Croat. Chem. Acta* **1981**, *54*, 53-59.

112. Godsil, C. D.. Matchings and Walks in Graphs. *J. Graph Theory* **1981**, *5*, 285-297.

113. Farrell, E. J. Matchings in Ladders. *Ars Comb.* **1978**, *6*, 153-161.

114. Farrell, E. J. Matchings in Complete Graphs and Complete Bipartite Graphs. *J. Comb. Inf. System Sci.* **1980**, *5*, 47-51.

115. Gutman, I.; Polansky, O. E. On the Matching Polynomial of the Graph G{$R_1,R_2,...$,R_n}. *Commun. Math. Chem. (MATCH)*, **1980**, *8*, 315-322.

116. Gutman, I. On Graph Theoretical Polynomials of Annulenes and Radialenes. *Z. Naturforsch.* **1980**, *35a*, 453-457.

117. Gutman, I. Characteristic and Matching Polynomials of some Compound Graphs. *Publ. Inst. Math. (Beograd)*, **1980**, *27*, 61-66.

118. Farrell, E. J. Matchings in Triangular Animals. *J. Comb. Inf. System Sci.* **1982**, *7*, 143-154.

119. Gutman, I. Characteristic and Matching Polynomials of Benzenoid Hydrocarbons. *J. Chem. Soc. Faraday Trans.* 2 **1983**, *79*, 337-345.

120. Gutman, I.; Farrell, E. J.; Wahid, S. A. On the Matching Polynomials of Graphs Containing Benzenoid Chains. *J. Comb. Inf. System Sci.* **1983**, *8*, 159-168.

121. Farrell, E. J.; Wahid, S. A. Matchings in Benzene Chains, *Discr. Appl. Math.* **1984**, *7*, 45-54.

122. Farrell, E. J.; Wahid, S. A. Matchings in Pentagonal Chains. *Discr. Appl. Math.* **1984**, *8*, 31-40.

123. Farrell, E. J.; Wahid, S. A. Matchings in Heptagonal and Octagonal Chains. *Caribb. J. Math.* **1985**, *3*, 45-57.

124. Farrell, E. J.; Wahid, S. A. Matchings in Nonagonal and Decagonal Chains. *Caribb. J. Math.* **1985**, *3*, 73-84.

125. Farrell, E. J.; Wahid, S. A., Matchings in Long Benzene Chains, *Commun. Math. Chem. (MATCH)* **1985**, *18*, 37-48.

126. Graovac, A.; Babić, D. Comment on ''Matching in Long Benzene Chains'' by E. J. Farrell and S. A. Wahid. *Commun. Math. Chem. (MATCH)*, **1985**, *18*, 49-53.

127. Babić, D.; Graovac, A.; Mohar, B.; Pisanski, T. The Matching Polynomial of a Polygraph. *Discr. Appl. Math.* **1986**, *15*, 11-24.

128. Strunje, M. On the Matching and the Characteristic Polynomial of the Graph of Truncated Octahedron. *Commun. Math. Chem. (MATCH)* **1986**, *20*, 283-287.

129. Gutman, I. El-Basil, S. Fibonacci Graphs. *Commun. Math. Chem. (MATCH)*, **1986**, *20*, 81-94.

130. Farrell, E. J. Counting Matchings in Graphs. *J. Franklin Inst.* **1987**, *324*, 331-339.

131. Randić, M.; Hosoya, H.; Polansky, O. E. On the Construction of the Matching Polynomial for Unbranched CataCondensed Benzenoids. *J. Comput. Chem.* **1989**, *10*, 683-697.

132. Farrell, E. J.; Wahid, S. A. A Note on the Characterization of Thistles by Their Matching Polynomials. *J. Comb. Math. Comb. Comput.* **1990**, *8*, 97-101.

133. Farrell, E. J.; Whitehead, E. G. On Matching and Chromatic Properties of Circulants. *J. Comb. Math. Comb. Comput.* **1990**, *8*, 79-88.

134. Gutman, I. Characteristic and Matching Polynomials of Some Bipartite Graphs. *Publ. Elektrotehn. Fak. (Beograd) Ser. Mat.* **1990**, *1*, 25-30.

135. Gutman, I.; Hosoya, H. Molecular Graphs with Equal Z-Counting and Independence Polynomials. *Z. Naturforsch.* **1990**, *45a*, 645-648.

136. Hosoya, H. Factorization and Recursion Relations of the Matching and Characteristic Polynomials of Periodic Polymer Networks. *J. Math. Chem.* **1991**, *7*, 289-305.

137. Farrell, E. J.; Guo, J. M.; Constantine, G. M. On Matching Coefficients. *Discr. Math.* **1991**, *89*, 203-209.

138. Farrell, E. J.; Kennedy, J. W.; Quintas, L. V.; Wahid, S. A. Graphs with Palindromic Matching and Characteristic Polynomials. *Vishwa Int. J. Graph Theory* **1992**, *1*, 59-76.

139. Balasubramanian, K. Matching Polynomials of Fullerene Clusters. *Chem. Phys. Lett.* **1993**, *201*, 306-314.

140. Gutman, I. A Contribution to the Study of Palindromic Graphs. *Graph Theory Notes New York* **1993**, *24*, 51-56.

141. Hosoya, H.; Harary, F. On the Matching Properties of Three Fence Graphs. *J. Math. Chem.* **1993**, *12*, 211-218.

142. Babić, D.; Ori, O. Matching Polynomial and Topological Resonance Energy of C_{70},. *Chem. Phys. Lett.* **1995**, *234*, 240-244.

143. Beezer, R. A.; Farrell, E. J. The Matching Polynomial of a Regular Graph. *Discr. Math.* **1995**, *137*, 7-18.

144. Gutman, I.; Hosoya, H.; Babić, D. Topological Indices and Graph Polynomials of some Macrocyclic Belt-Shaped Molecules. *J. Chem. Soc. Faraday Trans.* **1996**, *92*, 625-628.

145. Herndon, W. C.; Ellzey, M. L. Procedures for Obtaining Graph-Theoretical Resonance Energies. *J. Chem. Inf. Comput. Sci.* **1979**, *19*, 260-264.

146. Mohar, B.; Trinajstić, N. On Computation of the Topological Resonance Energy. *J. Comput. Chem.* **1982**, *3*, 28-36.

147. Hosoya, H.; Ohkami, N. Operator Technique for Obtaining the Recursion Formulas of Characteristic and Matching Polynomials as Applied to Polyhex Graphs. *J. Comput. Chem.* **1983**, *4*, 585-593.

148. El-Basil, S. Fibonacci Relations. On the Computation of some Counting Polynomials of Very Large Graphs. *Theor. Chim. Acta* **1984**, *65*, 199-213.

149. Hosoya, H.; Motoyama, A. An Effective Algorithm for Obtaining Polynomials for Dimer Statistics. Application of Operator Technique on the Topological Index to Two- and Three-Dimensional Rectangular and Torus Lattices. *J. Math. Phys.* **1985**, *26*, 157-167.

150. Ramaraj, R.; Balasubramanian, K. Computer Generation of Matching Polynomials of Chemical Graphs and Lattices. *J. Comput. Chem.* **1985**, *6*, 122-141.

151. Herndon, W. C.; Radhakrishnan, T. P.; Živković, T. P. Characteristic and Matching Polynomials of Chemical Graphs. *Chem. Phys. Lett.* **1988**, *152*, 233-238.

152. Hosoya, H.; Balasubramanian, K. Computational Algorithms for Matching Polynomials of Graphs from the Characteristic Polynomials of Edge-Weighted Graphs. *J. Comput. Chem.* **1989**, *10*, 698-710.

153. Balakrishnarajan, M. M.; Venuvanalingam, P. General Method for the Computation of Matching Polynomials of Graphs. *J. Chem. Inf. Comput. Sci.* **1994**, *34*, 1122-1126.

154. Schaad, L. J.; Hess, B. A.; Nation, J. B.; Trinajstić, N.; Gutman, I. On the Reference Structure for the Resonance Energy of Aromatic Hydrocarbons. *Croat. Chem. Acta* **1979**, *52*, 233-248.

155. Graovac, A. On the Construction of a Hermitean Matrix Associated with the Acyclic Polynomial of a Conjugated Hydrocarbon. *Chem. Phys. Lett.* **1981**, *82*, 248-251.

156. Gutman, I.; Graovac, A.; Mohar, B., On the Existence of a Hermitean Matrix whose Characteristic Polynomial is the Matching Polynomial of a Molecular Graph. *Commun. Math. Chem. (MATCH)*, **1982**, *13*, 129-150.

157. Mizoguchi, N. The Structure Corresponding to the Reference Polynomial of {N}Annulene in the Topological Resonance Energy Theory. *J. Am. Chem. Soc.* **1985**, *107*, 4419-4421.

158. Graovac, A.; Polansky, O. E. On Hermitean Matrices Associated with the Matching Polynomials of Graphs. Part I. On some Graphs whose Matching Polynomial is the Characteristic Polynomial of a Hermitean Matrix. *Commun. Math. Chem. (MATCH)* **1986**, *21*, 33-45.

159. Polansky, O. E.; Graovac, A. On Hermitean Matrices Associated with the Matching Polynomials of Graphs. Part II. Konstruktion von Gewichtmatrizen für beliebige, mit K_4 homöomorphe trizyklische Graphen, welche das Matching Polynom des Graphen erzeugen. *Commun. Math. Chem. (MATCH)*, **1986**, *21*, 47-79.

160. Graovac, A.; Polansky, O. E. On Hermitean Matrices Associated with the Matching Polynomials of Graphs. Part III. The Matching Polynomials of Bicyclic and Tricyclic Cata-Condensed Graphs are the Characteristic Polynomials of

Hermitean Matrices with Quaternionic Weights. *Commun Math. Chem.* (*MATCH*), **1986**, *21*, 81-91.

161. Polansky, O. E.; Graovac, A. On Hermitean Matrices Associated with the Matching Polynomials of Graphs. Part IV. Zur Existenz und Konstruktion von mit dem Matching Polynom Assozierten Gewichtmatrizen für Polyzyklische Graphen. *Commun. Math. Chem.* (*MATCH*), **1986**, *21*, 93-113.

162. Mizoguchi, N. Graphs Corresponding to Reference Polynomial or to Circuit Characteristic Polynomial. *J. Math. Chem.* **1993**, *12*, 265-277.

163. Farrell, E. J.; Wahid, S. A. D-Graphs 1; An introduction to Graphs whose Matching Polynomials are Determinants of Matrices. *Bull. Inst. Comb. Appl.* **1995**, *15*, 81-86.

164. Schwenk, A. J. On Unimodal Sequences of Graphical Invariants. *J. Comb. Theory B* **1981**, *30*, 247-250.

165. Godsil, C. D. Hermite Polynomials and a Duality Relation for Matchings Polynomials. *Combinatorica* **1981**, *1*, 257-262.

166. Godsil, C. D. Matching Behavior is Asymptotically Normal. *Combinatorica* **1981**, *1*, 369-376.

167. Gutman, I. A Note on Analogies between the Characteristic and the Matching Polynomial of a Graph. *Publ. Inst. Math.* (*Beograd*) **1982**, *31*, 27-31.

168. Gutman, I.; Zhang, F. On the Ordering of Graphs with Respect to Their Matching Numbers. *Discr. Appl. Math.* **1986**, *15*, 25-33.

169. Gutman, I. Some Analytical Properties of the Independence and Matching Polynomials. *Commun. Math. Chem.* (*MATCH*), **1992**, *28*, 139-150.

170. Gutman, I. Some Relations for the Independence and Matching Polynomials and Their Chemical Applications. *Bull. Acad. Serbe Sci. Arts* (*Cl. Math. Natur.*), **1992**, *105*, 39-49.

171. Gutman, I.; Cvetković, D. M. Relations between Graphs and Special Functions. *Coll. Sci. Papers Fac. Sci. Kragujevac* **1980**, *1*, 101-119.

172. Hosoya, H. Graphical and Combinatorial Aspects of some Orthogonal Polynomials. *Nat. Sci. Repts. Ochanomizu Univ.* **1981**, *31*, 127-138.

173. Godsil, C. D. Some Graphs with Characteristic Polynomials which are not Solvable by Radicals. *J. Graph Theory* **1982**, *6*, 211-214.

174. Heilbronner, E. Das Kompositions-Prinzip: Eine anschauliche Methode zur elektronen-Theoretischen Behandlung nicht oder niedrig Symmetrischer Molekeln im Rahmen der MO-Theorie. *Helv. Chim. Acta* **1953**, *36*, 170-188.

175. Heilbronner, E. Ein Graphisches Verfahren zur Faktorisierung der Sekulärdeterminante Aromatischer Ringsysteme im Rahmen der LCAO-MO Theorie. *Helv. Chim. Acta* **1954**, *37*, 913-921.

176. Schwenk, A. J. Computing the Characteristic Polynomial of a Graph. In: Bari, R. A.; Harary F., Eds.; *Graphs and Combinatorics*; Springer-Verlag: Berlin, 1974; pp 153-172.

177. Kaulgud, M. V.; Chitgopkar, V. H. Polynomial Matrix Method for the Calculation of π-Electron Energies for Linear Conjugated Polymers. *J. Chem. Soc. Faraday Trans. 2.* **1977**, *73*, 1385-1395.

178. Kaulgud, M. V.; Chitgopkar, V. H. Polynomial Matrix Method for the Calculation of Charge Densities and Bond Orders in Linear Conjugated Polymers. *J. Chem. Soc. Faraday Trans.* 2. **1977**, *74*, 951-957.

179. Gutman, I. Generalizations of a Recurrence Relation for the Characteristic Polynomials of Trees. *Publ. Inst. Math. (Beograd)*, **1977**, *21*, 75-80.

180. Schwenk, A. J.; Wilson, R. J.; On the Eigenvalues of a Graph. In: Beineke, L. W.; Wilson, R. J., Eds.; *Selected Topics in Graph Theory*; Academic Press: London, 1978; pp 307-336.

181. Gutman, I. Polynomial Matrix Method for the Estimation of π-Electron Energies of some Linear Conjugated Molecules. *J. Chem. Soc. Faraday Trans.* 2. **1980**, *76*, 1161-1169.

182. Jiang, Y. Graphical-Deleting Method for Computing Characteristic Polynomial of a Graph. *Sci. Sinica* **1982**, *25*, 681-689.

183. Gutman, I.; Shalabi, A. Edge Erasure Technique for Computing the Characteristic Polynomials of Molecular Graphs. *Commun. Math. Chem. (MATCH)* **1985**, *18*, 3-15.

184. Kiang, Y. S.; Tang, A. C. A Graphical Evaluation of Characteristic Polynomial of Hückel Trees. *Int. J. Quantum Chem.* **1986**, *29*, 229-240.

185. Krivka, P.; Jeričević, Ž.; Trinajstić, N. On the Computation of Characteristic Polynomial of a Chemical Graph. *Int. J. Quantum Chem. Quantum Chem. Symp.* **1986**, *19*, 129-147.

186. Rowlinson, P. A. A Deletion-Construction Algorithm for the Characteristic Polynomial of a Multigraph. *Proc. Roy. Soc. Edinburgh* **1987**, *A105*, 153-160.

187. Kolmykov, V. A. Several Relationships for Characteristic Polynomials of Graphs. *Vychisl. Sistemy* **1990**, *136*, 35-37 (in Russian).

188. John, P.; Sachs, H. Calculating the Characteristic Polynomial, Eigenvectors and Number of Spanning Trees of a Hexagonal System. *J. Chem. Soc. Faraday Trans.* **1990**, *86*, 1033-1039.

189. John, P. E. Calculating the Characteristic Polynomial and the Eigenvectors of a Weighted Hexagonal System (Benzenoid Hydrocarbons with Heteroatoms). *Commun. Math. Chem. (MATCH)*, **1994**, *30*, 153-169.

190. Rosenfeld, V. R.; Gutman, I. A New Recursion Relation for the Characteristic Polynomial of a Molecular Graph. *J. Chem. Inf. Comput. Sci.* **1996**, *36*, 527-530.

191. Zhao, H.; Wang, Y. Calculation of the Coefficients of the Characteristic Polynomial from its Subgraphs. *Int. J. Quantum Chem.* **1996**, *59*, 97-102.

192. John, P. E.; Schild, G. Calculating the Characteristic Polynomial and the Eigenvectors of a Tree. *Commun. Math. Chem. (MATCH)* **1996**, *34*, 217-237.

193. Clarke, F. H. A Graph Polynomial and Its Applications, *Discr. Math.* **1972**, *3*, 305-313.

194. Gutman, I.; Cvetković, D. M. The Reconstruction Problem for Characteristic Polynomials of Graphs. *Publ. Elektrotehn. Fak. (Beograd)*, *Ser. Mat. Fiz.* **1975**, *505*. 45-48.

195. Krivka, P.; Mallion, R. B.; Trinajstić, N. Chemical Graph Theory. Part VII. The Use of Ulam Sub-Graphs in Obtaining Characteristic Polynomials. *J. Mol. Struct. (Theochem)*, **1988**, *164*, 363-377.

196. Gutman, I.; Li, X.; Zhang, H. On a Formula Involving the First Derivative of the Characteristic Polynomial of a Graph. *Publ. Elektrotehn. Fak. (Beograd), Ser. Mat.* **1993**, *4*, 97-102.

197. Gutman, I. Graph Propagators. *Graph Theory Notes New York* **1990**, *19*, 26-30.

198. Gutman, I. Some Relations for Graphic Polynomials. *Publ. Inst. Math. (Beograd)*, **1986**, *39*, 55-62.

199. Türker, L. Graph Theoretical Approach to HMO Coefficient Polynomials of Certain Molecules. *METU J. Pure Appl. Sci.* **1980**, 13209-224.

200. Mladenov, I. M.; Kotarov, M. D.; Vassileva-Popova, J. G. Method for Computing the Characteristic Polynomial. *Int. J. Quantum Chem.* **1980**, *18*, 339-341.

201. Balasubramanian, K.; Randić, M. The Characteristic Polynomials of Structures with Pending Bonds. *Theor. Chim. Acta* **1982**, *61*, 307-323.

202. Randić, M. On Evaluation of the Characteristic Polynomial for Large Graphs. *J. Comput. Chem.* **1982**, *3*, 421-435.

203. Randić, M. On Alternative Form of the Characteristic Polynomial and the Problem of Graph Recognition. *Theor. Chim. Acta* **1983**, *62*, 485-498.

204. Hosoya, H.; Randić, M., Analysis of the Topological Dependency of the Characteristic Polynomial in Its Chebyshev Expansion. *Theor. Chim. Acta* **1983**, *63*, 473-495.

205. Kirby, E. C. The Characteristic Polynomial: Computer-Aided Decomposition of the Secular Determinant as a Method of Evaluation for Homo-Conjugated Systems. *J. Chem. Res.* **1984**, 4-5.

206. Balasubramanian, K. The Use of Frame's Method for the Characteristic Polynomials of Chemical Graphs. *Theor. Chim. Acta* **1984**, *65*, 49-58.

207. El-Basil, S. Characteristic Polynomials of Large Graphs. On Alternate form of Characteristic Polynomial, *Theor. Chim. Acta* **1984**, *65*, 191-197.

208. Balasubramanian, K. Computer Generation of the Characteristic Polynomial of Chemical Graphs. *J. Comput. Chem.* **1984**, *5*, 387-394.

209. Balasubramanian, K. Characteristic Polynomials of Organic Polymers and Periodic Structures. *J. Comput. Chem.* **1985**, *6*, 656-661.

210. Balasubramanian, K.; Randić, M. Spectral Polynomials of Systems with General Interactions. *Int. J. Quantum Chem.* **1985**, *28*, 481-498.

211. Randić, M. On the Characteristic Equations of the Characteristic Polynomial. *SIAM J. Algebr. Discr. Math.* **1985**, *6*, 145-162.

212. Brocas, J. Comments on Characteristic Polynomials of Chemical Graphs. *Theor. Chim. Acta* **1985**, *68*, 155-158.

213. Dias, J. R. Properties and Derivation of the Fourth and Sixth Coefficients of the Characteristic Polynomial of Molecular Graphs. New Graphical Invariants. *Theor. Chim. Acta* **1985**, *68*, 107-123.

214. Barakat, R. Characteristic Polynomials of Chemical Graphs via Symmetric Function Theory. *Theor. Chim. Acta* **1986**, *69*, 35-39.

215. Barysz, M.; Nikolić, S.; Trinajstić, N. A Note on the Characteristic Polynomial. *Commun. Math. Chem. (MATCH)* **1986**, *19*, 117-126.

216. Randić, M.; Baker, B.; Kleiner, A. F. Factoring the Characteristic Polynomial. *Int. J. Quantum Chem. Quantum Chem. Symp*. **1986**, *19*, 107-127.

217. Dias, J. R. Facile Calculations of Select Eigenvalues and the Characteristic Polynomial of Small Molecular Graphs Containing Heteroatoms. *Canad. J. Chem.* **1987**, *65*, 734-739.

218. Randić, M. On the Evaluation of the Characteristic Polynomial via Symmetric Function Theory. *J. Math. Chem.* **1987**, *1*, 145-152.

219. Kirby, E. C. The Factorization of Chemical Graphs and Their Polynomials: A Systematic Study of Certain Trees. *J. Math. Chem.* **1987**, *1*, 175-218.

220. Balasubramanian, K. Computer Generation of Characteristic Polynomials of Edge-Weighted Graphs, Heterographs, and Directed Graphs. *J. Comput. Chem.* **1988**, *9*, 204-211.

221. Kirby, E. C. A Note on some Coefficients of the Chebyshev Polynomial form of the Characteristic Polynomial. *J. Math. Chem.* **1988**, *2*, 83-87.

222. Jiang, Y.; Zhang, H. Moments and Characteristic Polynomials of Bipartite Hückel Graphs. *J. Math. Chem.* **1989**, *3*, 357-375.

223. Mohar, B. Computing the Characteristic Polynomial of a Tree. *J. Math. Chem.* **1989**, *3*, 403-406.

224. Balasubramanian, K. Recent Development in Tree-Pruning Methods and Polynomials for Cactus Graphs and Trees. *J. Math. Chem.* **1990**, *4*, 89-102.

225. Živković, T. P. On the Evaluation of the Characteristic Polynomial of a Chemical Graph. *J. Comput. Chem.* **1990**, *11*, 217-222.

226. Venuvanalingam, P.; Thangavel, P. Parallel Algorithm for the Computation of Characteristic Polynomials of Chemical Graphs. *J. Comput. Chem.* **1991**, *12*, 779-783.

227. Ivanciuc, O. Chemical Graph Polynomials. Part 2. The Propagation Diagram Algorithm for the Computation of the Characteristic Polynomial of Molecular Graphs. *Rev. Roum. Chim.* **1992**, *37*, 1341-1345.

228. John, P. E.; Rausch, W. Beschreibung eines Programms zur Berechnung des Charakteristischen Polynoms und der Eigenvektoren eines Graphen. *Commun. Math. Chem. (MATCH)* **1995**, *32*, 237-249.

229. Sachs, H. über Teiler, Faktoren und Charakteristische Polynome von Graphen, Teil I. *Wiss. Z. TH Ilmenau* **1966**, *11*, 7-12.

230. Sachs, H. über Teiler, Faktoren und Charakteristische Polynome von Graphen, Teil II. *Wiss. Z. TH Ilmenau* **1967**, *13*, 405-412.

231. McClelland, B. J. Graphical Method for Factorizing Secular Determinants of Hückel Molecular Orbital Theory. *J. Chem. Soc. Faraday Trans.* 2. **1974**, *70*, 1453-1456.

232. King, R. B. Symmetry Factoring of the Characteristic Equations of Graphs Corresponding to Polyhedra. *Theor. Chim. Acta* **1977**, *44*, 223-243.

233. Tang, A.; Kiang, Y. Graph Theory of Molecular Orbitals. II. Symmetrical Analysis and Calculation of MO Coefficients. *Sci. Sinica* **1977**, *20*, 595-612.

234. Sachs, H.; Stiebitz, M. Authorphism Group and Spectrum of a Graph. In: Hajnal, A.; Sós, V. T., Eds.; *Combinatorics*; North-Holland: Amsterdam, 1978; pp 657-670.

235. Zhang, Q.; Li, L.; Wang, N. Graphical Method of Hückel Matrix. *Sci. Sinica* **1979**, *22*, 1169-1184.

236. Yan, J. Symmetry Rules in the Graph Theory of Molecular Orbitals. *Adv. Quantum. Chem.* **1981**, *13*, 211-241.

237. Yan, J.; Wang, Z. The Characteristics of LCAO-MO's Secular Matrices for Symmetric Molecules and Their Applications. *Acta Chim. Sin.* **1981**, *39*, 17-22.

238. McClelland, B. J. Eigenvalues of the Topological Matrix. Splitting of Graphs with Symmetrical Components and Alternant Graphs. *J. Chem. Soc. Faraday Trans.* 2. **1982**, *78*, 911-916.

239. Dias, J. R. Characteristic Polynomials and Eigenvalues of Molecular Graphs with a Greater than Twofold Axis of Symmetry. *J. Mol. Struct. (Theochem)* **1988**, *165*, 125-148.

240. Datta, K. K.; Mukherjee, A. K. A Graph-Theoretical Method for Stepwise Factorization of Symmetric Graphs for Simultaneous Determination of Eigenvectors and Eigenvalues. *Proc. Indian Acad. Sci. (Chem. Sci.)*, **1989**, *101*, 143-154.

241. Rakshit, S. C.; Hazra, B. Extension of Graphical Method of Reduction of Topological Matrices of Chemical Graphs. *J. Indian Chem. Soc.* **1991**, *68*, 509-512.

242. Gineityte, V. Some Common Algebraic and Spectral Properties of Secular Polynomials of Graphs for Molecules of Similar Structure. *Lithuan. J. Phys.* **1992**, *32*, 175-182.

243. Hosoya, H.; Tsukano, Y. Efficient Way for Factorizing the Characteristic Polynomial of Highly Symmetrical Graphs such as the Buckminsterfullerene. *Fullerene Sci. Technol.* **1994**, *2*, 381-393.

244. Patra, S. M.; Mishra, R. K. Factors of Polyhedral Spectra: A Computational Approach. *J. Mol. Struct. (Theochem)* **1995**, *342*, 201-209.

245. Patra, S. M.; Mishra, R. K. Splitting the Characteristic Polynomial (CP) Using a Computational Technique to Obtain the Factors of the Mirror Plane and Two-~, Three-, and n-Fold Symmetric Graphs. *Int. J. Quantum Chem.* **1995**, *53*, 361-374.

246. Gineityte, V. Secular Polynomials for Chemical Graphs of Alkanes in Terms of Atoms and Bonds and Their Spectral Properties. *Int. J. Quantum Chem.* **1996**, *60*, 743-752.

247. Mallion, R. B.; Trinajstić, N.; Schwenk, A. J. Graph Theory in Chemistry - Generalization of Sachs' Formula. *Z. Naturforsch.* **1974**, *29a*, 1481-1484.

248. Aihara, J. General Rules for Constructing Hückel Molecular Orbital Characteristic Polynomials. *J. Am. Chem. Soc.* **1976**, *98*, 6840-6844.

249. Rigby, M. J.; Mallion, R. B.; Day, A. C. Comment on a Graph-Theoretical Description of Heteroconjugated Molecules. *Chem. Phys. Lett.* **1977**, *51*, 178-182.

250. Ivanciuc, O. Chemical Graph Polynomials. 1. The Polynomial Description of Generalized Chemical Graphs. *Rev. Roum. Chim.* **1988**, *33*, 709-717.

251. Rosenfeld, V. R.; Gutman, I. On the Graph Polynomials of a Weighted Graph. *Coll. Sci. Papers Fac. Sci. Kragujevac* **1991**, *12*, 49-57.

252. Collatz, L.; Sinogowitz, U. Spektren endlicher Graphen. *Abh. Math. Sem. Univ. Hamburg* **1957**, *21*, 63-77.

253. Derflinger, G.; Sofer, H. Die HMO-Koeffizienten der Linearen Polyacene in geschlossener Form. *Monatsh. Chem.* **1968**, *99*, 1866-1875.

254. Mowshowitz, A. The Characteristic Polynomial of a Graph. *J. Comb. Theory B* **1972**, *12*, 177-193.

255. Gutman, I. Electronic Properties of Möbius Systems. *Z. Naturforsch.* **1978**, *33a*, 214-216.

256. Gutman, I. Partial Ordering of Forests According to Their Characteristic Polynomials. In: Hajnal, A.; Sós, V. T., Eds.; *Combinatorics*; North-Holland: Amsterdam, 1978; pp 429-436.

257. Polansky, O. E.; Zander, M. Topological Effect on MO Energies. *J. Mol. Struct.* **1982**, *84*, 361-385.

258. Polansky, O. E.; Zander, M.; Motoc, I. Topological Effect on MO Energies. II. On the MO Energies of Structurally Related Aza-Arenes. *Z. Naturforsch.* **1983**, *38a*, 196-199.

259. Polansky, O. E. Topological Effects Displayed in Absorption and Photoelectron Spectra. *J. Mol. Struct.* **1984**, *113*, 281-298.

260. Godsil, C. D. Spectra of Trees. *Ann. Discr. Math.* **1984**, *20*, 151-159.

261. Gutman, I.; Graovac, A.; Polansky, O. E., On the Theory of S- and T-Isomers. *Chem. Phys. Lett.* **1985**, *116*, 206-209.

262. Graovac, A.; Gutman, I.; Polansky, O. E. An Interlacing Theorem in Simple Molecular-Orbital Theory. *J. Chem. Soc. Faraday Trans.* 2, **1985**, *81*, 1543-1553.

263. Polansky, O. E. Research Notes on the Topological Effect on Molecular Orbitals (TEMO); 1. On the Characteristic Polynomials of Topologically Related Isomers Formed within Three Different Models. *Commun. Math. Chem. (MATCH)*, **1985**, *18*, 111-166.

264. Polansky, O. E. Research Notes on the Topological Effect on Molecular Orbitals (TEMO); 2. Some Structural Requirements for the Central Subunit in a Particular Topological Model, *Commun. Math. Chem. (MATCH)*, **1985**, *18*, 167-216.

265. Polansky, O. E.. Research Notes on the Topological Effect on Molecular Orbitals (TEMO); 3. Einige neue Beziehungen zwischen den mit bestimmten Graphen assozierten Characteristischen Polynomen. *Commun. Math. Chem. (MATCH)*, **1985**, *18*, 217-248.

266. Gutman, I.; Zhang, F. On a Quasiordering of Bipartite Graphs. *Publ. Inst. Math. (Beograd)*, **1986**, *40*, 11-15.

267. Zhang, Y.; Zhang, F.; Gutman, I. On the Ordering of Bipartite Graphs with Respect to Their Characteristic Polynomials. *Coll. Sci. Papers Fac. Sci. Kragujevac* **1988**, *9*, 9-20.

268. Hosoya, H.; Balasubramanian, K. Exact Dimer Statistics and Characteristic Polynomials of Cacti Lattices. *Theor. Chim. Acta* **1989**, *76*, 315-329.

269. Balasubramanian, K. Characteristic Polynomials of Spirographs. *J. Math. Chem.* **1989**, *3*, 147-159.

270. Rowlinson, P. On the Characteristic Polynomials of Spirographs and Related Graphs. *J. Math. Chem.* **1991**, *8*, 345-354.

271. Kennedy, J. W. Palindromic Graphs. *Graph Theory Notes New York* **1992**, *22*, 27-32.

272. Balasubramanian, K. Characteristic Polynomials of Fullerene Cages. *Chem. Phys. Lett.* **1992**, *198*, 577-586.

273. Fowler, P. W. Comment on "Characteristic Polynomials of Fullerene Cages". *Chem. Phys. Lett.* **1993**, *203*, 611-612.

274. Babić, D.; Graovac, A.; Gutman, I. Comment on "Characteristic Polynomials of Fullerene Cages". *Chem. Phys. Lett.* **1993**, *206*, 584-585.

275. Zhang, H.; Balasubramanian, K. Moments and Characteristic Polynomials for Square Lattice Graphs. *J. Math. Chem.* **1993**, *12*, 219-234.

276. Sciriha, I. On the Coefficient of λ in the Characteristic Polynomial of Singular Graphs. *Util. Math.* **1997**, *52*, 97-111.

277. Sciriha, I. The Characteristic Polynomial of Windmills with an Application to the Line Graphs of Trees. *Graph Theory Notes New York* **1988**, *35*, 16-21.

278. Polansky, O. E.; Graovac, A. On the Expansion of the μ-Polynomial of a Simple Graph Partitioned into Subgraphs with at Least Two Components. *Commun. Math. Chem. (MATCH)*, **1982**, *13*, 151-166.

279. Polansky, O. E. Die Entwicklung der μ-Polynoms eines schlichten Graphen, entschprechend seiner Zerlegung an einem Knoten. *Commun. Math. Chem. (MATCH)*, **1985**, *18*, 71-81.

280. Gutman, I. Some Relations for the μ-Polynomial. *Commun. Math. Chem. (MATCH)*, **1986**, *19*, 127-137.

281. Gutman, I.; Polansky, O. E. μ-Polynomial of a Graph with an Articulation Point. *Commun. Math. Chem. (MATCH)* **1986**, *19*, 139-145.

282. Farrell, E. J. On a General Class of Graph Polynomials. *J. Comb. Theory B* **1979**, *26*, 111-122.

283. Farrell, E. J. A Note on the Circuit Polynomials and Characteristic Polynomials of Weels and Ladders. *Discr. Math.* **1982**, *39*, 31-36.

284. Farrell, E. J.; Grell, J. C. On Reconstructing the Circuit Polynomial of a Graph. *Caribb. J. Math.* **1984**, *1*, 109-119.

285. Farrell, E. J.; Grell, J. C. Some Analytical Properties of the Circuit Polynomial of a Graph. *Caribb. J. Math.* **1985**, *2*, 69-76.

286. Farrell, E. J.; Grell, J. C. Circuit Polynomials, Characteristic Polynomials and μ-Polynomials of some Polygonal Chains. *Commun. Math. Chem. (MATCH)* **1985**, *18*, 17-35.

287. Farrell, E. J.; Grell, J. C. Cycle Decomposition of Linear Benzene Chains. *Commun. Math. Chem. (MATCH)*, **1986**, *21*, 325-337.

288. Gutman, I. Comment on the Paper: Cycle Decomposition of Linear Benzene Chains. *Commun. Math. Chem. (MATCH)*, **1987**, *22*, 317-318.

289. Farrell, E. J. On F-Polynomials of Thistles. *J. Franklin Inst.* **1987**, *324*, 341-349.

290. Farrell, E. J. The Impact of F-Polynomials in Graph Theory. *Ann. Discr. Math.* **1993**, *55*, 173-178.

291. Farrell, E. J. The possible Applications of F-Polynomials in the Investigations of Phase Transitions. *Proc. Caribb. Acad. Sci.* **1995**, *6*, 15-25.

292. Knop, J. V.; Trinajstić, N. Chemical Graph Theory. II. On the Graph Theoretical Polynomials of Conjugated Structures. *Int. J. Quantum Chem. Quantum Chem. Symp.* **1980**, *14*, 503-520.

293. Gutman, I. Topological Properties of Benzenoid Systems. X. Note on a Graph-Theoretical Polynomial of Knop and Trinajstić. *Croat. Chem. Acta* **1982**, *55*, 309-313.

294. Seibert, J.; Trinajstić, N. Chemical Graph Theory. IV. On the Cyclic Polynomial, *Int. J. Quantum Chem.* **1983**, *23*, 1829-1841.

295. Klein, D. J.; Randić, M.; Babić, D.; Trinajstić, N. On Conjugated-Circuit Polynomials. *Int. J. Quantum Chem.* **1994**, *50*, 369-384.

296. Gutman, I.; Mizoguchi, N. Conjugated Systems with Identical Circuit Resonance Energies. *Bull. Chem. Soc. Japan* **1990**, *63*, 1083-1086.

297. Lepović, M.; Gutman, I.; Petrović, M.; Mizoguchi, N. Some Contributions to the Theory of Cyclic Conjugation. *J. Serb. Chem. Soc.* **1990**, *55*, 193-198.

298. Gutman, I.; Mizoguchi, N. A Property of the Circuit Characteristic Polynomial. *J. Math. Chem.* **1990**, *5*, 81-82.

299. Gutman, I. A Contribution to the Study of Real Graph Polynomials. *Publ. Elektrotehn. Fak. (Beograd) Ser. Mat.* **1992**, *3*, 35-40.

300. Lepović, M.; Gutman, I.; Petrović, M. A Conjecture in the Theory of Cyclic Conjugation and an Example Supporting Its Validity. *Commun. Math. Chem. (MATCH)*, **1992**, *28*, 219-234.

301. Herndon, W. C. On the Concept of Graph-Theoretical Individual Ring Resonance Energies, *J. Am. Chem. Soc.* **1982**, *104*, 3541-3542.

302. Gutman, I. On some Graphic Polynomials whose Zeros are Real. *Publ. Inst. Math. (Beograd)* **1985**, *37*, 29-32.

303. Ivanciuc, O. Chemical Graph Polynomials. Part 3. The Laplacian Polynomial of Molecular Graphs. *Rev. Roum. Chim.* **1993**, *38*, 1499-1508.

304. Kel'mans, A. K.; Chelnokov, V. M. A Certain Polynomial of a Graph and Graphs with an Extremal Number of Trees. *J. Comb. Theory B* **1974**, *16*, 197-214.

305. Lepović, M.; Gutman, I.; Petrović, M. On Canonical Graphs and the Laplacian Characteristic Polynomial. *Publ. Elektrotehn. Fak. (Beograd) Ser. Mat.* (submitted).

306. Godsil, C. D.; Gutman, I. Wiener Index, Graph Spectrum, Line Graph. ACH Models in Chem., (in press).

307. Gutman, I.; Harary, F. Generalizations of the Matching Polynomial. *Util. Math.* **1983**, *24*, 97-106.

308. Gutman, I. An Identity for the Independence Polynomials of Trees. *Publ. Inst. Math. (Beograd)* **1991**, *50*, 19-23.

309. Gutman, I. Topological Properties of Benzenoid Systems. Merrifield-Simmons Indices and Independence Polynomials of Unbranched Catafusenes. *Rev. Roum. Chim.* **1991**, *36*, 379-388.

310. Stevanović, D. Graphs with Palindromic Independence Polynomial. *Graph Theory Notes New York* **1998**, *34*, 31-36.

311. Motoyama, A.; Hosoya, H. King and Domino Polynomials for Polyomino Graphs. *J. Math. Phys.* **1977**, *18*, 1485-1490.

312. Balasubramanian, K.; Ramaraj, R. Computer Generation of King and Color Polynomials of Graphs and Lattices and Their Applications to Statistical Mechanics. *J. Comput. Chem.* **1985**, *6*, 447-454.

313. El-Basil, S. On Color Polynomials of Fibonacci Graphs. *J. Comput. Chem.* **1987**, *8*, 956-959.

314. Farrell, E. J. On a Class of Polynomials Associated with the Stars of a Graph and Its Application to Node Disjoint Decomposition of Complete Graphs and Complete Bipartite Graphs into Stars. *Canad. Math. Bull.* **1978**, *2*, 35-46.

315. Farrell, E. J.; De Matas, C. On Star Polynomials of Complements of Graphs. *Ark. Math.* **1988**, *26*, 185-190.

316. Farrell, E. J.; De Matas, C. On the Characterizing Properties of Star Polynomials. *Util. Math.* **1988**, *33*, 33-45.

317. Farrell, E. J.; De Matas, C. On Star Polynomials, Graphical Partitions and Reconstruction. *Int. J. Math. Math. Sci.* **1988**, *11*, 87-94.

318. Farrell, E. J. Star Polynomials: Some Possible Applications. *Proc. Caribb. Acad. Sci.* **1994**, *5*, 163-168.

319. Farrell, E. J. On a Class of Polynomials Associated with the Cliques in a Graph and Its Applications. *Int. J. Math. Math. Sci.* **1989**, *12*, 77-84.

320. Hoede, C.; Li, X. L. Clique Polynomials and Independent Set Polynomials of Graphs. *Discr. Math.* **1994**, *125*, 219-228.

321. Stevanović, D. Clique Polynomials of Threshold Graphs. *Publ. Elektrotehn. Fac. (Beograd) Ser Mat.* **1997**, *8*, 84-87.

322. Merrifield, R. E.; Simmons, H. E. *Topological Methods in Chemistry*; Wiley: New York, 1989.

323. Ohkami, N.; Motoyama, A.; Yamaguchi, T.; Hosoya, H.; Gutman, I. Graph-Theoretical Analysis of the Clar's Aromatic Sextet. Mathematical Properties of the Set of the Kekulé Patterns and the Sextet Polynomial for Polycyclic Aromatic Hydrocarbons. *Tetrahedron* **1981**, *37*, 1113-1122.

324. Gutman, I. Topological Properties of Benzenoid Systems. IX. On the Sextet Polynomial. *Z. Naturforsch.* **1982**, *37a*, 69-73.

325. El-Basil, S. A Novel Graph-Theoretical Identity for the Sextet Polynomial. *Chem. Phys. Lett.* **1982**, *89*, 145-148.

326. El-Basil, S.; Gutman, I. Combinatorial Analysis of Kekulé Structures of Non-Branched Cata-Condensed Benzenoid Hydrocarbons. Proof of some Identities for the Sextet Polynomial. *Chem. Phys. Lett.* **1983**, *94*, 188-192.

327. Ohkami, N.; Hosoya, H. Topological Dependency of the Aromatic Sextets in Polycyclic Benzenoid Hydrocarbons. Recursive Relations of the Sextet Polynomial. *Theor. Chim. Acta* **1983**, *64*, 153-170.

328. Gutman, I.; El-Basil, S. Topological Properties of Benzenoid Systems. XXIV. Computing the Sextet Polynomial. *Z. Naturforsch.***1984**, *39a*, 276-281.

329. El-Basil, S.; Trinajstić, N. Application of the Reduced Graph Model to the Sextet Polynomial. *J. Mol. Struct. (Theochem)*, **1984**, *110*, 1-14.

330. El-Basil, S. An Easy Combinatorial Algorithm for the Construction of Sextet Polynomials of Cata-Condensed Benzenoid Hydrocarbons. *Croat. Chem. Acta* **1984**, *57*, 1-20.

331. El-Basil, S. On a Graph-Theoretical Approach to Kekulé Structures. Novel Identities of Sextet Polynomials and a Relation to Clar's Sextet Theory. *Croat. Chem. Acta* **1984**, *57*, 47-64.

332. Gutman, I.; Cyvin, S. J. A New Method for the Calculation of the Sextet Polynomial of Unbranched Catacondensed Benzenoid Hydrocarbons. *Bull. Chem. Soc. Japan* **1989**, *62*, 1250-1252.

333. Ohkami, N. Graph-Theoretical Analysis of the Sextet Polynomial. Proof of the Correspondence between the Sextet Patterns and Kekulé Patterns. *J. Math. Chem.* **1990**, *5*, 23-42.

334. Zhang, H. The Clar Covering Polynomials of S,T-Isomers. *Commun. Math. Chem. (MATCH)* **1993**, *29*, 189-197.

335. John, P. E. Calculating the Cell Polynomial of Catacondensed Polycyclic Hydrocarbons. *J. Chem. Inf. Comput. Sci.* **1994**, *34*, 357-359.

336. Ohkami, N. Hosoya, H. Wheland Polynomial. I. Graph-Theoretical Analysis of the Contribution of the Excited Resonance Structures to the Ground State of Acyclic Polyenes. *Bull. Chem. Soc. Japan* **1979**, *52*, 1624-1633.

337. Randić, M.; Hosoya, H.; Ohkami, N.; Trinajstić, N. The Generalized Wheland Polynomial, *J. Math. Chem.* **1987**, 1, 97-122.

338. Balakrishnarajan, M. M.; Venuvanalingam, P. Learning Approach for the Computation of Generalized Wheland Polynomials of Chemical Graphs. *J. Chem. Inf. Comput. Sci.* **1994**, *34*, 1113-1117.

339. John, P. Multilineare Auzählpolynome für Anzahlen von Linearfaktoren in katakondensierten hexagonalen Systemen. *Commun. Math. Chem. (MATCH)* **1991**, *26*, 137-154.

340. Gutman, I. Some Properties of the Wiener Polynomial. *Graph Theory Notes New York* **1993**, *25*, 13-18.

341. Dobrynin, A. A. Graphs with Palindromic Wiener Polynomials. *Graph Theory Notes of New York,* **1994**, *27*, 50-54.

342. Dobrynin, A. A. Construction of Graphs with a Palindromic Wiener Polynomial. *Vychisl. Sistemy* **1994**, *151*, 37-54 (in Russian).

343. Lepović, M.; Gutman, I. A Collective Property of Trees and Chemical Trees. *J. Chem. Inf. Comput. Sci.* **1998**, *38*, 823-826.

344. Gutman, I.; Estrada, E.; Ivanciuc, O. Some Properties of the Wiener Polynomial of Trees.

345. Graph Theory Notes New York **1999**, *36*, 7-13
346. Gutman, I. Hosoya Polynomial and the Distance of the Total Graph of a Tree
347. Publ. Elektrotehn. Fak. (Beograd) Ser. Mat. (in press).

Chapter 4

TOPOLOGICAL INDICES

A single number, representing a chemical structure, in graph-theoretical terms, is called a topological descriptor. Being a structural invariant it does not depend on the labeling or the pictorial representation of a graph. Despite the considerable loss of information by the *projection* in a single number of a structure, such descriptors found broad applications in the correlation and prediction of several molecular properties[1,2] and also in tests of similarity and isomorphism.[3,4]

When a topological descriptor correlates with a molecular property, it can be denominated as molecular index or topological index (*TI*).

Randić[5] has outlined some desirable attributes for the topological indices in the view of preventing their *hazardous proliferation.*

List of desirable attributes for a topological index

1. Direct structural interpretation
2. Good correlation with at least one property
3. Good discrimination of isomers
4. Locally defined
5. Generalizable to *higher* analogues
6. Linearly independent
7. Simplicity
8. Not based on physico-chemical properties
9. Not trivially related to other indices
10. Efficiency of construction
11. Based on familiar structural concepts
12. Show a correct size-dependence
13. Gradual change with gradual change in structures

Only an index having a direct and clear structural interpretation can help to the interpretation of a complex molecular property. If the index correlates with a single molecular property it could indicate the *structural composition* of that property.[5] If it is sensible to gradual structural changes (e.g. within a set of isomers) then the index could give information about the molecular shape. If it is locally defined, the index could describe *local contributions* to a given property. If the index can be generalized to higher analogues or it can be built up on various bases (e.g. on various matrices[4,6]) it could offer a larger pool of descriptors for the regression analysis.

Among the molecular properties, good correlation with the structure was found for: thermodynamic properties (e.g. boiling points, heat of combustion, enthalpy of formation, etc.), chromatographic retention indices, octane number and various biological properties.

Thus, a topological index converts a chemical structure into a single number, useful in QSPR/QSAR studies. More than a hundred of topological descriptors were proposed so far and tested for correlation with physico-chemical (QSPR) or biological activity (QSAR) of the molecules.

In the construction of a *TI*, two stages can be distinguished:[7] (a) the assignment stage and (b) the operational stage.

(a) In the *assignment stage*, the topological information is encoded as local invariants (*LOI*s - *LO*cal *I*nvariants). The *LOI*s can be pure topological or weighted by chemical properties (e.g. atomic properties, when the chemical nature of vertices is needed).

(b) In the *operational stage*, the *LOI*s are mathematically *operated* for producing global (molecular) invariants as single number descriptors of a structure. They are referred to as topological indices. The operational stage may (or may not) encompass topological information. Often the operation is the simple addition. It is a *natural* operation, since many molecular properties are considered to be additive. There exist more sophisticated global invariants (e.g. the eigenvalues), as a result of complex matrix operations. In such cases, the two stages are indistinguishable.

When a *TI* shows one and the same value for two or more structures, it is said that *TI* is *degenerated*. The degeneracy may appear both in the assignment and the operational stages.[7] The *assignment degeneracy* appears when non-equivalent subgraphs (i.e. vertices) receive identical *LOI*s or when non-isomorphic graphs show the same ordered *LOI*s (e.g. the same distance degree sequence, DDS_i). The *operational degeneracy* is seldom encountered (e.g. when simple operations act on weakly differentiated *LOI*s) and leads to the same value of *TI* for non-isomorphic graphs which do not show assignment degeneracy.

The *discriminating sensitivity* of a *TI* is a measure of its ability to distinguish among nonisomorphic graphs by distinct numerical values. An evaluation of this sensitivity, s, on a fixed set, M, of nonisomorphic graphs can be achieved by formula $s = (m - m_i)/m$, where $m = |M|$ and m_i is the number of graphs undistinguished by *TI* within the set M.

A criterion used in the classification of topological indices is that of the matrix which supplies the topological information[1] (the *info* matrix), in the assignment stage. When the *LOI*s are assigned on combined matrices (see the Schultz index) or the two stages are indistinguishable, such a classification becomes difficult. Other criterion would take into account the mathematical operations involved in achieving a certain index. In the following, these two criteria will be used in an attempt of classifying the main topological indices.

4.1. INDICES BASED ON ADJACENCY MATRIX

4.1.1. The Index of Total Adjacency

The simplest TI is the half sum of entries in the adjacency matrix $\mathbf{A}$:

$$A = (1/2)\sum_i \sum_j [\mathbf{A}]_{ij} \tag{4.1}$$

It was called the *total adjacency index*[8] and is equal to the number of edges, Q, in graph. Figure 4.1 shows the matrix $\mathbf{A}$ and the corresponding index for 2-Methylbutane, $G_{4.1}$.

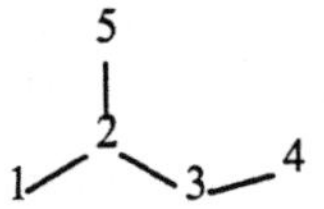

	1	2	3	4	5
1	0	1	0	0	0
2	1	0	1	0	1
3	0	1	0	1	0
4	0	0	1	0	0
5	0	1	0	0	0

$A(G_{4.1}) = 4$

$G_{4.1}$

Figure 4.1. Adjacency matrix and A index for the graph $G_{4.1}$.

4.1.2. The Indices of Platt, F, Gordon-Scantlebury, N2, and Bertz, B1

Platt[9,10] has introduced the total adjacency of edges in a graph, as the F index

$$F = \sum_i \sum_j [\mathbf{EA}]_{ij} = 2\sum_i \binom{\delta_i}{2} = 2N_2 = 2B_1 \tag{4.2}$$

where $\mathbf{EA}$ is the Edge Adjacency matrix. This index is twice the Gordon - Scantlebury[11] index, N_2, defined as the number of modes in which the acyclic fragment C-C-C may be superposed on a molecular graph

$$N_2 = \sum_i (P_2)_i \tag{4.3}$$

This index equals the number of all paths of length 2, P_2, in graph. The F index is also twice the Bertz index,[12] B_1 , defined as the number of edges in the line graph $L_1(G)$. The last one can be calculated combinatorially from the vertex degree, δ_i . For the graph $G_{4.2}$, the calculus is given in Figure 4.2.

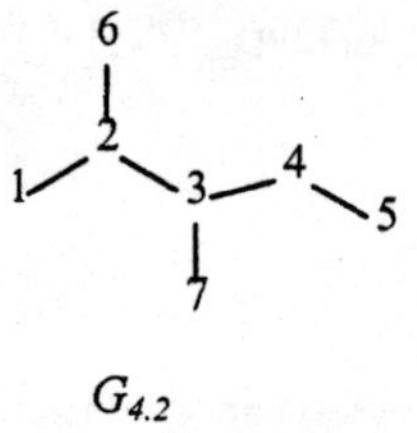

$$B_1(G_{4.2}) = 2\binom{3}{2} + \binom{2}{2} = 7$$

$G_{4.2}$

Figure 4.2. Bertz index B_1, for the graph $G_{4.2}$.

4.1.3. The Indices of ZAGREB Group

First TIs based on adjacency matrix (i.e. based on connectivity) were introduced by the Group from Zagreb[13,14]

$$M_1 = \sum_i \delta_i^2 \tag{4.4}$$

$$M_2 = \sum_{(ij)\in E(G)} \delta_i \delta_j \tag{4.5}$$

where δ_i, δ_j - are the vertex degrees for any two adjacent vertices. For the graph $G_{4.2}$ one calculates:

$M_1 = 4 \times 1^2 + 2 \times 3^2 + 2^2 = 26;$ $M_2 = 3(1 \times 3) + (3 \times 3) + (2 \times 3) + (1 \times 2) = 26$

4.1.4. The Randić Index, χ

The χ index was introduced by Randić[15] for characterizing the branching in graphs:

$$\chi = \sum_{(ij)\in E(G)} (\delta_i \delta_j)^{-1/2} \tag{4.6}$$

For the graph $G_{4.3}$, the calculus is illustrated in Figure 4.3.

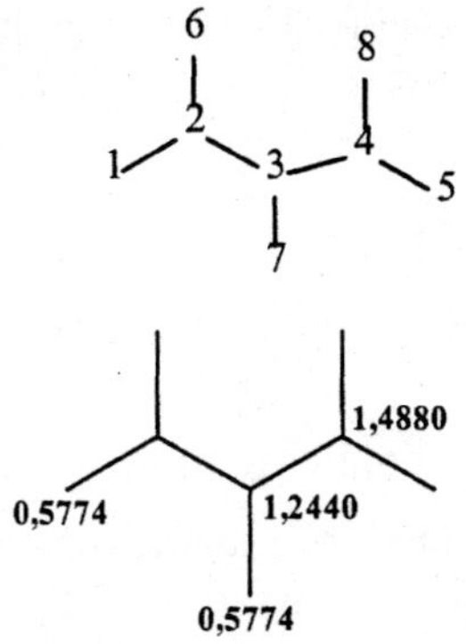

(a) $\chi = 5(1 \times 3)^{-1/2} + 2(3 \times 3)^{-1/2} = 3.5535$
(b) $\chi = (1/2)[5 \times 0.5774 + 2 \times 1.4880 + 1.2440]$
$= 3.5535$

$G_{4.3}$

$G_{4.3} \{\chi_i\}$

Figure 4.3. Randić Index χ for the graph $G_{4.3}$: (a) on edge and (b) on vertex calculation

χ is a connectivity index (i.e. an index calculated on edge by using the vertex degrees - see Figure 4.3 (a)) and, due to its extreme popularity, it was called *the connectivity index*. Its relatedness with M_2 is immediate.

Diudea et al.[16] have defined χ on vertex

$$\chi_i = \sum_{j(i)\in E(G)} (\delta_i \delta_j)^{-1/2} \tag{4.7}$$

$$\chi = \frac{1}{2}\sum_i \chi_i \tag{4.8}$$

Such a definition was used in connection with some fragmental descriptors (see Sect. 4.7). For $G_{4.3}$ the calculus is given in Figure 4.3 (b).

The χ values decrease as the branching increases within a set of alkane isomers. They increase by the number of atoms in the molecular graph.

This index was shown to correlate with various physico-chemical (e.g. enthalpy of formation, molar refraction, van der Waals areas and volumes, chromatographic retention index etc.) and biological properties.[17]

4.1.5. Extensions of χ Index

4.1.5.1. Kier and Hall Extensions

Kier and Hall[17] have generalized the χ index , considering the edge as the simplest path (i.e. the path of length 1) and extending the summation over all paths of length e:

$$^e\chi = \sum_{p_e}(\delta_i\delta_j..\delta_{e+1})^{-1/2} \tag{4.9}$$

where $\delta_i\delta_j..\delta_{e+1}$ are the vertex degrees along the path p_e. Indices $^e\chi$ are used as a family of structurally related topological indices[5,18] Other subgraph based connectivities have been developed.[17]

The authors have also extended the validity of χ to heteroatom-containing molecules. They introduced the δ_i^v valencies (see Sect. 8.3.2.1) in the construction of the analogous index χ^v:

$$\chi^v = \sum_{(ij)\in E(G)}(\delta_i^v\delta_j^v)^{-1/2} \tag{4.10}$$

The δ_i^v values and other electronic and topological considerations are assembled in the *electrotopological state index*.[19,20]

4.1.5.2. Estrada Extensions

Estrada[21] changed the vertex degree δ_i with the *edge degree*, $\delta(e_i)$, eq 4.6 becoming

$$\varepsilon = \sum_r [\delta(e_i)\delta(e_j)]_r^{-1/2} \tag{4.11}$$

where the summation runs over all r-pairs of adjacent edges. It is obvious that the ε index of a graph equals the connectivity index of the corresponding line graph: $\varepsilon(G) = \chi(L_l(G))$.

The ε index was extended to heteroatom-containing molecules[22] (with the Pauling[23] k_{C-X} parameter for calculating the weighted edge degrees $\delta^W(e_i)$) and to a 3D-descriptor[24] counting an *electron charge density connectivity*. These indices showed good correlating ability with the molar volume[21,22] (of alkanes and a mixed set of ethers and halogeno-derivatives) and with the boiling points of alkenes.[24]

4.1.5.3. *Razinger Extension*

Razinger[25] introduced the vertex degrees of higher rank ^eW_i, in construction of χ^eW index[6]

$$\chi^eW = \sum_{(ij)\in E(G)}(^ew_i\,^ew_j)^{-1/2} \tag{4.12}$$

Diudea[16] has defined this index on vertex (see also eqs 4.7 and 4.8):

$$\chi^e W_i = \sum_{j(i,j) \in E(G)} ({}^e w_i \, {}^e w_j)^{-1/2} \qquad (4.13)$$

$$\chi^e W = \frac{1}{2} \sum_i \chi^e W_i \qquad (4.14)$$

Values $\chi^e W_i$ and $\chi^e W$ for $G_{4.3}$ are shown in Figure 4.4. Indices $\chi^e W$ are also used as a family of topological indices.

Graphs $G_{4.3} \{{}^e W_i\}$.

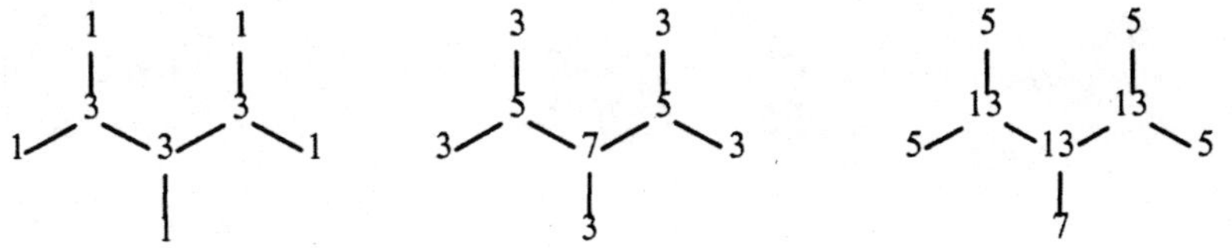

Graphs $G_{4.3} \{\chi^e W_i\}$.

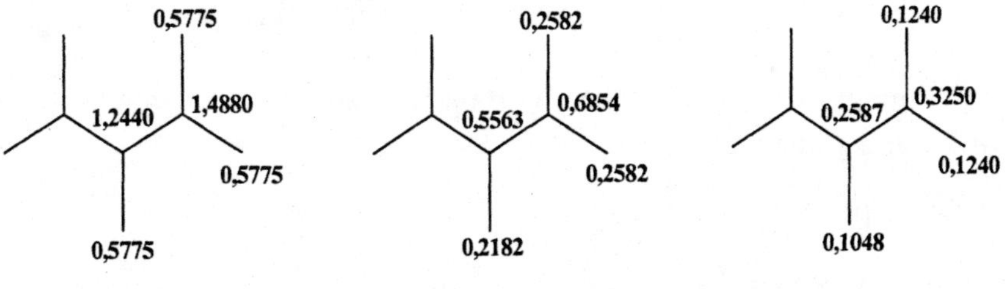

$$\chi^1 W = 3.5534 \qquad \chi^2 W = 1.5890 \qquad \chi^3 W = 0.7548$$

Figure 4.4. Indices $\chi^e W_i$ and $\chi^e W$ (e=1-3) for the graph $G_{4.3}$.

Indices $\chi^e W$ can be weighted cf. the algorithms ${}^e W_M$ (eqs 2.5 - 2.8) or ${}^e E_M$ (eqs 8.39 - 8.41) thus accounting for the chemical nature of atoms and edges.

4.1.5.4. Diudea Extensions

Diudea and Silaghi[26] used the group electronegativity valencies, denoted *EVG* (see Sect 8.3.2.1 - eqs 8.30 - 8.32) in constructing a χ analogue index, *DS* (on vertex defined):

$$DS_i = \sum_{j(i)\in E(G)}(EVG_iEGV_j)^{-1/2} \tag{4.15}$$

$$DS = \sum_i DS_i \tag{4.16}$$

The DS index showed excellent correlating ability.[26] It is exemplified in Figure 4.5 for the graph $G_{4.3}$.

(a)

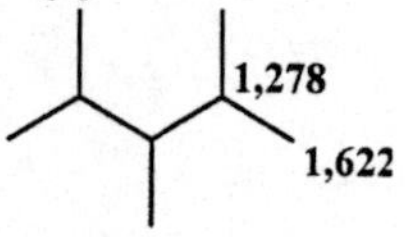

1,278
1,622

$G_{4.3}$ {EVG_i}

$DS = 5 \times 0.6945 + 2 \times 2.1714 + 2.2593 = 10.0745$

(b)

0,9716
0,9575

$G_{4.3}$ {EC_i}

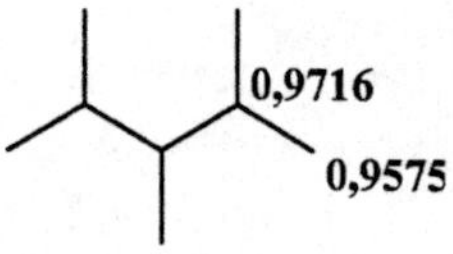

$EC_N = 5 \times 1.0368 + 2 \times 3.1028 + 3.0952 = 14.4848$
$EC_P = 5 \times 0.9645 + 2 \times 2.9006 + 2.9077 = 13.5314$

Figure 4.5. (a) DS and (b) EC indices for the graph $G_{4.3}$.

A similar index, $EC_{P/N}$, was built up by using the EC electronegativities[27] (see Sect 8.3.2.1 and Table 8.2)

$$EC_{P/N,i} = \sum_{j(i)\in E(G)}(EC_iEC_j)^{\pm 1/2} \tag{4.17}$$

$$EC_{P/N} = \sum_i EC_{P/N,i} \tag{4.18}$$

The subscript symbol is P for +1/2 and N for -1/2. Values of these indices are given in Figure 4.5 (b) for the graph $G_{4.3}$.

Excellent correlations of the EC indices with some physico-chemical and biological properties of aliphatic alcohols, amines and halogeno-derivatives were obtained.[27]

4.2. INDICES BASED ON WIENER, DISTANCE AND DETOUR MATRICES

4.2.1 Wiener-Type Indices

Wiener proposed in 1947 the first structural index in connection to some studies on the thermodynamic properties of hydrocarbons (i.e. acyclic structures).[28-31] W index, (called by its author the *path number* and eventually referred to as the Wiener index/number), has later become one of the central subjects which focused the attention of theoretical chemists. Extensive studies have been performed for finding correlations with various physico-chemical and biological properties. In this respect, the reader can consult three recent reviews.[32-34]

4.2.1.1. Main Definitions

In acyclic structures, the *Wiener index*, W,[28] and its extension, the *hyper-Wiener index*, WW,[35] can be defined as

$$W = W(G) = \sum_e N_{i,e} \, N_{j,e} \tag{4.19}$$
$$WW = WW(G) = \sum_p N_{i,p} \, N_{j,p} \tag{4.20}$$

where N_i and N_j denote the number of vertices lying on the two sides of the edge e or path p, respectively, having the endpoints i and j. Eq 4.19 follows the method of calculation given by Wiener himself:[28] "Multiply the number of carbon atoms on one side of any bond by those on the other side; W is the sum of these values for all bonds".

The *edge contributions*, $N_{i,e} \, N_{j,e}$ and the *path contributions*, $N_{i,p} \, N_{j,p}$, to the global index are just the entries in the Wiener matrices,[36,37] $\mathbf{W}_e$ and $\mathbf{W}_p$, from which W and WW can be calculated by

$$W = (1/2) \sum_i \sum_j [\mathbf{W}_e]_{ij} \tag{4.21}$$

$$WW = (1/2) \sum_i \sum_j [\mathbf{W}_p]_{ij} \tag{4.22}$$

The indices W and WW (calculated according to eqs 4.19-4.22) count all *external* paths passing through the two endpoints, i and j, of all edges and paths, respectively, in an acyclic graph.

Other main definitions[38,39] of the Wiener-type indices are based on the distance matrices, $\mathbf{D}_e$ and $\mathbf{D}_p$

$$W = (1/2) \sum_i \sum_j [\mathbf{D}_e]_{ij} \tag{4.23}$$
$$WW = (1/2) \sum_i \sum_j [\mathbf{D}_p]_{ij} \tag{4.24}$$

Recall that Wiener[28] has calculated the path number W "as the sum of the distance between any two carbon atoms in the molecule, in terms of carbon-carbon bonds." In other words, W is given as the sum of elements above the main diagonal of the distance matrix, as shown by Hosoya.[38] In opposition to the edge/path contribution definitions, (see eqs 4.19-4.22), relations (4.23) and (4.24) are valid both for acyclic and cycle-containing structures.

It is useful to indicate in the symbols of the Wiener-type indices the matrix on which they are calculated, by a subscript letter (see below).

The hyper Wiener index WW_{Dp} (calculated according to eq 4.24) counts all *internal* paths existing between the two endpoints, i and j, of all paths of a graph.[39] In acyclic graphs, the WW_{Wp} values (cf. eq 4.22) are identical to WW_{Dp} values (cf. eq 4.24), by virtue of the equality of the sum of all internal and external paths, with respect to all pairs of vertices, i and j. [40]

4.2.1.2. Other Definitions

Attempts have been made to express W in terms of *edge contributions* and to extend such definitions (4.19) to cycle-containing structures[41 - 44]

$$W = \sum_e \sum_g 1/\#_g (G) \tag{4.25}$$

where $\#_g(G)$ is the number of different geodesics between the endpoints of g and sum is over all geodesics containing e and next over all e in G.

Klein, Lukovits and Gutman[40] have decomposed the hyper-Wiener number of trees by a relation that can be written as:

$$WW = (Tr(D_e^2)/2 + W)/2 \tag{4.26}$$

where $Tr(D_e^2)$ is the trace of the squared distance matrix. Relation (4.26) is used as a definition for the hyper-Wiener index of cycle-containing graphs.

Expansion of the right-hand side of eq 4.24, by taking into account the definition of $\mathbf{D}_p$ matrix,[39, 40] results in a new decomposition (i.e. a new definition) of the hyper-Wiener index WW

$$WW = \sum_{i<j}[\mathbf{D}_P]_{ij} = \sum_{i<j}\binom{[\mathbf{D}_e]_{ij}+1}{2} = \sum_{i<j}[\mathbf{D}_e]_{ij} + \sum_{i<j}\binom{[\mathbf{D}_e]_{ij}}{2} \tag{4.27}$$

The first term is just the Wiener index, W. The second term is the *non-Wiener* part of the hyper-Wiener index, or the contributions of $\mathbf{D}_p$ when $|p|>1$. It is denoted by W_Δ

$$W_\Delta = \sum_{i<j}\binom{[\mathbf{D}_e]_{ij}}{2} \tag{4.28}$$

Thus, the hyper-Wiener index can be written as

$$WW = W + W_\Delta \tag{4.29}$$

W_Δ is related to $Tr(D_e^2)$ by[6]

$$W_\Delta = (Tr(D_e^2)) - 2W_e)/4 \tag{4.30}$$

W_Δ is correlated 0.99975 with WW in the set of octanes. Wiener indices express the *expansiveness* of a molecular graph.[28, 32] Values of W and WW indices for a set of acyclic and cyclic octanes are presented in Table 4.1 (see also Table 4.5). These values decrease as the branching increases within a set of isomers (see entries 1-18).

Table 4.1 Wiener-Type Indices of a Set of Acyclic and Cyclic Octanes.

No.	Graph	W	WW	w	ww
1	P_8	84	210	84	210
2	$2MP_7$	79	185	79	185
3	$3MP_7$	76	170	76	170
4	$4MP_7$	75	165	75	165
5	$3EP_6$	72	150	72	150
6	$25MP_6$	74	161	74	161
7	$24MP_6$	71	147	71	147
8	$23MP_6$	70	143	70	143
9	$34MP_6$	68	134	68	134
10	$3E2MP_5$	67	129	67	129
11	$22M2P_6$	71	149	71	149
12	$33M2P_6$	67	131	67	131
13	$234M3P_5$	65	122	65	122
14	$3E3MP_5$	64	118	64	118
15	$224MP_5$	66	127	66	127
16	$223MP_5$	63	115	63	115
17	$233MP_5$	62	111	62	111
18	$2233MP_4$	58	97	58	97
19	$112MC_5$	56	92	106	278
20	$113MC_5$	58	100	104	266
21	IPC_5	62	114	106	286
22	PC_5	67	135	111	315
23	$11MC_6$	59	103	119	337
24	$12MC_6$	60	106	124	362
25	$13MC_6$	61	110	123	355
26	$14MC_6$	62	115	122	349
27	EC_6	64	122	124	368
28	C_8	64	120	160	552
29	$123MC_5$	58	99	109	290
30	$1M2EC_5$	61	110	111	307
31	$1M3EC_5$	63	119	109	294
32	MC_7	61	109	142	451

M = Methyl; E = Ethyl; P = Propyl; IP = Isopropyl;
P_N = path of length N; C_N = N-membered cycle.

4.2.1.3. Wiener and Hyper-Wiener Indices of Some Particular Graphs

4.2.1.3.1. Path and Tree Graphs

In *path graphs*, P_N, the combinatorial analysis lead to the following relation for the Wiener index[45-47]

$$W(P_N) = \binom{N}{2} + \binom{N}{3} = \binom{N+1}{3} \tag{4.31}$$

In *trees*, T_N, the branching introduced by the vertices r, of degree $\delta_r > 2$, will lower the value of W, as given by the Doyle-Graver formula[45, 48]

$$W(T_N) = \binom{N}{2} + \binom{N}{3} - \sum_r \sum_{1 \le i < j < k \le \delta_r} n_i n_j n_k \tag{4.32}$$

where $n_1, n_2, ..., n_{\delta_r}$ are the number of vertices in branches attached to the vertex r; $n_1 + n_2 + ... + n_{\delta_r} + 1 = N$, and summation runs as follows: first summation over all branching points in graph and the second one over all $\binom{\delta_r}{3}$ triplet products around a branching point. In eqs 4.31 and 4.32, the first term appears to be the *size* term while the second (and the third) one (ones) gives (give) account for the *shape* of a structure.[47]

A relation similar to (4.31) can be written for the hyper-Wiener index of the path graph[47]

$$WW(P_N) = \binom{N+1}{3} + \binom{N+1}{4} = \binom{N+2}{4} \tag{4.33}$$

In such graphs, WW can be written (cf eq 4.29) as[6]

$$WW(P_N) = W(P_N) + W_\Delta(P_N) = W_\Delta(P_{N+1}) \tag{4.34}$$

which, iteratively, becomes

$$W_\Delta(P_{N+1}) = W(P_1) + W(P_2) + ... + W(P_{N-1}) + W(P_N) + W_\Delta(P_1) \tag{4.35}$$

and, keeping in mind that $W_\Delta(P_1) = 0$, one obtains[39]

$$W_\Delta(P_N) = W(P_1) + W(P_2) + ... + W(P_{N-1}) \tag{4.36}$$

This relation expresses the fact that, in n-alkanes, W_Δ can be calculated from the Wiener index of the preceding homologues. By substituting $W_\Delta(P_N)$ (eq 4.36) in eq 4.34, one obtains the expression for the hyper-Wiener number, WW

$$WW(P_N) = W(P_1) + W(P_2) + ... + W(P_{N-1}) + W(P_N) \tag{4.37}$$

This relation was also reported by Lukovits.[49]

The relations (4.31) and (4.33) lead to the simpler relations[49-52]
$$W(P_N) = N(N^2 -1)/6 \tag{4.38}$$
$$WW(P_N) = N(N-1)(N+1)(N+2)/24 \tag{4.39}$$

4.2.1.3.2. Cycle Graphs

In *cycle graphs*, C_N, the following combinatorial relations were found[39,52]

$$W(C_N) = N\binom{(N-z+2)/2}{2} - (1-z)\frac{N^2}{4} \tag{4.40}$$

$$W_\Delta(C_N) = N\sum_{i=1}^{(N-z-2)/2}\binom{i+1}{2} - (1-z)\frac{N}{2}\binom{N/2}{2} \tag{4.41}$$

$$WW(C_N) = N\sum_{i=1}^{(N-z)/2}\binom{i+1}{2} - (1-z)\frac{N}{2}\binom{N/2+1}{2} \tag{4.42}$$

where $z = N \bmod 2$.
Expansion of the above relations lead to[34,52,53]

$$W(C_N) = N(N^2 - z)/8 \tag{4.43}$$
$$W_\Delta(C_N) = (N^2 - z)(N^2 - 3N + 2 - 2z)/48 \tag{4.44}$$
$$WW(C_N) = (N-z)(N-z+1)(N-z+2)(N+3z)/48 \tag{4.45}$$

From these relations, the following recurrences are straightforward[39]

$$W(C_{N+1}) = W(C_N) + \frac{W(C_N)}{N} + \binom{N+1}{2}\binom{N+z}{2} = W(C_N) + W(e+1) \tag{4.46}$$

$$W_\Delta(C_{N+1}) = W_\Delta(C_N) + \frac{W_\Delta(C_N)}{N} + \binom{N+1}{2}\binom{(N+z)/2}{2} = W_\Delta(C_N) +$$

$$W_\Delta(e+1) \tag{4.47}$$

$$WW(C_{N+1}) = WW(C_N) + \frac{WW(C_N)}{N} + \binom{N+1}{2}\binom{(N+z)/2+1}{2} = WW(C_N) + WW(\epsilon$$

(4.48)

from which it is easily seen that

$$WW(e+1) = W(e+1) + W_\Delta(e+1)$$

(4.49)

Quantities $WW(e+1)$, $W(e+1)$ and $W_\Delta(e+1)$ represent the bond contribution of the newly introduced edge (i.e. the $(e+1)^{th}$ edge, where $e = |N|$).

From the above relations one obtains[39]

$$WW(C_{N+1}) = WW(C_N) + WW(e+1) = W(C_{N+1}) + W_\Delta(C_{N+1}) = W(C_N) + W_\Delta(C_N) +$$
$$W(e+1) + W_\Delta(e+1)$$

(4.50)

4.2.1.3.3. Spiro-Graphs

A spiro-graph is obtained from simple rings by fusing a single vertex of one ring with a single vertex of another ring, for giving a single vertex (of degree four) in the resulted coalesced graph.[54] The process can be repeated, thus resulting in spiro-chains.[55]

For rings larger than three vertices, the construction of spiro-graphs have to take into account all the possibilities of connection. Thus, for four- and five-membered ring *1,2-* and *1,3-* structures are considered whereas for six-membered rings, a third *1,4-* structure is taken into account (see Figure 4.6).

The formulas (Table 4.2) for evaluating the Wiener, W, and the hyper-Wiener, WW, indices were derived on the ground of **LC** matrices,[4,55] by the aid of the MAPLE V Computer Algebra System (release 2). For other graph - theoretical aspects in spiro-graphs see.[54,56]

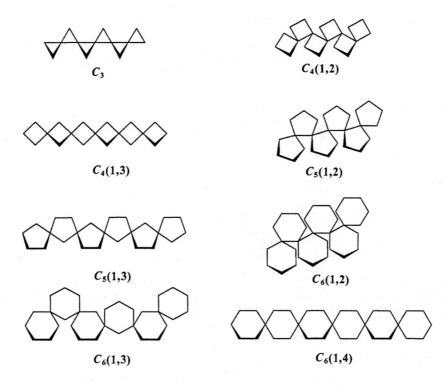

C_3

$C_4(1,2)$

$C_4(1,3)$

$C_5(1,2)$

$C_5(1,3)$

$C_6(1,2)$

$C_6(1,3)$

$C_6(1,4)$

Figure 4.6 Spiro-chains with three- (C_3) to six- (C_6) membered cycles.

Table 4.2. Formulas for W and WW Indices of Spiro-Chains

Spiro-Chain	W Index	WW Index
Three-membered cycles	$n(2n^2 + 6n + 1)/3$	$n^2(n^2 + 6n + 11)/6$
Four-membered cycles		
1,3- Spiro-chains	$n(3n^2 + 3n + 2)$	$n(3n^3 + 7n^2 + 4n + 6)/2$
1,2- Spiro-chains	$n(3n^2 + 15n - 2)/2$	$n(3n^3 + 26n^2 + 85n - 34)/8$
Five-membered cycles		
1,3-spiro-chains	$n(4n + 1)(4n + 5)/3$	$2n(n + 1)(2n + 1)(2n + 3)/3$
1,2-spiro-chains	$n(8n^2 + 48n - 11)/3$	$2n(n^3 + 10n^2 + 38n - 19)/3$
Six-membered cycles		
1,4-spiro-chains	$n(25n^2 + 15n + 14)/2$	$n(75n^3 + 110n^2 + 29n + 122)/8$
1,3-spiro-chains	$n(25n^2 + 60n - 4)/3$	$n(25n^3 + 105n^2 + 173n - 51)/6$
1,2-spiro-chains	$n(25n^2 + 195n - 58)/6$	$n(25n^3 + 310n^2 + 1547n - 847)/24$

n = number of cycles in chain

Table 4.3 lists values of W and WW in spiro-chains.

Table 4.3. Wiener, W, and Hyper-Wiener, WW, Indices
of Spiro-Chains with Three- (C_3) to Six- (C_6) Membered Cycles.

n	C_3	C_4 (1,2)	C_4 (1,3)	C_5 (1,2)	C_5 (1,3)	C_6 (1,2)	C_6 (1,3)	C_6 (1,4)
W index								
2	14	40	40	78	78	144	144	144
3	37	105	114	205	221	376	401	426
4	76	212	248	412	476	748	848	948
5	135	370	460	715	875	1285	1535	1785
6	218	588	768	1130	1450	2012	2512	3012
7	329	875	1190	1673	2233	2954	3829	4704
8	472	1240	1744	2360	3256	4136	5536	6936
WW index								
2	18	66	66	140	140	305	305	305
3	57	201	243	424	504	904	1044	1209
4	136	457	652	952	1320	1979	2614	3399
5	275	885	1440	1820	2860	3695	5470	7745
6	498	1545	2790	3140	5460	6242	10167	15342
7	833	2506	4921	5040	9520	9835	17360	27510
8	1312	3846	8088	7664	15504	14714	27804	45794

From Table 4.3 one can see that the values of W and WW increase as the type of connection between the cycles of a spiro-chain change from *1,2* to *1,3* and *1,4*. This fact is in agreement with the idea that the Wiener-type indices express the expansiveness[28] of structures: a spiro-chain is as more expanded as the connectivity of its cycles involves more bonds. Conversely, a spiro-chain is as more branched as the number of separating bonds is lower.[55]

4.2.1.4. Detour Extensions of the Wiener and Hyper-Wiener Indices

Extension of the distance based definitions of W and WW by changing the notion *distance* by that of *detours*, lead to the detour-analogues, w,[57-60] and ww.[59] Thus, these indices are calculated as the half sum of entries in the corresponding detour matrices

$$w = (1/2)\sum_i \sum_j [\Delta_e]_{ij} \qquad\qquad (4.51)$$

$$ww = (1/2)\sum_i \sum_j [\Delta_p]_{ij} \qquad\qquad (4.52)$$

Values of these indices for a set of acyclic and cyclic octanes are included in Table 4.1.

Formulas for calculating the detour indices in simple cycles have been derived by Lukovits[59]

$$w \quad = N(3N^2 - 4N + z) / 8 \qquad\qquad (4.53)$$
$$ww = N(7N^3 - 3N^2 - 10N + 3z(N+1)) / 48 \qquad\qquad (4.54)$$
$$z = N \bmod 2$$

The *detour* and *hyper-detour* indices have promoted very interesting correlating searches.[59,60]

4.2.1.5. Walk Numbers, ^eW_M : Wiener Indices of Higher Rank

The ^eW_M *algorithm*[16,39] (see Sect. 2.1 - eqs. 2.5-2.8) supplies *global walk numbers,* ^eW_M , as the half-sum of the local numbers $^eW_{M,i}$

$$^eW_M = (1/2)\sum_i {}^eW_{M,i} = (1/2)\sum_i \sum_j [\mathbf{M}^e]_{ij} \qquad\qquad (4.55)$$

The subscript M denotes the matrix on which the algorithm runs, thus giving walk degrees weighted by the property collected by that matrix. The algorithm eludes the raising at power e of the matrix $\mathbf{M}$ (see the last member of eq 4.55). When $\mathbf{M}$ is the distance matrix (or other matrix involving distances or paths), the walk number ^eW_M is just a Wiener-type index of rank e:[39] $^eW_{De}$ denotes a *Wiener* (*Hosoya*) number; $^eW_{We}$ represents a *Wiener* (*Wiener*) number; $^eW_{Wp}$ denotes a *hyper-Wiener* (*Randić*) number and so on. Walk numbers of rank 2 are listed in Table 4.4 for the octane isomers.[39] The degenerate values are shaded.

^eW_M numbers can be calculated by the aid of the walk matrix, $\mathbf{W}_{(M1,M2,M3)}$ (see sect. 2.14).

Walk numbers, ^eW_M , are useful in discriminating nonisomorphic isomers. Usually a rank of two suffices in discriminating i.e. the octane isomers.[39] Special graphs, need however a rank higher than two (see Chap. 8).

Walk numbers, as the classical W index, showed good correlation with octane numbers. ON.[39]

Table 4.4. Walk Numbers, $^{e}W_M : e = 2$, of Octanes.

Graph	$^2W_{De}$	$^2W_{We}$	$^2W_{Dp}$	$^2W_{Wp}$
P_8	1848	2100	12726	12054
$2MP_7$	1628	2000	9711	9829
$3MP_7$	1512	1892	8256	8338
$4MP_7$	1476	1848	7830	7815
$3EP_6$	1360	1740	6412	6460
$25M_2P_6$	1420	1900	7171	7825
$24M_2P_6$	1312	1792	6023	6536
$23M_2P_6$	1280	1748	5772	6163
$34M_2P_6$	1208	1684	5050	5426
$3E2MP_5$	1172	1640	4646	4992
$22M_2P_6$	1316	1808	6277	6779
$33M_2P_6$	1176	1664	4878	5221
$234M_3P_5$	1096	1648	4076	4700
$3E3MP_5$	1072	1564	3916	4222
$224M_3P_5$	1128	1708	4406	5165
$223M_3P_5$	1032	1600	3653	4220
$233M_3P_5$	1000	1564	3402	3917
$2233M_4P_4$	868	1516	2521	3169

M = Methyl; E = Ethyl; P_N = path of length N; C_N = N-membered cycle.

4.2.1.6. Tratch Extension of the Wiener Index

Tratch et al.[61] proposed an *extended distance matrix*, **E**, which in the case of trees, we considered it as a *distance-extended Wiener* matrix and denoted by $\mathbf{D\_W}_p$ (See Sect. 2.12). The half sum of its entries supplies a *distance-extended Wiener* index, denoted here D_WW

$$D\_WW = (1/2)\sum_i\sum_j[\mathbf{D\_W}_p]_{ij} = (1/2)\sum_i\sum_j d_{ij}N_i N_j \qquad (4.56)$$

where d_{ij} is the distance between the vertices i and j whereas N_i, N_j have the same meaning as in eq 4.19. The extension by the distance of indices other than the Wiener index was used by Diudea[62] as a basis in the construction of 2D- and 3D- distance-extended Cluj indices (See Chap. 6 and Chap. 7).

4.2.1.7. Other Extensions of the Wiener Index

When $\mathbf{D}_e$ in eq 4.23 is changed by the 3D-Distance matrix (actually the $\mathbf{G}$ matrix - see Sect. 2.5) a *3D-Wiener index*, ^3W, is obtained.[63,64]

Marjanović and Gutman[65] derived an index counting all shortest paths in a graph. The index is identical to W in trees but is different in cycle-containing graphs.

Lukovits[66] proposed the *all path index, P*, with the hope of an unitary description for both acyclic and cycle-containing molecular graphs.

Gupta et al.[67] defined a *superpendentic index*, based on the *pendent matrix*, which is a submatrix of De obtained by retaining only the columns corresponding to pendent vertices (i.e. vertices of degree 1).

Szeged and Cluj indices are the most important extensions of the Wiener index. They will be detailed in Chap. 5 and 6, respectively.

4.2.2. Balaban Index, J

Balaban[68] applied a Randić formula on the *distance sum DS_i* (i.e. row sum in the distance matrix) for defining the J index

$$J = \frac{Q}{\mu + 1} \sum_{(ij) \in E(G)} (DS_i DS_j)^{-1/2} \tag{4.57}$$

where Q is the number of edges, DS_i and DS_j denote the distance sum of vertices i and j, respectively and μ is the cyclomatic number (i.e. the number of rings in graph). The factor before sum takes into account the variation of edge number with the cyclicity, thus being a normalizing factor. This index shows an extremely low degeneracy (in alkanes, the first degenerated pair appears in dodecanes),[69] a good variation with the molecular branching and a satisfactory correlational ability. Figure 4.7 illustrates the calculus of j for the graph $G_{4.3}$.

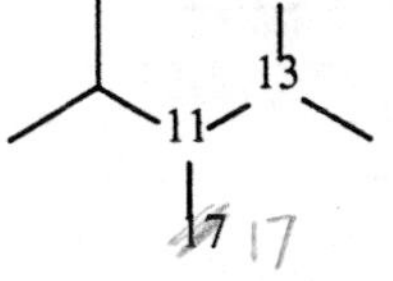

$$J = 7\,[\,4(19\mathrm{x}13)^{-1/2} + 2(11\mathrm{x}13)^{-1/2} + (11\mathrm{x}17)^{-1/2}\,] = 3.4642$$

$G_{4.3}\{DS_i\}$

Figure 4.7. Balaban index J for the graph $G_{4.3}$

Various extensions of J index, taking into account the chemical nature of vertices and edges[6] were proposed and tested for correlating ability.

4.3. INDICES BASED ON RECIPROCAL MATRICES

The reciprocal matrix **RM** of a square matrix **M** was defined in Sect. 2.13. The half sum of entries in such a matrix is an index, referred to as a *Harary-type index* (name given in honor of Frank Harary)[52, 70,71]

$$H_M = H_M(G) = (1/2)\sum_i \sum_j [\mathbf{RM}]_{ij} \tag{4.58}$$

the subscript M being the identifier for the matrix **M**.

The original *Harary index*, H_{De}, is constructed on the *reciprocal distance matrix*, **RD**$_e$.[70,71] The entries in this matrix suggest the interactions between the atoms of a molecule, which decrease as their mutual distances increase. Table 4.5 lists values H_{De} for octanes. One can see that they increase with the branching (in the opposite to the values of Wiener index) within the set of isomers and no degeneracy is encountered. This index was tested[6] for correlations with boiling points and van der Waals areas of octanes.

By analogy to H_{De}, Diudea[52] proposed the H_{We} and H_{Wp} indices, on the reciprocal Wiener matrices. These indices correlate excellent with the octane number, ON (r = 0.992; s = 3.313; F = 409.967 for H_{Wp})[52] H_{We} shows the same degenerate pairs (marked by italics) as the Wiener index, within this set of octane isomers (see Table 4.5).

In path graphs, a composition relation of the form: $H_{Wp}(P_N) = H_{We}(P_1) + H_{We}(P_2) + ... + H_{We}(P_{N-1}) + H_{We}(P_N)$ holds. It is analogous to that found by Lukovits[49] for the hyper Wiener index WW (see eq. 4.37). Despite the equality $WW_{Dp} = WW_{Wp}$, a similar composition relation for $H_{Dp}(P_N)$ was not found.

Table 4.5. Wiener-Type and Harary-Type Indices of Octanes.

Graph	W	WW	$W_{W(A,De,1)}$	H_{De}	H_{We}	H_{Dp}	H_{Wp}	$H_{W(A,De,1)}$
P_8	84	210	256	13.7429	0.6482	10.56429	5.8593	7.4281
$2MP_7$	79	185	253	14.1000	0.7077	10.86191	7.8938	7.4450
$3MP_7$	76	170	*209*	14.2667	0.7244	10.98095	8.5244	7.6542
$4MP_7$	75	165	208	14.3167	0.7286	11.01429	8.6897	7.6562
$3EP_6$	72	150	172	14.4833	0.7452	11.13333	9.2952	7.8529
$25M_2P_6$	74	161	207	14.4667	0.7673	11.16667	10.1784	7.5312
$24M_2P_6$	*71*	147	194	14.6500	*0.7839*	11.30000	10.8923	7.6650
$23M_2P_6$	70	143	181	14.7333	0.7881	11.36667	11.0992	7.8140
$34M_2P_6$	68	134	*167*	14.8667	0.8006	11.46667	11.6339	7.9382
$3E2MP_5$	*67*	129	161	14.9167	*0.8048*	11.50001	11.7881	7.9500
$22M_2P_6$	*71*	149	208	14.7667	*0.7839*	11.43333	10.9589	7.5250
$33M_2P_6$	*67*	131	179	15.0333	*0.8048*	11.63333	11.8548	7.7762
$234M_3P_5$	65	122	*167*	15.1667	0.8476	11.73333	13.7587	7.8996
$3E3MP_5$	64	118	145	15.2500	0.8214	11.79999	12.5714	8.0202
$224M_3P_5$	66	127	*209*	15.1667	0.8435	11.76667	13.5768	7.4805
$223M_3P_5$	63	115	164	15.4167	0.8601	11.96667	14.4018	7.8850
$233M_3P_5$	62	111	147	15.5000	0.8643	12.03334	14.5976	7.9971
$2233M_4P_4$	58	97	139	16.0000	0.9196	12.50000	17.4196	7.9643

Another Harary-type index is $H_{W(A,De,1)}$. It is calculated on the *restricted random walk* matrix of Randić,[72] which is identical to the $\mathbf{RW}_{(A,De,1)}$ matrix[52] (see Sect. 2.14). Values of this index, for octanes, are listed in Table 4.5, along with the corresponding $W_{(A,De,1)}$ values. Other Harary-type indices are exemplified in Table 4.5. As a general feature, the Harary-type indices possess more powerful discriminating ability than the corresponding Wiener-type indices (see Table 4.5).

Formulas[52] for calculating the Harary-type indices given in Table 4.5 and the corresponding Wiener-type indices, for path, cycles and stars, are listed in Table 4.6. Formulas for Harary-detour indices[73] are also included. Some interrelating formulas are given in Table 4.7.

Table 4.6. Formulas for Wiener- and Harary-Type Indices of Paths, Cycles and Stars

Index		Sum	Final Relation	Examples
Paths				
1	W_{De}	$\sum_{i=1}^{N-1}(N-i)i$	$\dfrac{1}{6}N(N-1)(N+1)$	$N=11$ 220
2	H_{De}	$\sum_{i=1}^{N-1}(N-i)i^{-1}$	$-N+N\Psi(N)+1+N\gamma$	$N=11$ 22.219
3	W_{We}	$\sum_{i=1}^{N-1}(N-i)i$	$\dfrac{1}{6}N(N-1)(N+1)$	$N=11$ 220
4	H_{We}	$\sum_{i=1}^{N-1}(N-i)^{-1}i^{-1}$		$N=11$ 0.533
5	WW_{Dp}	$\sum_{i=1}^{N-1}(N-i)\binom{i+1}{2}$	$\dfrac{1}{24}N(N-1)(N+2)(N+1)$	$N=11$ 715
6	H_{Dp}	$\sum_{i=1}^{N-1}(N-i)\binom{i+1}{2}^{-1}$	$-2\Psi(N+1)+2N-2\gamma$	$N=11$ 15.96
7	WW_{Wp}	$\sum_{i=1}^{N-1}\sum_{j=1}^{N-i}ij$	$\dfrac{1}{24}N(N-1)(N+2)(N+1)$	$N=11$ 715
8	H_{Wp}	$\sum_{i=1}^{N-1}\sum_{j=1}^{N-i}i^{-1}j^{-1}$		$N=11$ 7.562
	Cycles			
9	W_{De}	$N\sum_{i=1}^{\left(\frac{N-z}{2}\right)}i-(1-z)\left(\dfrac{N-z}{2}\right)\left(\dfrac{N-z}{2}\right)$	$\dfrac{1}{8}N(N^2-z)$	$N=11;165$ $N=12;216$
10	H_{De}	$N\sum_{i=1}^{\left(\frac{N-z}{2}\right)}i^{-1}-(1-z)\left(\dfrac{N-z}{2}\right)\left(\dfrac{N-z}{2}\right)^{-1}$	$N\left(\Psi\left(\dfrac{1}{2}N-\dfrac{1}{2}z+1\right)+\gamma\right)-1+z$	$N=11;25.117$ $N=12;\ 28.4$
11	$w_{\Delta e}$		$N(3N^2-4N+z)/8$	Ref. 57

12 $H_{\Delta e}$

$$H_{\Delta_e} = z\left[N\sum_{i=1}^{(N-1)/2}(N-i)^{-1}\right] + (1-z)\left[N\sum_{i=1}^{(N-2)/2}(N-i)^{-1} + \frac{N}{2}\left(\frac{N}{2}\right)^{-1}\right]$$

$$H_{\Delta_e} = -zN\Psi\left(\frac{1-N}{2}\right) - N\Psi\left(\frac{-N}{2}\right) + N\Psi(1-N) + 1 + zN\Psi\left(\frac{-N}{2}\right) - z$$

13 WW_{Dp}

$$N\sum_{i=1}^{\left(\frac{N-z}{2}\right)}\binom{i+1}{2} - (1-z)\left(\frac{N-z}{2}\right)\left(\frac{\frac{N-z}{2}}{2}+1\right) \quad \frac{1}{48}(N-z)(N-z+1)(N-z+2)(N+\quad$$ $N=11;\ 385$ $N=12;\ 546$

14 H_{Dp}

$$N\sum_{i=1}^{\left(\frac{N-z}{2}\right)}\binom{i+1}{2}^{-1} - (1-z)\left(\frac{N-z}{2}\right)\left(\frac{\frac{N-z}{2}}{2}+1\right)^{-1} \quad 2\frac{(N^2-Nz+2z-2)}{(N-z+2)}$$ $N=11;$ $18.333;$ $N=12;\ 20.286$

15 $ww_{\Delta p}$ $N(7N^3-3N^2-10N+3z(N+1))/48$ Ref. 57

16 $H_{\Delta p}$

$$H_{\Delta p} = zN\sum_{i=1}^{(N-1)/2}\left[(N-i+1)(N-i)/2\right]^{-1} +$$

$$(1-z)\left[N\sum_{i=1}^{(N-2)/2}\left[N-i+1)(N-i)/2\right]^{-1} + \frac{N}{2}\left[\left(\frac{N}{2}+1\right)\left(\frac{N}{2}\right)/2\right]^{-1}\right]$$

$$H_{\Delta p} = 2(N^2+N-2z)/(N+1)(N+2)$$

17 $W_{(A,De,1)}$

$$\frac{1}{2}N\sum_{i=1}^{\left(\frac{N-z}{2}-1\right)}(2^i+2^i) + \frac{1}{2}\frac{N}{2-z}\left(2^{\frac{N-z}{2}}+2^{\frac{N-z}{2}}\right) \qquad N\left(\frac{3\times 2^{\left(\frac{N-z}{2}\right)}-2^{\left(\frac{N-z}{2}\right)}z-4+2z}{2-z}\right)$$

$N=11;\ 682$
$N=12;\ 1128$

18 $H_{W(A,De,1)}$

$$\frac{1}{2}N\sum_{i=1}^{\left(\frac{N-z}{2}-1\right)}(2^{-i}+2^{-i}) + \frac{1}{2}\frac{N}{2-z}\left(2^{\frac{N-z}{2}}+2^{\frac{N-z}{2}}\right) \qquad N\left(\frac{-3\times 2^{\left(\frac{-N+z}{2}\right)}-2x2^{\left(\frac{-N+z}{2}\right)}z+2-z}{2-z}\right)$$

$N=11;\ 10.656$
$N=12;\ 11.719$

Stars

19 W_{De}	$N-1+\sum_{i=1}^{N-2}i\times 2$		$(N-1)^2$	$N=11;\ 100$
20 H_{De}	$N-1+\sum_{i=1}^{N-2}i\times 2^{-1}$		$\frac{1}{4}(N+2)(N-1)$	$N=11;\ 32.5$
21 W_{We}	$(N-1)(1\times(N-1))$		$(N-1)^2$	$N=11;\ 100$
22 H_{We}	$(N-1)(1\times(N-1)^{-1})$		1	$N=11;\ 1$
23 WW_{Dp}	$N-1+\sum_{i=1}^{N-2}i\times 3$		$\frac{1}{2}(N-1)(3N-4)$	$N=11;\ 145$

24 H_{Dp}	$N-1+\sum_{i=1}^{N-2} ix3^{-1}$	$\frac{1}{6}(N+4)(N-1)$	$N=11; 25$
25 H_{Wp}	$(N-1)(1x(N-1)^{-1})+\sum_{i=1}^{N-2} ix1x1$	$\frac{1}{2}(N^2-3N+4)$	$N=11; 46$
26 $W_{(A,De,1)}$	$\frac{1}{2}(N-1)(1+(N-1))+\sum_{i=1}^{N-2} i(N-1)$	$\frac{1}{2}(N-1)(N^2-2N+2)$	$N=11; 505$
27 $H_{W(A,De,1)}$	$\frac{1}{2}(N-1)(1+(N-1)^{-1})+\sum_{i=1}^{N-2} i(N-1)^{-1}$	$N-1$	$N=11; 10$

$y = N \mod 4$; $z = N \mod 2$; $\gamma(x) = int(exp(-t)*t^\wedge(x-1)$, $t = 0,..$infinity; $\psi(x) =$ diff(ln($\gamma(x),x$)); $\psi(n,x) = $ diff($\psi(x), x\$n$); $\psi(0,x) = \psi(x)$;

Table 4.7. Interrelating Formulas

	Graph	Relation		Graph	Relation
1	Paths	$WW = \frac{1}{4}(N+2)W$	8		$H_{De} = \frac{1}{4}\frac{N+2}{N-1}W$
2		$WW = W + \binom{N+1}{4}$	9		$H_{Dp} = \frac{1}{3}\frac{N+4}{3N-4}WW$
3	Stars	$WW = \frac{1}{2}\frac{3N-4}{N-1}W$	10		$H_{We} = W - N(N-2)$
4		$H_{Dp} = \frac{2}{3}\frac{N+4}{N+2}H_{De}$	11		$H_{Wp} = \frac{N^2}{2} - \frac{WW}{N-1}$
5		$H_{We} = N - W^{1/2}$	12		$H_{W(A,D,1)} = W_{(A,D,1)} + \frac{2}{N^2-2N+2}$
6		$H_{W(A,D,1)} = N - H_{We} = W^{1/2}$	13		$H_{W(A,D,1)} = \frac{6H_{Dp}}{N+4}$
7		$H_{Wp} = H_{We} + \binom{N-1}{2}$	14		$H_{W(A,D,1)} = \frac{2WW}{3N-4}$

The Harary-type indices found interesting applications in correlating studies and particularly in discriminating sets of isomers.[74]

4.4. INDICES BASED ON COMBINATION OF MATRICES

4.4.1. Schultz-Type Indices

Among the modifications of the Wiener index, the *molecular topological index*,[75] MTI, (or the Schultz number) was extensively studied[76-88] by virtue of its relatedness with the Wiener index. It is defined as

$$MTI = MTI(G) = \sum_i [\deg(A + D_e)]_i \tag{4.59}$$

where A and D_e are the adjacency and the distance matrices, respectively and $\deg = (deg_1, deg_2,..., deg_N)$ is the vector of vertex degrees of the graph.

In matrix form (see also Sect. 2.15), this index can be written as

$$MTI = \mathbf{uSCH}_{(A,A,D_e)}\mathbf{u}^T = \mathbf{uA(A + D}_e)\mathbf{u}^T = \mathbf{uA^2u^T} + \mathbf{uAD_eu^T} = I_{A^2} + I_{A,D_e} \tag{4.60}$$

where I_{A^2} is the sum of squared degrees (it is exactly M_1, the Zagreb Group first index) and I_{ADe} is *the true Schultz index* (i.e. the non-trivial part[87] of MTI). In trees, MTI and $I_{A,De}$ are related to the classical Wiener index by[85-87]

$$MTI = 4W + 2P_2 - (N-1)(N-2) \tag{4.61}$$

$$I_{A,De} = 4W - N(N-1) \tag{4.62}$$

where P_2 the number of paths of length 2 (or Platt,[9] or Gordon-Scantlebury,[11] or also Bertz[12] number)

$$P_2 = \sum_{i=1}^{N}\binom{v_i}{2} \tag{4.63}$$

Schultz and co-workers proposed some weighting schemes in accounting for the chemical nature and stereochemistry of molecular graphs.[76-83] Estrada[89] calculated an index based on distances of edges in a graph. It is the MTI index of the corresponding line graph (see also the ε index, Sect. 4.1).

In the extended Schultz matrices,[90] (see Sect. 2.15) a decomposition similar to eq 4.60 leads to

$$I(SCH_{(M_1,A,M_3)}) = \mathbf{uSCH}_{(M_1,A,M_3)}\mathbf{u}^T = \mathbf{uM_1(A+M_3)u^T} = \mathbf{uM_1Au^T} + \mathbf{uM_1M_3u^T} = I_{M_1,A} + I_{M_1,M} \tag{4.64}$$

where none of the indices $I_{M_1,A}$ and I_{M_1,M_3} are trivial. Values of $I_{M_1,A}$ and the global indices $I(SCH_{(M_1,A,M_3)})$, calculated as the sum of all entries in the Schultz matrix of sequence $(\mathbf{D}_e, \mathbf{A}, \mathbf{D}_e)$, and $(\mathbf{W}_p, \mathbf{A}, \mathbf{W}_p)$, are listed, for octanes, in Table 4.8. They were tested for correlating ability.[88,91,92]

Table 4.8. $I_{M_1,A}$ and $I(SCH_{(M_1,A,M_3)})$ Indices of Octane Isomers.

Graph	$I_{D_e,A}$	$I_{W_e,A}$	$I(SCH_{(D_e,A,D_e)})$	$I(SCH_{(W_e,A,W_e)})$
P_8	280	322	3976	4522
$2MP_7$	260	324	3516	4324
$3MP_7$	248	318	3272	4102
$4MP_7$	244	316	3196	4012
$3EP_6$	232	306	2952	3786
$25M_2P_6$	240	326	3080	4126
$24M_2P_6$	228	320	2852	3904
$23M_2P_6$	224	318	2784	3814
$34M_2P_6$	216	314	2632	3682
$3E_2MP_5$	212	308	2556	3588
$22M_2P_6$	228	330	2860	3946
$33M_2P_6$	212	322	2564	3650
$234M_3P_5$	204	320	2396	3616
$3E_3MP_5$	200	314	2344	3442
$224M_3P_5$	208	332	2464	3748
$223M_3P_5$	196	326	2260	3526
$233M_3P_5$	192	324	2192	3452
$2233M_4P_4$	176	338	1912	3370

An extended Schultz index is *walk matrix* calculabe by[39,90]

$$I(SCH_{(M_1,A,M_3)}) = \mathbf{u}SCH_{(M_1,A,M_3)}\mathbf{u}^T = \mathbf{u}W_{(M_1,1,(A+M_3))}\mathbf{u}^T = \mathbf{u}W_{(M_1,1,A)}\mathbf{u}^T + \mathbf{u}W_{(M_1,1,M_3)}\mathbf{u}$$

$$(4.65)$$

In the Schultz matrix, one of the square matrices, say M_3, may be a non-symmetric one. In such a case, $\mathbf{SCH}_{(M_1,A,M_3)}$ is no more a symmetric matrix. Two composite indices can be now calculated: one is $I(SCH_{(M_1,A,M_3)})$, and the second is calculated similar to the *hyper-Cluj* indices (see Chap. 6)

$$I2(SCH_{(M_1,A,M_3)}) = \sum_p [(\mathbf{SCH}_{(M_1,A,M_3)})]_{ij}[(\mathbf{SCH}_{(M_1,A,M_3)})]_{ji} \qquad (4.66)$$

The summation goes over all paths $p(ij)$. As symmetric matrices: $\mathbf{A}$, $\mathbf{D}$, Δ and as unsymmetric matrices, Cluj and Szeged matrices are most often used. (see Chap. 5 and 6) Such composite indices are very discriminating among non-isomorphic isomers and therefore useful in this respect (see Chap. 8).

4.4.2. QXR, Local Vertex Invariants

Filip, Balaban and Ivanciuc[7] have combined matrices and vectors to produce novel vectors. The model provides sets of *LOVIs* (local vertex invariants) by solving the following system:

$$Q \times X = R \qquad (4.67)$$

where:

Q is a topological matrix derived from the adjacency matrix and R is a column vector which encodes either a topological property (vertex degree, distance sum etc.) or chemical property (atomic number, Z).

X is the column vector of *LOVIs* (i.e. solutions of eq 4.67);

The numerical solutions will depend on the specification of Q and R. In the case: $Q = A + P \times I$ ($P = Z$; I is the unity matrix) and $R = V$ (vertex valency vector), eq 4.67 becomes:

$$(A + Z \times I) \times X = V \qquad (4.68)$$

AZV-LOVIs are thus obtained, which Figure 4.8 illustrates for some simple graphs:[7,93]

Graph	$A + Z \times I$	$(A + Z \times I) \times X = V$	*LOVI*

Graph 1 (star, vertices 3—4—2 with 5):

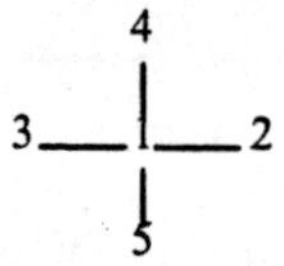

```
6 1 1 1 1   6x₁ + x₂ + x₃ + x₄ + x₅  = 4   x₁ = 0.6250
1 6 0 0 0   x₁ + 6x₂                  = 1   x₂ = 0.0625
1 0 6 0 0   x₁      + 6x₃             = 1   x₃ = 0.0625
1 0 0 6 0   x₁            + 6x₄       = 1   x₄ = 0.0625
1 0 0 0 6   x₁                  + 6x₅ = 1   x₅ = 0.0625
```

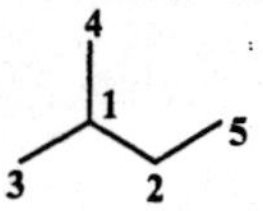

$6x_1 + x_2 + x_3 + x_4 + x_5 = 4$, $x_1 = 0.6250$
$x_1 + 6x_2 = 1$, $x_2 = 0.0625$
$x_1 + 6x_3 = 1$, $x_3 = 0.0625$
$x_1 + 6x_4 = 1$, $x_4 = 0.0625$
$x_1 + 6x_5 = 1$, $x_5 = 0.0625$

Graph 2:

$6x_1 + x_2 + x_3 + x_4 = 3$, $x_1 = 0.4281$
$x_1 + 6x_2 + x_5 = 2$, $x_2 = 0.2409$
$x_1 + 6x_3 = 1$, $x_3 = 0.0953$
$x_1 + 6x_4 = 1$, $x_4 = 0.0953$
$x_2 + 6x_5 = 1$, $x_5 = 0.1265$

```
6 1 1 1 0   6x₁ + x₂ + x₃ + x₄       = 3   x₁ = 0.4281
1 6 0 0 1   x₁ + 6x₂          + x₅   = 2   x₂ = 0.2409
1 0 6 0 0   x₁      + 6x₃            = 1   x₃ = 0.0953
1 0 0 6 0   x₁            + 6x₄      = 1   x₄ = 0.0953
0 1 0 0 6       x₂               + 6x₅ = 1 x₅ = 0.1265
```

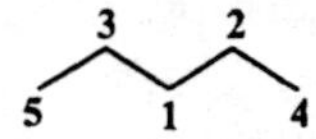

Graph 3:

```
6 1 1 0 0   6x₁ + x₂ + x₃        = 2   x₁ = 0.2424
1 6 0 1 0   x₁ + 6x₂ + x₄        = 2   x₂ = 0.2727
1 0 6 0 1   x₁      + 6x₃ + x₅    = 2   x₃ = 0.2727
0 1 0 6 0       x₂       + 6x₄    = 1   x₄ = 0.1212
0 0 1 0 6           x₃       + 6x₅ = 1  x₅ = 0.1212
```

Figure 4.8. Calculation of *AZV-LOVIs* .

The **AZV**-invariants have been used for constructing several global invariants (the so called triplet *TI*s) with very low degeneracy and good correlation ability.

4.5. INDICES BASED ON POLYNOMIAL COEFFICIENTS

4.5.1. Hosoya Index, Z

Hosoya[38] first introduced the notion of *topological index*, and proposed the Z index, as :

$$Z = Z(G) = \sum_{k=0}^{[N/2]} a(G,k) \qquad (4.69)$$

where $a(G,k)$ is the number of *k-matchings* of G, N is the number of vertices of G and $[N/2]$ is the integer part of the real $N/2$ value. By definition $a(G,0) = 1$ and $a(G,1)$ is equal to the number of edges of G. The calculation of the Z index for the graph $G_{4.3}$ is illustrated in Figure 4.9.

$a(G_{4.3},2) = 12$

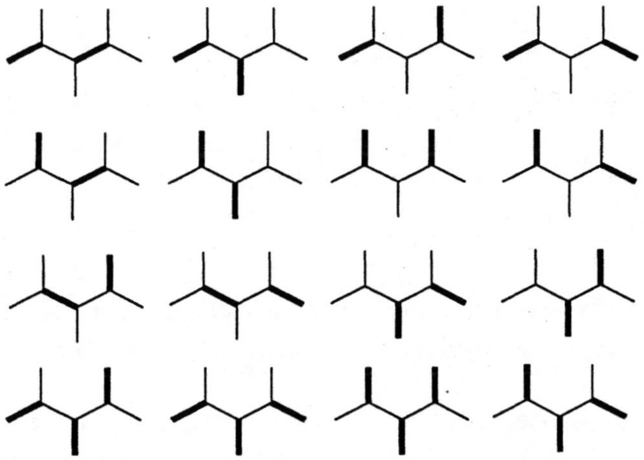

$a(G_{4.3},3) = 4$

$a(G_{4.3},4) = 0$

$Z(G_{4.3}) = 1 + 7 + 12 + 4 + 0 = 24$

Figure 4.9 Calculation of the Hosoya index, Z, for the graph $G_{4.3}$.

The a(G,k)'s are the coefficients of the *Z-counting polynomial*, *Q(G;x)*, of G (see also Chap. 3)

$$Q(G) = Q(G;x) = \sum_{k=0}^{[N/2]} a(G,k)x^k \tag{4.70}$$

The Z index is equal to the sum of the coefficients of the $Q(G;x)$ polynomial
$$Z = Q(G;1) \tag{4.71}$$

In larger graphs, the polynomial can be computed by a recursive relation[94]
$$Q(G;x) = Q(G-e;x) + xQ(G-[e];x) \tag{4.72}$$

where $G\text{-}e$ and $G\text{-}[e]$ are the spanning subgraphs of G resulted by removing an edge and the edge together with its incident edges, from G, respectively.

From an edge-weighted graph, Plavšić et al.[95] derived a weighted Z^* index, thus making possible the consideration of the chemical nature of molecular bonds.

Randić proposed the Hosoya matrix (see Sect. 2.10) as a basis for calculating two indices, called *path numbers* ^1Z and ^2Z. Plavšić et al.[96] demonstrated analytical relationships between these path numbers and the classical index Z.

Plavšić et al.[95] also extended the definition of the Z matrix to cycle-containing graphs

$$[Z]_{ij} = \left(\sum_{k \, p_{ij} \in D_{ij}(G)} Z(G - {}^k p_{ij}) \right) / |D_{ij}(G)| \tag{4.73}$$

where the non-diagonal entries are taken as the arithmetic mean among all $Z(G - {}^k p_{ij})$ numbers resulting by removing each time a geodesic (i,j) of G and $|D_{ij}(G)|$ is the cardinality of the set of these geodesics. Of course, the edge-weighted scheme was also extended. The invariants of the generalized Z matrix[95,97]

$$^kZ = \sum [^kZ]_{ij} \tag{4.74}$$

may offer a family of descriptors favoring a simpler interpretation of the regression equations than a set of ad-doc descriptors. Correlating studies were performed both on Z and ^kZ indices.[95]

Invariants calculated as *sum of polynomial coefficients* (cf. eqs 4.70 and 4.71), for the characteristic polynomial of various topological matrices were reported by Ivanciuc et al.[98] Some of them showed good selectivity among isomeric structures.

4.6. INDICES BASED ON EIGENVALUES AND EIGENVECTORS

Eingenvalues and eigenvectors, resulting by complex manipulations on square matrices, often enable a structural interpretation.

4.6.1. Eigenvalues

The first eigenvalue of the adjacency matrix, $\lambda_1(\mathbf{A})$, (see Chap. 3) has been proposed by Lovasz and Pelikan[99] as a measure of molecular branching. In the case of *distance-distance matrix*,[100] the leading eigenvalue $\lambda_1(\mathbf{DD})$ indicates the molecular folding. The corresponding eigenvalue of the Wiener matrix, $\lambda_1(\mathbf{W}_p)$, can be accepted as an alternative to $\lambda_1(\mathbf{A})$ in evaluating the branching of alkanes.[37] Eigenvalues (minimum or maximum values) of various topological matrices are often used in correlating studies.[98]

The eigenvalues of the Laplacian matrix, $\lambda_i(\mathbf{La})$, are used in calculating the number of spanning trees, $t(G)$, in graph[101] (see Sect. 2.2).

Mohar[102] defined two topological indices, TI_1 and TI_2 on the ground of Laplacian spectrum

$$(TI)_1 = 2N \log(Q/N) \sum_{i=2}^{N} (1/\lambda_i) \tag{4.75}$$

$$(TI)_2 = 4/(N\lambda_2) \tag{4.76}$$

In trees, TI_1 is related to the Wiener index

$$(TI)_1 = 2\log(Q/N)W \tag{4.77}$$

Table 4.9. shows eigenvalues of matrices $\mathbf{A}$, $\mathbf{D}_e$, $\mathbf{W}_p$ and $\mathbf{La}$, Mohar indices and their correlation with $\lambda_1(\mathbf{A})$ within the set of octanes.

Table 4.9. Eigenvalues of $\mathbf{A}$, $\mathbf{D}_e$, $\mathbf{W}_p$ and $\mathbf{La}$ and Mohar Indices of Octane Isomers.

Graph	$\lambda_1(\mathbf{A})$	$\lambda_1(\mathbf{D}_e)$	$\lambda_1(\mathbf{W}_p)$	$\lambda_2(\mathbf{La})$	$(TI)_1$	$(TI)_2$
P_8	1.879	21.836	57.170	0.1522	-9.7826	3.2843
$2MP_7$	1.950	20.479	52.612	0.1667	-9.1627	2.9991
$3MP_7$	1.989	19.763	48.406	0.1864	-8.8148	2.6825
$4MP_7$	2.000	19.542	46.661	0.1981	-8.6988	2.5245
$3EP_6$	2.029	18.779	42.204	0.2434	-8.3508	2.0542
$25M_2P_6$	2.000	19.111	47.724	0.1864	-8.5828	2.6825
$24M_2P_6$	2.042	18.396	43.419	0.2137	-8.2349	2.3399
$23M_2P_6$	2.074	18.181	42.059	0.2243	-8.1189	2.2293
$34M_2P_6$	2.095	17.676	39.290	0.2509	-7.8869	1.9930
$3E2MP_5$	2.101	17.419	37.428	0.3065	-7.7709	1.6315
$22M_2P_6$	2.112	18.413	44.471	0.2023	-8.2349	2.4721
$33M_2P_6$	2.157	18.308	38.533	0.2538	-7.7709	1.9702
$234M_3P_5$	2.136	16.808	37.025	0.2679	-7.5390	1.8660
$3E3MP_5$	2.189	16.670	34.142	0.3820	-7.4230	1.3090
$224M_3P_5$	2.149	17.034	39.141	0.2384	-7.6549	2.0969
$223M_3P_5$	2.206	16.315	34.994	0.2888	-7.3070	1.7313
$233M_3P_5$	2.222	16.068	33.468	0.3187	-7.1910	1.5690
$2233M_4P_4$	2.303	14.937	30.331	0.3542	-6.7271	1.4114
r vs $\lambda_1(\mathbf{A})$	1.000	0.964	0.958	0.867	0.979	0.903

The next step on this way was the *Quasi-Wiener* index,[101,103,104] W^*, defined as

$$W^* = N\sum_{i=2}^{N}\frac{1}{\lambda_i} \qquad (4.78)$$

where λ_i, $i = 2, 3, ..., N$ denote the positive eigenvalues of the Laplacian matrix. In acyclic structures, $W^* = W$, but in cycle-containing graphs the two quantities are different. In benzenoid molecules, a linear (but not particularly good) correlation between these indices was found.[105]

Klein and Randić[106] have considered the so-called *resistance distances* between the vertices of a graph, by analogy to the resistance between the vertices of an electrical network (superposable on the considered graph and having unit resistance of each edge). The sum of resistance distances is a topological index which was named the *Kirchhoff index*.[106,107] It satisfies the relation[106,108]

$$Kf = N\,Tr(\mathbf{La^*}) \qquad (4.79)$$

where $Tr(\mathbf{La^*})$ is the trace of the Moore-Penrose generalized inverse[109,110] of the Laplacian matrix. Further, Gutman and Mohar have demonstrated the identity of the quasi-Wiener and the Kirchhoff numbers for any graph.[108]

4.6.2. Eigenvectors

Huckel, in his theory, has equated the eigenvectors of the adjacency matrix for weighting of atomic orbitals in the construction of molecular orbitals. Balaban et al.[111] and Diudea et al.[112] have used the eigenvectors for defining several topological indices which showed good correlation with some physico-chemical properties and interesting intra- and inter-molecular ordering.

4.7. INDICES BASED ON GRAPHICAL BOND ORDER

4.7.1 X'/X Indices

The molecular descriptors X'/X are defined, according to the concept of *graphical bond orders*,[5,113] by the scenario

$$X_e = X(G - e) \qquad (4.80)$$

$$X' = \sum_e X_e \qquad (4.81)$$

$$X = X(G) \qquad (4.82)$$

$$X'/X = (\sum_e X_e)/X \tag{4.83}$$

where: - $G\text{-}e$ denotes the spanning subgraph (connected or not) resulted by cutting
the edge e;

$X(G\text{-}e)$ is a given topological index calculated on that subgraph
(if the subgraph is disconnected, the index is calculated by summing the values
on fragments, viewed as individual graphs);

- X_e is the edge contribution to the global index X', which is normalized by dividing it by $X(G)$

- the overall index X'/X is a bond-additive molecular characteristic. The quotient X_e/X (referred to as *graphical bond order*) can be used for weighting the adjacency matrix.

The algorithm is quite general and allows one to include molecular properties as source data in the construction of X'/X descriptors.

Figure 4.10 illustrates the construction of some indices X'/X for the graph $G_{4.3}$ ($234M_3P_5$).

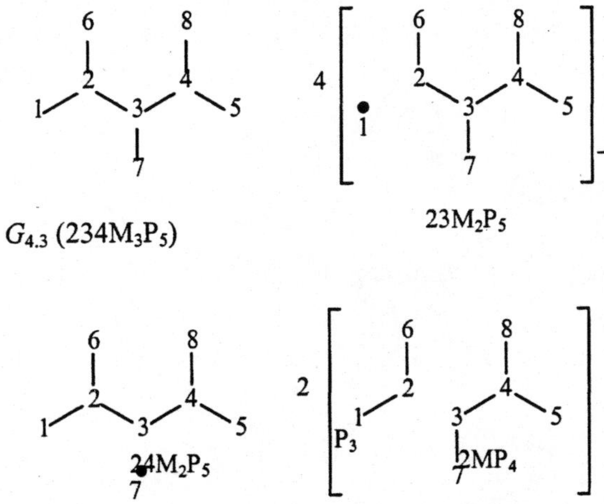

G	P	χ	J	W
P_3	3	1.4142	1.6330	4
$2MP_4$	10	2.2700	2.5396	18
$23M_2P_5$	21	3.1807	3.1440	46
$24M_2P_5$	21	3.1259	2.9532	48
$234M_3P$	28	3.5534	3.4643	65

$P'/P = [4 \times 21 + 21 + 2\times(3 + 10)]/28$ $\qquad = 4.6786$

$\chi'/\chi = [4 \times 3.1807 + 3.1259 + 2\times(1.4142 + 2.2700)]/3.5534 = 6.5338$

$J'/J = [4 \times 3.1440 + 2.9532 + 2\times(1.6330 + 2.5396)]/3.4643 = 6.8916$

$W'/W = [4 \times 46 + 48 + 2\times(4 + 18)]/ 65$ $\qquad = 4.2462$

Figure 4.10. Construction of some X'/X indices for the graph $G_{4.3}$ ($234M_3P_5$).

The most important invariant of this type is P'/P which, in trees, is related to the Wiener index by[114]

$$P'/P = (N\text{-}1) - [2/(N^2 -N)]W \tag{4.84}$$

4.7.2. SP Indices

SP (Subgraph Property) indices[115,116] represent an alternative to the X'/X descriptors and are basically *graphical bond orders*.[113]

They are constructed by the following algorithm:

(i) For each edge $e \in E(G)$, one defines, two subgraphs, $S_{L,e}$ and $S_{R,e}$, which collect the vertices lying to the left and to the right of edge e;

(ii) Subgraph properties, $P(S_{L,e})$ and $P(S_{R,e})$, are calculated by summing the vertex contributions, P_i (taken as *LOVIs* from the global property $P(G) = \Sigma_i P_i$) of all vertices i belonging to the given subgraph:

$$P(S_{L,e}) = \sum_{i \in S_{L,e}} P_i \qquad (4.85)$$

$$P(S_{R,e}) = \sum_{i \in S_{R,e}} P_i \qquad (4.86)$$

By dividing by $P(G)$, normalized values $P'(S_e)$, in the range 0 to 1, are obtained.

(iii) The edge contributions SP_e are evaluated as products to the left and to the right of edge e of the normalized values $P'(S_e)$. Since $P'(S_e)$ is taken as a part of the graph property, it is obvious that only one normalized value is needed:

$$SP_e = P'(S_{L,e})\, P'(S_{R,e}) = P'(S_{L,e})(\, 1 - P'(S_{L,e})) \qquad 4.87)$$

(iv) The global SP index is calculated by summing the edge contributions for all edges in graph:

$$SP = SP(G) = \sum_e SP_e \qquad (4.88)$$

The SP descriptors represent an extension of the procedure for calculating the Wiener index W. It is easily seen that, if $P = |V| = N$, then $SN = W/N^2$. Since this algorithm works with *LOVIs*, it becomes obvious the necessity of topological indices *defined on vertex* as a source for novel SP indices. Table 4.10 illustrates the construction of SP indices for some *properties* $(P_i = N_i; {}^1W_i; \chi_i)$.

The SP indices show a higher discriminating power in comparison to the original indices. The $S\chi^e W$ indices offer a family of indices, suitable in correlating studies (e.g. they supplied the best two variable model for the boiling point of octanes.[115,116]

Table 4.10. Construction of *SP* Indices for the Graph $G_{4.3}$
(see Figure 4.10; $P_i = N_i$; ^1W_i; χ_i; and $P(G)=\Sigma_i P_i$)

vertex	N_i	^1W_i	χ_i
1	1	1	0.57735
2	1	3	1.48803
3	1	3	1.24402
4	1	3	1.48803
5	1	1	0.57735
6	1	1	0.57735
7	1	1	0.57735
8	1	1	0.57735
$P(G)$	8	14	7.10684

SN_e:

$$4[\ 1/8 \times 7/8\ ] = 4[\ 0.125 \times 0.875\ ] = 4 \times 0.10938$$
$$1[\ 1/8 \times 7/8\ ] = \quad[\ 0.125 \times 0.875\ ] = \quad 0.10938$$
$$2[\ 3/8 \times 5/8\ ] = 2[\ 0.375 \times 0.625\ ] = 2 \times 0.23438$$
$$SN = \sum_e SN_e = W/N^2 = 65/(8\times8) = 1.01563$$

S^lW_e:

$$4[\ 1/14 \times 13/14\ ]= 4[\ 0.07143 \times 0.92857\ ]\ = 4 \times 0.06633$$
$$[\ 1/14 \times 13/14\ ] = [\ 0.07143 \times 0.92857\ ] = \quad 0.06633$$
$$2[\ 5/14 \times\ 9/14\ ] = 2[\ 0.35714 \times 0.64286\ ] = 2x\ 0.22959$$
$$S^lW = \sum_e S^lW_e = 0.79082$$

$S\chi_e$:

$$4[\ 0.57735 \times 6.529486\ ]/(\ 7.10684\ )^2 = 4 \times 0.07464$$
$$[\ 0.57735 \times 6.529486\ ]/(\ 7.10684\ )^2 = \quad 0.07464$$
$$2[\ 2.64273 \times 4.464106\ ]/(\ 7.10684\ )^2 = 2 \times 0.23358$$
$$S\chi = \sum_e S\chi_e = 0.84035$$

4.8. INDICES BASED ON LAYER MATRICES

Layer matrices **LM** have been used to derive two types of indices: (i) indices of *centrality* $C(LM)$ and (ii) indices of *centrocomplexity* $X(LM)$.[4,16] The X-type indices originated the famous super-index EAID.[117]

4.8.1. Indices of centrality

Indices of centrality[4,16] $C(LM)$ look for the center of the graph and are defined as

$$C(LM)_i = [\sum_{j=1}^{ecc_i} ([\mathbf{LM}]_{ij})^{j/d}]^{-1} \qquad (4.89)$$

$$C(LM) = \sum_i C(LM)_i \qquad (4.90)$$

where d is a specified topological distance ($eg.$ $d = 10$).

4.8.2. Indices of centrocomplexity

These indices express the location vs. a vertex of high complexity (eg. a vertex of high degree or a heteroatom).[4] They are defined as

$$X(LM)_i = [\sum_{j=0}^{ecc_i} [\mathbf{LM}]_{ij} 10^{-zj}]^{\pm 1} t_i \qquad (4.91)$$

$$X(LM) = \sum_i X(LM)_i \qquad (4.92)$$

where z denotes the number of bits (of the integer part) of $max([\mathbf{LM}]_{ij})$ in G and t_i is a weighting factor for heteroatom specification. For the graph $G_{4.3}$, matrices $\mathbf{L^1W}$ and $\mathbf{LDS}$ and the corresponding indices are shown in Table 4.11.

Table 4.11. Layer Matrices $\mathbf{L^1W}$ and $\mathbf{LDS}$ and the C and X Indices (with $d = 10$ and $t = 1$, Respectively) for the Graph $G_{4.3}$.

	$\mathbf{L^1W}$	$C(L^1W)_i$	$X(L^1W)_i$	$\mathbf{LDS}$	$C(LDS)_i$	$X(LDS)_i$
1	1 3 4 4 2	0.1897	1.3442	19 13 30 30 38	0.0968	19.13303038
2	3 5 4 2 0	0.2684	3.5420	13 49 30 38 00	0.1555	13.49303800
3	3 7 4 0 0	0.3945	3.7400	11 43 76 00 00	0.2608	11.43760000
4	3 5 4 2 0	0.2684	3.5420	13 49 30 38 00	0.1555	13.49303800
5	1 3 4 4 2	0.1897	1.3442	19 13 30 30 38	0.0968	19.13303038
6	1 3 4 4 2	0.1897	1.3442	19 13 30 30 38	0.0968	19.13303038
7	1 3 6 4 0	0.2461	1.3640	17 11 26 76 00	0.1458	17.11267600
8	1 3 4 4 2	0.1897	1.3442	19 13 30 30 38	0.0968	19.13303038
Global Index:		1.9365	17.5648		1.1052	132.06847352

These indices were useful in studies of intramolecular ordering of subgraphs of various size as well as of intermolecular ordering (see Chap. 8).

4.8.3. EATI Super-Indices

Two super-indices, $EATI_1$ and $EATI_2$, were recently proposed by Hu and Xu on weighted molecular graphs.[117,118] They follow the scenario described below.

(1). *Set the weight of atoms.*

For $EATI_1$ the δ_i^v valencies of Kier and Hall[17] are used (see Sect. 8.3.2.1). The index $EATI_2$ benefits of a more elaborated weight, based on layer matrices: the *connectivity valence matrix* **CVM** (whose element $[\mathbf{CVM}]_{ij}$ is the sum of δ_i^v for all vertices belonging to the jth layer) and the *bond matrix* **B** (whose element $[\mathbf{B}]_{ij}$ is defined as the sum of conventional bond orders - see below - of bonds connecting two subsequent layers). The weight of a vertex is given by the function

$$S_i = [\mathbf{CVM}]_{i1} + \sum_{j=1}^{ecc_i}[\mathbf{CVM}]_{i(j+1)}[\mathbf{B}]_{ij} \times 10^{-j} \tag{4.93}$$

Recall that both these layer matrices originate in the $\mathbf{L^1W}$ matrix[16,119] (see Chap. 2) and eq 4.93 follows from eq 4.91.

(2). *Set up the adjacency matrix, according to the conventional bond orders.*

$$[\mathbf{A}]_{ij} = \begin{cases} 0 & \text{no connection} \\ 1 & \text{a single bond} \\ 2 & \text{a double bond} \\ 3 & \text{a triple bond} \\ 1.5 & \text{an aromatic bond} \end{cases} \tag{4.94}$$

(3). *Set up the extended adjacency matrix,* **EA.**

$$.[\mathbf{EA}]_{ij} = \begin{cases} \dfrac{\sqrt{(Radii)_{ij}}}{6} & i=j \\ \dfrac{(\sqrt{[\mathbf{A}]_{ij}})w_{ij}}{6} & i \neq j \end{cases} \tag{4.95}$$

where $(Radii)_i$ is the covalent radii (in angstroms) of the ith atom and w_{ij} is the weighting for an edge (which is different for the two indices)

$$w_{ij} = \frac{1}{\sqrt{\delta_i^v \delta_j^v}} \qquad \text{for } EATI_1 \qquad\qquad (4.96)$$

$$w_{ij} = \sqrt{\frac{S_i}{S_j}} + \sqrt{\frac{S_j}{S_i}} \qquad \text{for } EATI_2 \qquad\qquad (4.97)$$

(4). *Evaluate a new matrix* **EA*** *as the sum of* **EA** *powers.*

$$\mathbf{EA}^* = \sum_{k=0}^{N-1} (\mathbf{EA})^k \qquad\qquad (4.98)$$

where N is the number of vertices in the molecular graph. When $k = 0$, then $(\mathbf{EA})^0$ is the unity matrix.

(5). *Calculate the topological indices.*

$$EATI_1 = \sum_{i=1}^{N} ([\mathbf{EA}^*]_{ii})^2 \qquad\qquad (4.99)$$

$$EATI_2 = \sum_{i=1}^{N} [\mathbf{EA}^*]_{ii} \qquad\qquad (4.100)$$

The algorithm is illustrated in Figure 4.11 for calculating $EATI_2$ for furane, $G_{4.4}$

$EATI_1$ was tested for selectivity on over 610,000 structures and also good correlating ability was found.[118] $EATI_2$ (also called $EAID$[117]) was particularly tested for selectivity (more than 4 bilion structures were investigated) and no degeneracy appeared. It is the most powerful index designed so far and is a candidate for CAS Registry Numbers.

4.9. INFORMATION INDICES

The following well-known principle[120] is generally used in construction of information indices. Let X be a set consisting of n elements. By assuming some equivalence criterion the elements are shared into m equivalence classes (i.e. subsets) X_i, each class having n_i elements and $n = \sum_{i=1}^{m} n_i$. The probability for an element to belong to the ith subset is $p_i = n_i / n$. The mean information content IC, corresponding to one element of the considered set is given by the Shanon formula[121]

$$H = -\sum_{i=1}^{n} p_i \log p_i \qquad\qquad (4.101)$$

In chemical literature Bonchev and Trinajstić[51,122,123] and the group of Basak[124,125] have pioneered this topic. Various schemes and various topological quantities were used so far in devising information indices. For early work along these lines see the book.[122]

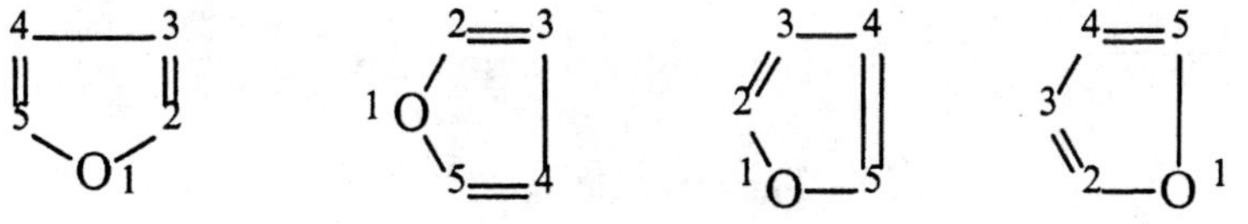

$G_{4.4}$ $G_{4.4}\{1\}$ $G_{4.4}\{2,5\}$ $G_{4.4}\{3,4\}$
Layer Partitions

Layers			CVM			B		S_i	(Radii)$_i$
1	(2,5)	(3,4)	6	6	6	2	4	7.44	0.702
2	(1,3)	(4,5)	3	9	6	3	2	5.82	0.772
3	(2,4)	(1,5)	3	6	9	3	3	5.07	0.772
4	(3,5)	(1,2)	3	6	9	3	3	5.07	0.772
5	(1,4)	(2,3)	3	9	6	3	2	5.82	0.772

A					EA					EA*				
0	1	0	0	1	.1396	.3358	0	0	.3358	1.612	.719	.419	.419	.719
1	0	2	0	0	.3358	.1464	.4725	0	0	.719	1.847	1	.437	.369
0	2	0	1	0	0	.4725	.1464	.3333	0	.419	1	1.875	.776	.437
0	0	1	0	2	0	0	.3333	.1464	.4725	.419	.437	.776	1.875	1
1	0	0	2	0	.3358	0	0	.4725	.1464	.719	.369	.437	1	1.874

$$EATI_2 = 9.021$$

Figure 4.11. Calculation of EATI2 index for the graph $G_{4.4}$

Recently, Ivanciuc et al.[126] have defined an information invariant on the vertex distance degree sequence, DDS (see Sect. 2.16)

$$u_i = -\sum_{j=1}^{ecc_i} \frac{jDDS_{ij}}{ds_i} \log \frac{j}{ds_i} \qquad (4.102)$$

where $ds_i = \sum_{j=1}^{N} d_{ij}$, j is the magnitude of distances within DDS and $ecc_i = \max j$. The above mean local information on the magnitude of distances was next modified in the view of constructing four global indices, U, V, X and Y, on the formula of J index (see Sect.4.2.2).

A similar descriptor, the Information Distance Index, IDI, was introduced by Konstantinova[127]

$$IDI_i = -\sum_{j=1}^{N} \frac{d_{ij}}{ds_i} \log \frac{d_{ij}}{ds_i} \qquad (4.103)$$

where the ratio $\dfrac{d_{ij}}{ds_i}$ is the probability for an accidentally chosen vertex to be at the

distance d_{ij} from the vertex i. Then the global IDI takes the form

$$IDI = IDI(G) = \sum_{i=1}^{N} IDI(i) \qquad (4.104)$$

The selectivity of information indices is rather high. Thus, IDI was tested on the set of all unbranched hexagonal chains with three to ten rings (2562 structures).[74,127] No degeneracy appeared within this set. However, Ivanciuc et al.[126] found some pairs of trees with degenerate U, V, X and Y indices.

4.10. OTHER TOPOLOGICAL INDICES

4.10.1. Stereochemical Descriptors

Descriptors for *cis/trans* stereoisomers or staggered alkane rotamers were introduced by the *three-dimensional Wiener* index, ^3W. In addition to a net improvement of selectivity, a good correlating ability with alkane properties was reported.[63,64]

A *three dimensional $\chi\chi$ connectivity* descriptor was developed by Randić[128] on the ground of topographic matrices.

Estrada[24] included a corrected electron charge density (calculated with MOPAC) in a connectivity-type index, Ω, also able to differentiate *cis/trans* isomers of alkenes.

The *cis/trans connectivity descriptor* χ_{ct} of Pogliani[129] is based on an extension of the concept of connectivity. It considers virtual ring fragments formed by embedded *cis* structures and appears to accurate describe some physico-chemical properties of olefins.

Other description of stereoisomers is obtained by using the GAI index,[130] (based on quantum chemistry considerations), the topographic indices,[131] the stereoisomeric and optical topological indices[132-135] or the complex numbers in the characterization of a plane.[136]

Enantiomers are more difficult to be described topologically, some results being reported by Schultz[82] and by Galvez et al.[137]

The *shape descriptors* ^kS proposed by Randić,[138] are constructed on a distance topographic matrix $\mathbf{T}$, whose elements are raised to the power k. Next, the average row sum of the $^k\mathbf{T}$ matrix

$$^kR = (1/N)\sum_i\sum_j [^k\mathbf{T}]_{ij} \qquad (4.105)$$

is normalized by division by $k!$

$$^kS = {}^kR / k! \qquad (4.106)$$

The *molecular shape profile S* is obtained as a sequence

$$S = {}^0S, {}^1S, ..., {}^kS \tag{4.107}$$

where the leading term is just the number of atoms in molecule. The factorial in the denominators ensures the convergence of the series. The maximum k value will equal the number of atoms on the molecular periphery. The consideration of atomic profiles may offer an alternative route to graph center (see Chap. 8). The individual ${}^kS$ descriptors can be used in correlating studies.[139,140]

The *kappa indices* ${}^Lk$, of Kier[141-144] are also related to as shape descriptors. A *flexibility index* φ for molecules was derived from the kappa shape attributes.[145]

Actually there exists a net trend for a more appropriate description of the 3D-structures. A bright illustration of this trend the reader can find in two recent monographs by Balaban.[146,147]

4.10.2. ID Numbers

In the design of a topological index, the two main attributes: (i) the correlating ability and (ii) the selectivity (i.e. the discriminating power among structures) are not reached at the same time. For the second purpose, an appropriate weighting scheme would discriminate the subgraphs (i.e. atoms, edges, paths) that are topologically (and chemically) different. It this respect, the so called *molecular identification numbers ID* were designed.

Randić[148] defined the molecular *ID* number as follows:
The mapping from the edge set $E(G)$ to the real numbers, $f: E(G) \rightarrow R$ is

$$f(e_{ij}) = (\delta_i \delta_j)^{-1/2} \tag{4.108}$$

The mapping from the path set $P(G)$ to the real numbers, $f^*: P(G) \rightarrow R$ takes the form

$$f^*(p) = \prod_{}^{m} f(e_{ij}) \tag{4.109}$$

for a path p consisting of m edges. The *connectivity ID* number of a molecular graph having N vertices is

$$ID = ID(G) = N + \sum_p f^*(p) \tag{4.110}$$

where summation goes over all distinct paths in G.

The selectivity of the connectivity ID number, CID was tested for all alkanes with up to 20 carbon atoms.[149] The smallest pair of trees with the same ID numbers was found at $N = 15$ (a duplicate among some 7500 graphs). Recall that the first degenerate pair for the Balaban number J appears at $N = 12$.

The *prime ID* number,[150] PID, changed the weighting scheme (4.108) by $(p_i p_j)^-$
(1/2)

where p_i and p_j are successive prime numbers, thus reducing the chance of accidental duplication of ID numbers. Indeed, for PID, Szymanski et al.[151] found the first duplicate at $N = 20$, meaning one pair among 618,050 graphs.

Despite the failure of the ID numbers to be unique, such a high resolution among structures is of interest in chemical data storage.

The race for a higher selective index continued with BID (Balaban's ID number),[152] WID (the walk ID number),[153] $MINID$,[154] $MINSID$,[154] (both based on self-returning weighted distances in the graph), τ number,[155] SID (the self-returning walk ID number)[156] and $EAID$[117] (see Sect. 4.8.3). The last two indices, for which no degeneracy was found so far, are constructed in a very close manner (both are self-returning walks but the weighting scheme differs, however).

A 3-dimensional *molecular identification number*, MID, which can be obtained from the Cartesian coordinates, differentiates the alkane rotamers.[157]

* * *

The main application of TIs in chemistry is the evaluation and prediction of physico-chemical properties. It is related to as $QSPR$ (Quantitative Structure-Property Relationship). The correlating studies make use of TIs as mathematical molecular descriptors.

The physical meaning of topological indices is a more or less proved percept.[1, 158-162] The majority of TIs show a *size*-dimension (e.g. W, the connectivity indices, etc.) but some of them emphasize a *shape*-component (e.g. k and other shape descriptors).[163]

Despite TIs do not offer a causal explanation of the molecular properties, they condense important structural features into a single numerical parameter that parallels some properties. The structural insight is often permitted, such that TIs are useful tools for looking at *fragmental components* of a certain property. Some aspects about the correlating studies are detailed in Chap. 9.

For the computation of TIs some commercial programs, such as POLLY,[164] MOLCONN[165] or CODESSA[166,167] are available.

REFERENCES

1.   A.T. Balaban, A. T.; Moţoc, I.; Bonchev, D.; Mekenyan, O. Topological Indices for

2.   Structure - Activity Correlations, *Top. Curr. Chem.* **1993**, *114*, 21-55.
3.   Rouvray, D. H. The Challange of Characterizing Branching in Molecular Species
4.   Discr. Appl. Math. **1988**, 19, 317-338.
5.   Randić, M. Design of Molecules with Desired Properties. A Molecular Similarity Approach to Property Optimization, in *Concepts and Applications of Molecular Similarity*, M.A. Johnson and G.M. Maggiora, Eds. John Wiley & Sons, Inc., 1990, Chap. 5, pp.77-145
6.   Diudea, M. V. Layer Matrices in Molecular Graphs, *J. Chem. Inf. Comput. Sci.*
7.   **1994**, *34*, 1064-1071.
8.   Randić, M. Generalized Molecular Descriptors, *J. Math. Chem.* **1991**, 7, 155-168.
9.   Diudea, M. V.; Ivanciuc, O. *Molecular Topology*, Comprex Cluj, 1995 (in Romanian).
10.   Filip, P.; Balaban, T. S.; Balaban, A. T. A New Approach for Devising Local Graph
11.   Invariants, Derived Topological Indices with Low Degeneracy and Good Correlation Ability, *J. Math. Chem.* **1987**, *1*, 61-83.
12.   Bonchev, D.; Mekenyan, O.; Frische, I. An Approach to the Topological Modeling of
13.   Crystal Growth, *Cryst. Growth* , **1980**, *49*, 90-96.
14.   Platt, J. R. Influence of Neighbor Bonds on Additive Bond Properties in Parafins,
15.   *J. Chem. Phys.* **1947**, *15*, 419-420.
16.   Platt, J. R. Prediction of Isomeric Differences in Parafin Properties, *J. Phys. Chem.*
17.   **1952**, *56*, 328-336.
18.   Gordon, M.; Scantlebury, G. R. Nonrandom Polycondensation: Statistical Theory of the Substitution Effect, *Trans. Faraday Soc.* **1964**, *60*, 604-621.
19.   Bertz, S. H. Branching in Graphs and Molecules, *Discr. Appl. Math.***19** (1988) 65-83.
20.   Gutman, I.; Trinajstić, N. Graph Theory and Molecular Orbitals. Total pi-Electron Energy of Alternant Hydrocarbons, *Chem. Phys. Lett.* **1972**, *17*, 535-538.
21.   Gutman, I.; Rusćić; Trinajstić, N.; Wilkox, Jr., C. F. Graph Theory and Molecular
22.   Orbitals. XII. Acyclic Polyenes, *J. Chem. Phys.* **1975**, *62*, 3339-3405.
23.   Randić, M. On Characterization of Molecular Branching, *J. Amer. Chem.Soc.*
24.   **1975**, *97*, 6609-6615.
25.   Diudea, M. V.; Topan, M. I.; Graovac, A. Layer Matrix of Walk Degrees,
26.   J. Chem. Inf. Comput. Sci. **1994**, 34, 1072-1078.
27.   Kier, L. B.; Hall, L. H. Molecular Connectivity in Chemistry and Drug Research
28.   Acad. Press, 1976.
29.   Randić, M. Search for Optimal Molecular Descriptors, *Croat. Chem. Acta.*
30.   **1991**, *64*, 43-54.
31.   Hall, L. H.; Kier, L. B. Electrotopological State Indices for Atom Types: A Novel
32.   Combination of Electronic, Topological, and Valence State Information,
33.   J. Chem. Inf. Comput. Sci. **1995**, 35, 1039-1045.
34.   Kier, L. B.; Hall, L. H. *The Electrotopological State*, Academic Press, New York, 1999.
35.   Estrada E. Edge Adjacency Relationships and a Novel Topological Index Related

36.   to Molecular Volume, *J. Chem. Inf. Comput. Sci.* **1995**, *35*, 31-33.
37.   Estrada E. Edge Adjacency Relationships in Molecular Graphs Containing Heteroatoms: A New Topological Index Related to Molar Volume, *J. Chem. Inf. Comput. Sci.* **1995**, *35*, 701-707.
38.   Daudel, R.; Lefevre, R.; Moser, C. Quantum Chemistry. Methods and Applications,
39.   Ed. Rev. pp. 70-80.
40.   Estrada, E. Three-Dimensional Molecular Descriptors Based on Electron Charge Density Weighted Graphs, *J. Chem. Inf. Comput. Sci.* **1995**, *35*, 708-713.
41.   Razinger, M. Discrimination and Ordering of Chemical Structure by the Number of
42.   Walks, *Theor. Chim. Acta*, **1986**, *70*, 365 - 378.
43.   Diudea, M. V.; Silaghi-Dumitrescu, I. Valence Group Electronegativity as a Vertex
44.   Discriminator, *Rev. Roum. Chim.* **1989**, *34*, 1175.
45.   Diudea, M. V.; Kacso', I. E.; Topan, M. I. A QSPR/QSAR Study by Using New Valence Group Carbon-Related Electronegativities, *Rev. Roum. Chim.* **1996**, *41*, 141-157.
46.   Wiener, H. Structural Determination of Paraffin Boiling Points, *J. Amer. Chem. Soc.*
47.   **1947**, *69*, 17-20.
48.   Wiener, H. Correlation of Heats of Isomerization, and Differences in Heats of Vaporization of Isomers, among the Paraffin Hydrocarbons, *J. Amer. Chem. Soc.* **1947**, *69*, 2636-2638.
49.   Wiener, H. Vapor Pressure-Temperature Relationship among the Branched Paraffin
50.   Hydrocarbons, *J. Phys. Chem.* **1948**, *52*, 425-430.
51.   Wiener H. Relation of the Physical Properties of the Isomeric Alkanes to Molecular Structure, *J. Phys. Chem.* **1948**, *52*, 1082-1089.
52.   Diudea, M. V.; Gutman, I. Wiener-Type Topological Indices, *Croat. Chem. Acta.*
53.   **1998**, *71* 21-51.
54.   Gutman, I.; Yeh, Y. N, Lee, S. L.; Luo, Y. L. Some Recent Results in the Theory of the Wiener Index, *Indian J. Chem.* **1993**, *32A*, 651-661.
55.   Nikolić, S.; Trinajstić, N.; Mihalić, Z. The Wiener Index: Development and Applications, *Croat. Chem. Acta*, **1995**, *68*, 105-129.
56.   Randić, M. Novel Molecular Description for Structure-Property Studies,
57.   Chem. Phys. Lett. **1993**, 211, 478-483.
58.   Randić, M.; Guo, X.; Oxley, T.; Krishnapriyan, H. Wiener Matrix: Source of Novel Graph Invariants, *J. Chem. Inf. Comput. Sci.* **1993**, *33*, 709-716.
59.   Randić, M.; Guo, X.; Oxley, T.; Krishnapriyan, H.; Naylor, L. Wiener Matrix Invariants, *J. Chem. Inf. Comput. Sci.* **1994**, *34*, 361-367.

60.   Hosoya, H. Topological index. A Newly Proposed Quantity Characterizing the Topological Nature of Structural Isomers of Saturated Hydrocarbons, *Bull. Chem. Soc. Japan.* **1971**, *44*, 2332-2339.
61.   Diudea, M. V. Walk Numbers ^eW_M : Wiener Numbers of Higher Rank, *J. Chem. Inf. Comput. Sci.* **1996**, *36*, 535-540.

62. Klein, D. J.; Lukovits, I.; Gutman, I. On the Definition of the Hyper-Wiener Index for Cycle-Containing Structures, *J. Chem. Inf. Comput. Sci.* **1995**, *35*, 50-52.

63. Lukovits, I.; Linert, W. A Novel Definition of the Hyper-Wiener Index for Cycles, *J. Chem. Inf. Comput. Sci.* **1994**, *34*, 899-902.

64. Lukovits, I.; Gutman, I. Edge-Decomposition of the Wiener Number,

65. Commun.Math.Comput.Chem.(MATCH), **1994**, 31, 133-144.

66. Lukovits, I. An Algorithm for Computation of Bond Contribution of the Wiener Index, *Croat. Chem. Acta*, **1995**, *68*, 99-103.

67. Zhu, H.Y.; Klein, D. J.; Lukovits, I. Extensions of the Wiener Number, *J. Chem. Inf. Comput. Sci.* **1996**, *36*, 420-428.

68. Gutman, I. Calculating the Wiener Number: The Doyle-Graver Method, *J. Serb. Chem. Soc.* **58** (1993) 745-750.

69. 46. Gutman, I. A New Method for Calculation of the Wiener Number in Acyclic Molecules, *J. Mol. Struct. (Theochem)* **1993**, *285*, 137-142.

70. Diudea, Wiener and Hyper-Wiener Numbers in a Single Matrix,

71. J. Chem. Inf. Comput. Sci. **1996**, 36, 833-836.

72. Doyle, J.K.; Graver, J. E. Mean Distance in a Graph, *Discr. Math.* **1977**, *17*, 147-154.

73. Lukovits, I. Formulas for the Hyper-Wiener Index of Trees, *J. Chem. Inf. Comput. Sci.* **1994**, *34*, 1079-1081.

74. Gutman, I. A Formula for the Wiener Number of Trees and Its Extension to Graphs

75. Containing Cycles, *Graph Theory Notes of New York,* **1994**, *26*, 9-15.

76. Bonchev, D.; Trinajstić, N. Information Theory, Distance Matrix and Molecular Branching, *J. Chem. Phys.* **1977**, *67*, 4517-4533.

77. Diudea, M. V. Indices of Reciprocal Property or Harary Indices, *J. Chem. Inf. Comput. Sci.* **1997**, *37*, 292-299.

78. Diudea; M. V.; Minailiuc, O.; Katona, G.; Gutman, I. Szeged Matrices and Related

79. Numbers, Commun. Math. Comput. Chem. (MATCH), **1997**, 35, 129-143.

80. Balasubramanian, K. Recent Developments in Tree-Pruning Methods and Polynomials for Cactus Graphs and Trees, *J. Math. Chem.* **1990**, *4*, 89-102.

81. Diudea, M. V.; Katona, G.; Minailiuc, O.; Pârv, B. Wiener and Hyper-Wiener Indices in Spiro-Graphs, *Izvest. Akad. Nauk, Ser. Khim.* **1995**, *9*,1674-1679; *Russ. Chem. Bull.* **1995**, *44*, 1606-1611 (Eng).

82. Zefirov, N. S.; Kozhushkov, S. I.; Kuznstsova, T. S.; Kokoreva, O. V.; Lukin, K. A.; Ugrak, B. I.;.Tratch, S. S. Triangulanes: Stereoisomerism and General Method of Synthesis, *J. Am. Chem. Soc.* **1990**, *112*, 7702-7707.

83. Amić, D.; Trinajstić, N. On the Detour Matrix, *Croat. Chem. Acta*, **1995**, *68*, 53-62.

84. Ivanciuc, O.; Balaban, A. T. Design of Topological Indices.Part. 8. Path Matrices and Derived Molecular Graph Invariants, *Commun. Math. Comput. Chem. (MATCH)*, **1994**, *30*, 141-152.

85. Lukovits, I. The Detour Index, *Croat. Chem. Acta*, **1996**, *69*, 873-882.

86. Lukovits, I.; Razinger, M. On Calculation of the Detour Index, *J. Chem. Inf. Comput. Sci.* **1997**, *37*, 283-286.

87. Tratch, S. S.; Stankevich, M. I.; Zefirov, N. S. Combinatorial Models and Algorithms

88. in Chemistry. The Expanded Wiener Number- a Novel Topological Index,

89. J. Comput. Chem. **1990**, 11, 899-908.

90. Diudea, Cluj Matrix CJ_u : Source of Various Graph Descriptors, *Commun. Math. Comput. Chem. (MATCH)*, **1997**, *35*, 169-183.

91. Bogdanov, B.; Nikolić, S.; Trinajstić, N. On the Three-Dimensional Wiener Number,

92. *J. Math. Chem.* **1989**, *3*, 299-309.

93. Bogdanov, B.; Nikolić, S.; Trinajstić, N. On the Three-Dimensional Wiener Number.

94. A Comment, *J. Math. Chem.* **1990**, *5*, 305-306.

95. Marjanović, M.; Gutman I. A distance-Based Graph Invariant Related to the Wiener and Szeged Numbers, *Bull. Acad. Serbe Sci. Arts (Cl. Math. Natur.)* **1997**, *114*, 99-108.

96. Lukovits, I. An All-Path Version of the Wiener Index, *J. Chem. Inf. Comput. Sci.*

97. **1998**, *38*, 125-129

98. Gupta, S.; Singh, M.; Madan, A. K. Superpendentic Index: A Novel Topological Descriptor for Predicting Biological Activity, *J. Chem. Inf. Comput. Sci.* **1998**, *38*, 450-456.

99. Balaban, A. T. Highly Discriminating Distance-Based Topological Index, *Chem. Phys. Lett.* **1982**, *89*, 399-404.

100. Balaban, A. T.; Quintas, L. V. The Smallest Graphs, Trees and 4-Trees with Degenerate Topological Index J, *Commun. Math. Comput. Chem. (MATCH)*, **1983**, *14*, 213-233.

101. Ivanciuc, O.; Balaban, T. S.; Balaban, A. T.; Reciprocal Distance Matrix, Related Local Vertex Invariants and Topological Indices, *J. Math Chem.* **1993**, *12*, 309-318.

102. Plavšić, D.; Nikolić, S.; Trinajstić, N.; Mihalić, Z. On the Harary Index for the

103. Characterization of Chemical Graphs, *J. Math. Chem.* **1993**, *12*, 235-250.

104. Randić, M. Restricted Random Walks on Graphs, *Theor. Chim. Acta*, **1995**, *92*, 97-106.

105. Diudea, M. V.; Katona, G.; Lukovits, I.; Trinajstić, N. Detour and Cluj-Detour Indices, *Croat. Chem. Acta* **1998**, *71*, 459-471.

106. Konstantinova, E. V.; Diudea, M. V. The Wiener Polynomial Derivatives and

107. Other Topological Indices in Chemical Research, *Croat. Chem. Acta*, (in press).

108. Schultz, H. P.; Schultz, E. B.; Schultz, T. P. Topological Organic Chemistry. 2. Graph

109. Theory, Matrix Determinants and Eigenvalues, and Topological Indices of Alkanes, *J. Chem. Inf. Comput. Sci.* **1990**, *30*, 27-29.

110. Schultz, H. P.; Schultz, T. P. Topological Organic Chemistry.3. Graph Theory,

111. Binary and Decimal Adjacency Matrices, and Topological Indices of Alkanes,

112. J. Chem. Inf. Comput. Sci. **1991**, 31, 144-147.

113. 77. Schultz, H. P.; Schultz, E. B.; Schultz, T.P. Topological Organic Chemistry. 4. Graph Theory, Matrix Permanents, and Topological Indices of Alkanes, *J. Chem. Inf. Comput. Sci.* **1992**, *32*, 69-72.

114. 78. Schultz, H. P.; Schultz, T. P. Topological Organic Chemistry. 5. Graph Theory,

115. Matrix Hafnians and Pfaffnians, and Topological Indices of Alkanes, *J. Chem. Inf. Comput. Sci.* **1992**, *32*, 364-366.

116. 79. Schultz, H. P.; Schultz, T. P., Topological Organic Chemistry. 6. Theory and

117. Topological Indices of Cycloalkanes, *J. Chem. Inf. Comput. Sci.* **1993**, *33*, 240-244.

118. Schultz, H. P.; Schultz, E. B.; Schultz, T.P. Topological Organic Chemistry.7. Graph

119. Theory and Molecular Topological Indices of Unsaturated and Aromatic Hydrocarbons, *J. Chem. Inf. Comput. Sci.* **1993**, *33*, 863-867.

120. Schultz, H. P.; Schultz, E. B.; Schultz, T. P., Topological Organic Chemistry. 8.

121. Graph Theory and Topological Indices of Heteronuclear Systems, *J. Chem. Inf. Comput. Sci.* **1994,** *34*, 1151-1157.

122. Schultz, H. P.; Schultz, E. B.; Schultz, T. P., Topological Organic Chemistry.9.

123. Graph Theory and Molecular Topological Indices of Stereoisomeric Compounds, *J. Chem. Inf. Comput. Sci.* **1995,** *35*, 864-870.

124. Schultz, H. P.; Schultz, E. B.; Schultz, T. P., Topological Organic Chemistry.10.

125. Graph Theory and Topological Indices of Conformational Isomers,

126. J. Chem. Inf. Comput. Sci. **1996**, 36, 996-1000.

127. Mihalić, Z.; Nikolić, S.; Trinajstić, N. Comparative Study of Molecular Descriptors

128. Derived from the Distance Matrix, *J. Chem. Inf. Comput. Sci.* **1992**, *32*, 28-37.

129. Klein, D. J.; Mihalić, Z.; Plavšić, D.; Trinajstić, N. Molecular Topological Index:

130. A Relation with the Wiener Index, *J. Chem. Inf. Comput. Sci.* **1992**, *32*, 304-305.

131. Plavšić, D.; Nikolić, S.; Trinajstić, N.; Klein, D. J. Relation Between the Wiener Index and the Schultz Index for Several Classes of Chemical Graphs, *Croat. Chem. Acta*, **1993**, *66*, 345-353.

132. Gutman, I. Selected Properties of the Schultz Molecular Topological Index,

133. J. Chem. Inf. Comput. Sci. **1994**, 34, 1087-1089.

134. Estrada, E.; Rodriguez, L. Matrix Algebraic Manipulation of Molecular Graphs. 2. Harary and MTI-Like Molecular Descriptors, *Commun.Math.Comput.Chem.(MATCH)*, **1997**, *35*, 157-167.

135. Estrada, E.; Gutman, I. A Topological Index Based on Distances of Edges of Molecular Graphs, *J. Chem. Inf. Comput. Sci.* **1996**, *36*, 850-853.

136. Diudea, M. V.; Randić, M. Matrix Operator, $W_{(M1,M2,M3)}$ and Schultz-Type Numbers,

137. J. Chem. Inf. Comput. Sci. **1997**, 37, 1095-1100.

138. Diudea, M. V. Novel Schultz Analogue Indices, *Commun. Math. Comput. Chem.*

139. *(MATCH)*, **1995**, *32*, 85-103.

140. Diudea, M. V.; Pop, C. M. A Schultz-Type Index Based on the Wiener Matrix,

141. *Indian J. Chem.* **1996**. *35A*. 257-261.

142. Balaban, T. S.; Filip, P.; Ivanciuc, O. Computer Generation of Acyclic Graphs Based on Local Vertex Invariants and Topological Indices. Derived Canonical Labeling and Coding of Trees and Alkanes, *J. Math. Chem.* **1992**, *11*, 79-105.

143. Hosoya, H. Topological Index as a Common Tool for Quantum Chemistry, Statistical

144. Mechanics, and Graph Theory, in *Mathematics and Computational Concepts in Chemistry*, Ed. Trinajstić, N., Horwood: Chichester, England, 1986, Chap. 11, pp. 110-123.

145. Plavšić, D.; Šošić, M.; Đaković, Z.; Gutman, I.; Graovac, A. Extension of the Z Matrix to Cycle-containing and Edge-Weighted Molecular Graphs, *J. Chem. Inf. Comput. Sci.* **1997**, *37, 529-534.*

146. Plavšić, D.; Šošić, M.; Landeka, I.; Gutman, I.; Graovac, A. On the Relation between the Path Numbers ^1Z and ^2Z and the Hosoya Z Index, *J. Chem. Inf. Comput. Sci.* **1996**, *36,* 1118-1122.

147. Randic, M. Hosoya Matrix - A Source of New molecular Descriptors, *Croat. Chem. Acta*, **1994**, *34*, 368-376.

148. Diudea, M. V.; Ivanciuc, O.; Nikolić, S.; Trinajstić, N. Matrices of Reciprocal Distance, Polynomials and Derived Numbers, *Commun. Math. Comput. Chem.*

149. *(MATCH)*, **1997**, *35*, 41-64.

150. Lovasz, L.; Pelikan, J. On the Eigenvalues of Trees, *Period. Math. Hung.* **1973**, *3*, 175-182.

151. Randić, M.; Kleiner, A. F.; De Alba, L. M. Distance-Distance Matrices for Graph

152. Embeded on 3-Dimensional Lattices, *J. Chem. Inf. Comput. Sci.* **1994**, *34*, 277-286.

153. Trinajstić, N.; Babić, D.; Nicolić, S.; Plavšić, D.; Amić, D.; Mihalić, Z.

154. The Laplacian Matrix in Chemistry, *J. Chem. Inf. Comput. Sci.* **1994**, *34*, 368-376.

155. Mohar, B. *MATH/CHEM/COMP* 1988, Ed. Graovac, A. Elsevier, Amsterdam,

156. 1989, pp. 1-8.

157. Mohar, B.; Babić, D.; Trinajstić, N. A Novel Definition of the Wiener Index for Trees, *J. Chem. Inf. Comput. Sci.* **1993**, *33*, 153-154.

158. Gutman, I.; Lee, S.L.; Chu, C.H.; Luo, Y.L. Chemical Applications of the Laplacian Spectrum of Molecular Graphs: Studies of the Wiener Number, *Indian J. Chem.*, **1994**, *33A*, 603-608.

159. Marković, S.; Gutman, I.; Bančević, Ž. Correlation between Wiener and quasi-Wiener Indices in Benzenoid Hydrocarbons, *J. Serb. Chem. Soc.* **1995**, *60*, 633-636.

160. Klein, D. J.; Randić, M. Resistance Distance, *J. Math. Chem.* **1993**, *12*, 81-95.

161. Bonchev, D.; Balaban, A. T.; Liu, X.; Klein, D. J. Molecular Cyclicity and Centricity of Polycyclic Graphs. I. Cyclicity Based on Resistance Distances or Reciprocal Distances, *Int. J. Quantum Chem.* **1994**, *50*, 1-20.

162. Gutman; I.; Mohar, B. The Quasi-Wiener and the Kirshhoff Indices Coincide, *J. Chem. Inf. Comput. Sci.* **1996**, *36*, 982-985.

163. Campbell, S. L.; Meyer, C. D. Generalized Inverses of Linear Transformations,

164. Pitman. London. 1979.

165. Ben-Israel; A.; Greville, T. N. E. *Generalized Inverses - Theory and Applications*, Wiley, New York, 1974.

166. Balaban, A. T.; Ciubotariu, D.; Medeleanu, M. Topological Indices and Real Number Vertex Invariants Based on Graph Eigenvalues or Eigenvectors, *J. Chem. Inf. Comput. Sci.* **1991**, *31*, 517-523.

167. Diudea, M. V.; Pop, C. M.; Katona, G.; Dobrynin, A. A. Dual Descriptors in the Calculation of the Wiener Numbers, *J. Serb. Chem. Soc.* **1997**, *62*, 241-250.

168. Randić, M.; Mihalić, Z.; Nikolić, S.; Trinajstić, N. Graphical Bond Orders: Novel
169. Structural Descriptors, *J. Chem. Inf. Comput. Sci.* **1994**, *34*, 403-409.

170. Plavšić, D.; Šošić, M.; Landeka, I.; Trinajstić, N. On the Relation between the P'/P Index and the Wiener Number, *J. Chem. Inf. Comput. Sci.* **1996**, *36*, 1123-1126.

171. Diudea, O.M. Minailiuc, G. Katona, Novel Connectivity Descriptors Based
172. on Walk Degrees, *Croat. Chem. Acta*, **1996**, *69*, 857-871.

173. Diudea, M. V.; Minailiuc, O. M.; Katona, G. SP Indices: Novel Connectivity
174. Descriptors, *Rev. Roum. Chim.* **1997**, *42*, 239-249.

175. Hu, C. Y.; Xu, L. On Highly Discriminating Molecular Topological Index,
176. J. Chem. Inf. Comput. Sci. **1996**, 36, 82-90.

177. Guo, M.; Xu, l.; Hu, C.Y.; Yu, S.M. Study on Structure-Activity Relationship of Organic Compounds - Applications of a New Highly Discriminating Topological Index, *Commun. Math. Comput. Chem. (MATCH)*, **1997**, *35*, 185-197.

178. Diudea, M.V.; Minailiuc, O.; Balaban, A. T. Molecular Topology. IV. Regressive Vertex Degrees (New Graph Invariants) and Derived Topological Indices, *J. Comput. Chem.* **1991**, *12*, 527-535.

179. Bonchev, D. Information Indices for Atoms and Molecules, *Commun. Math. Comput. Chem. (MATCH)*, **1979**, *7*, 65-113.

180. Shannon, C.E.; Weaver, W. *Mathematical Theory of Communications*; University of Illinois, Urbana, 1949.

181. Bonchev, D. Information Theoretic Indices for Characterization of Chemical Structures, Res. Stud. Press, Chichester, 1983; Letchworth, 1993.

182. Trinajstić, N. *Chemical Graph Theory*, 2nd Ed., CRC Press, Boca Raton, Florida,
183. 1992, pp. 225-273.

184. Basak, S. C.; Magnuson, V. R. Determining Structural similarity of Chemicals Using Graph-Theoretical Indices, *Discrette Appl. Math. 1988*, *19*, 17-44.

185. Raychaudhury, C.; Ray, S. K.; Roy, A. B.; Ghosh, J. J.; Basak, S. C. Discrimination of Isomeric Structures Using Information Theoretic Indices, *J. Comput. Chem.* **1984**, *5*, 581-588.

186. Ivanciuc, O.; Balaban, T. S.; Balaban, A. T. Chemical Graphs with Degenerate Topological Indices Based on Information on Distances, *J. Math. Chem.* **1993**, *14*, 21-33.

187. Konstantinova, E.V. The Discrimination Ability of Some Topological and Information Indices for Graphs of Unbranched Hexagonal Systems, *J. Chem. Inf. Comput. Sci.* **1996**, *36*, 54-57.

188. Randić, M. On Characterization of Three-Dimensional Structures, *Int. J. Quantum*
189. Chem. Quantum Biol. Symp. **1988**, 15, 201-208.

190. Pogliani, L. On a Graph Theoretical Characterization of Cis/Trans Isomers,
191. J. Chem. Inf. Comput. Sci. **1994**, 34, 801-804.
192. Xu, L.; Wang, H. Y.; Su, Q. A Newly Proposed Molecular Topological Index for the Discrimination of CIS/TRANS Isomers and for the Studies of QSAR/QSPR.
193. *Comput. Chem.* **1992**, *16*, 187-194.
194. Zefirov, N. S.; Tratch, S. S. Some Notes on Randić-Razinger's Approach to Characterization of Molecular Shapes, *J. Chem. Inf. Comput. Sci.* **1997**, *37*, 900-912.
195. Pyka, A. A New Optical Topological Index (I_{opt}) for Predicting the Separation of D and L Optical Isomers by TLC. Part III. *J. Planar Chrom.* **1993**, *6*, 282-288.
196. Pyka, A. A New Stereoisomeric Topological Index (I_{STI}) for Predicting the Separation of Stereoisomers in TLC. Part VII. *J. Planar Chrom.* **1994**, *7*, 389-393.
197. Pyka, A. New Topological Indices for the Study of Isomeric Compounds. *J. Serb. Chem. Soc.* **1997**, *62*, 251-259.
198. Gutman, I.; Pyka, A. New Topological Indices for Distinguishing between Enantiomers and Stereoisomers: A Mathematical Analysis. *J. Serb. Chem. Soc.* **1997**, *62*, 261-265.
199. Lekishvilli, G. On the Characterization of molecular Structure: 1. Cis-Trans Isomerism, *J. Chem. Inf. Comput. Sci.* **1997**, *37*, 924-928.
200. Julian-Ortiz, J. V.; Alapent, C. G.; Rios-Santaorina, I.; Garcia-Domenech, R. J.; Galvez, J. Prediction of Properties of Chiral Compounds by Molecular Topology,
201. Molec. Graph. Model. **1998**, 16, 14-18.
202. Randić, M. Molecular Shape Profiles, *J. Chem. Inf. Comput. Sci.* **1995**, *35*, 373-382.
203. Randić, M. Molecular Profiles. Novel Geometry-Dependent Molecular Descriptors,
204. *New J. Chem.* **1995**, *19*, 781-791.
205. Randić, M. Quantitative Structure-Property Relationship. Boiling Points of Planar Benzenoids, *New J. Chem.* **1996**, *20*, 1001-1009.
206. Kier, L. B. Shape Index from Molecular Graphs, *Quant. Struct. Act. Relat.* **1985**, *4*, 109-116.
207. Kier, L. B. Shape Indexes of Orders One and Three from Molecular Graphs,
208. Quant. Struct. Act. Relat. **1986**, 5, 1-7.
209. Kier, L. B. Distinguishing Atom Differences in a Molecular Graph Shape Index,
210. Quant. Struct. Act. Relat. **1986**, 5, 7-12.
211. Kier, L. B. Inclusion of Symmetry as a Shape Attribute in Kappa Index Analysis,
212. Quant. Struct. Act. Relat. **1987**, 6, 8-12.
213. Kier, L. B. An Index of Molecular Flexibility from Kappa Shape Attributes,
214. Quant. Struct. Act. Relat. **1989**, 8, 221-224.
215. Balaban, A. T., Ed. From Chemical Topology to Three-Dimensional Geometry,
216. Plenum Publishing Corporation, New York, 1997.
217. Devillers, J.; Balaban, A. T. *Topological Indices and Related Descriptors in QSAR and QSPR*, Gordon and Breach, Reading, UK, 1999.
218. Randić, M. On Molecular Identification Numbers, *J. Chem. Inf. Comput. Sci.*
219. **1984**. *24*. 164-175.

220. Szymanski, K.; Muller, W. R.; Knop, J. V.; Trinajstić, N. On M. Randić's Identification Numbers, *J. Chem. Inf. Comput. Sci.* **1985**, *25*, 413-415.

221. Randić, M. Molecular ID Numbers: By Design, *J. Chem. Inf. Comput. Sci.*

222. **1986**, *26*, 134-136.

223. Szymanski, K.; Muller, W. R.; Knop, J. V.; Trinajstić, N. Molecular ID Numbers,

224. Croat. Chem. Acta **1986**, 59, 719-723.

225. Balaban, A. T. Numerical Modelling of Chemical Structures: Local Graph Invariants and Topological Indices, in Graph Theory and Topology, Eds. King, R.b.; Rouvray, D. H. Elsevier, Amsterdam, 1987, pp. 159-176.

226. Knop, J. V.; Muller, W. R.; Szymanski, K.; Trinajstić, N. On the Identification Numbers for Chemical Structures, Int. J. Quantum Chem.: Quantum Chem. Symp. **1986**, *20*, 173-183.

227. Ivanciuc, O.; Balaban, A. T. Design of Topological Indices. Part 3. New Identification Numbers for Chemical Structures: MINID and MINSID, *Croat. Chem. Acta*, **1996**, *69*, 9-16.

228. Hall, L. H.; Kier, L. B. Determination of Topological Equivalence in Molecular Graphs from the Topological State, *Quant. Struct.-Act. Relat.* **1990**, *9*, 115-131.

229. Muller, W. R.; Szymanski, K.; Knop, J. V.; Mihalić, Z.; Trinajstić, N. The Walk ID

230. Number Revisited, *J. Chem. Inf. Comput. Sci.* **1993**, *33*, 231-233.

231. Randić, M.; Jerman-Blazić, B.; Trinajstić, N. Development of 3-Dimensional Molecular Descriptors, *Comput. Chem.* **1990**, *14*, 237-246.

232. Randić, M. Topological Indices, in *Encyclopedia of Computational Chemistry*, Eds. Schleyer, P. v. R.; Allinger, N. L.; Clark, T.; Gasteiger, J.; Kollman, P. A.; Schaefer III, H. F.; Schreiner, P. R., Wiley, Chichester, 1998, pp. 3018-3032.

233. Stankevich, M. I.; Stankevich, I. V.; Zefirov, S. N. Topological Indexes in Organic Chemistry, *Usp. Khim.* **1988**, *57*, 337-366; *Russ. Chem. Rev. (Engl. Transl.)*, **1988**, *57*, 191-208.

234. Baskin, I. I.; Skvortsova, M. I.; Stankevich, I. V.; Zefirov, N. S. On Basis of Invariants of Labeled Molecular Graphs, *J. Chem. Inf. Comput. Sci.* **1995**, *35*, 527-531.

235. Randić, M. On Characterization of Chemical Structure, *J. Chem. Inf. Comput. Sci.*

236. **1997**, *37*, 672-687.

237. Randić, M. On Characterization of Cyclic Structure, *J. Chem. Inf. Comput. Sci.*

238. **1997**, *37*, 1063-1071.

239. Mezey, P. Shape in Chemistry: An Introduction to Molecular Shape and Topology,

240. VCH Publishers, New York, 1993.

241. Basak, S. C.; Harris, D. K.; Magnuson, V, R. *POLLY* 2.3 (Copyright of the University of Minnesota, 1988).

242. Hall, L. H. *MOLCONN* and *MOLCONN2*, Hall Associates Consulting, 2 Davis Street, Quincy, MA 02170.

243. Katritzky, A. R.; Lobanov, V. S.; Karelson, M. *CODESSA Manual*, University of Florida, Gainesville, Fl, 1995.

244. Karelson, M.; Lobanov, V. S.; Katritzky, A. R. Quantum-Chemical Descriptors

245. in QSAR/QSPR Study, *Chem. Rev.* **1996**, *96*, 1027-1043.

Chapter 5

SZEGED INDICES

5.1. INTRODUCTION

5.1.1. Historical Remarks

In the 1990s a country which once was called Yugoslavia (or more officially: Socialist Federal Republic of Yugoslavia) fragmented into a number of newly formed states, followed by a sequence of bloody civil wars conducted by unimaginable cruelty. A part of the former country retained the name Yugoslavia. In 1993 and early 1994 an unprecedented inflation ravaged this new Yugoslavia, reaching at its peaks a daily value of 50%. In order to escape from the social, economic, political, cultural, ethical etc. collapse in his country, one of the present authors moved to Szeged, a city in Hungary lying on the very border of Hungary and Yugoslavia. There he happily spent a couple of years as a Visiting Professor at the Institute of Physical Chemistry of the *Attila JÓzsef* University.

Soon after settling in Szeged he produced a paper which eventually appeared in 1994.1 In this work a novel distance-based graph invariant has been introduced, and some of its basic properties established. Neither in the paper[1] nor in three subsequent articles[2-4] was any name given to that invariant. As research in this direction continued to expand, the lack of a name became quite annoying and in 1995 the invariant was named the Szeged index and denoted by Sz.[5,6] This name and symbol seem nowadays to be universally accepted.

A remarkably vigorous research followed these early works on the Szeged index. In just four years more than 30 papers on Sz were produced.[7-37] The review[38] covers, among other distance-based topological indices, also the theory of the Szeged index. The review[39] is devoted solely to the mathematical properties of Sz. Quite a few mathematicians and chemists, belonging to different and independent research groups, were involved in the elaboration of the theory and application of the Szeged index. The authors who worked on Sz are from Austria, Croatia, France, Hong Kong, Hungary, India, Romania, Russia, Slovenia and Yugoslavia.

5.1.2. Towards the Szeged Index Concept

The basic idea behind the Szeged index is the following. For a long time it is known[40] that the identity:

$$W = W(G) = \sum_e n_{1,e} \, n_{2,e} \tag{5.1}$$

is satisfied for the Wiener index W of a molecular graph G, provided that G is acyclic. (For details of the notation used in eq 5.1 see the subsequent section and also Sect. 4.2.1.1) This formula is rather attractive, because it can be understood as a decomposition of the Wiener index into contributions coming from individual chemical bonds of the respective molecule. Indeed, $n_{1,e}n_{2e}$ can be in a natural way interpreted as the increment associated with the carbon-carbon bond which in G is represented by the edge e.

Formula (5.1) is, however, not valid for graphs containing cycles. The efforts to modify the right-hand side of eq 5.1 and to find an edge-decomposition of the Wiener index, applicable to all molecular graphs were successful,[41-45] but resulted in rather perplexed and difficult-to-apply expressions. In the work[1] a seemingly awkward idea was put forward: the complications with the generalization of eq 5.1 to graphs containing cycles could be overcome by using the right-hand side of this equation as the definition of a new graph invariant. If so, then the right-hand side of eq 5.1 is automatically applicable to all graphs and, in addition, in the case of acyclic systems the newly introduced invariant coincides with the Wiener index.

Such a definition happened to be a lucky hit: the new graph invariant - the Szeged index - possesses interesting and nontrivial properties and, furthermore, is of some relevance for chemical applications.

5.2. DEFINITION OF THE SZEGED INDEX

Let G be a molecular graph, which means that G is necessarily connected (in fact, G may be any connected graph). Let $V(G)$ and $E(G)$ be the vertex and edge sets of G. Denote by i and j two vertices of G, which we formally write as $i,j \in V(G)$. If i and j are adjacent vertices, then the edge between them is denoted by (i,j). Then we write $(i,j) \in E(G)$. Sometimes, however, we denote the edge (i,j) simply as e. One should therefore bear in mind that throughout this chapter, the endpoints of the edge denoted by e are the vertices denoted by i and j.

Let $u,v \in V(G)$. Recall that the distance between the vertices u and v is equal to the number of edges in a shortest path connecting u and v, and is denoted by $d_{u,v}$.

As usual, the number of elements of a set S will be denoted by $|S|$. Hence, $|V(G)|$ and $|E(G)|$ stand for the number of vertices and edges, respectively, of the graph G.

Definition 5.1. Let G be a connected graph and $e = (x,y)$ its edge. The sets $N_{1,e}$ and $N_{2,e}$ are then defined as

$$N_{1,e} = \{v|v \in V(G) , d_{v,i} < d_{v,j}\} \tag{5.2}$$
$$N_{2,e} = \{v|v \in V(G) , d_{v,j} < d_{v,i}\} \tag{5.3}$$

Further,

$$n_{1,e} = |N_{1,e}| \tag{5.4}$$
$$n_{2,e} = |N_{2,e}| \tag{5.5}$$

When misunderstanding is avoided, we denote $N_{1,e}$ and $n_{1,e}$ simply by N_1 and n_1. Similarly, $N_{2,e}$ and $n_{2,e}$ will sometimes be written as N_2 and n_2, respectively.

In words: $n_{1,e}$ counts the vertices of G lying closer to one endpoint i of the edge e than to its other endpoint j. The meaning of $n_{2,e}$ is analogous. Vertices equidistant to x and y are not counted. Because the quantities n_1 and n_2 play the crucial role in the theory of the Szeged index, we illustrate Definition 5.1 by the graphs $G_{5.1}$ and $G_{5.2}$, depicted in Figure 5.1.

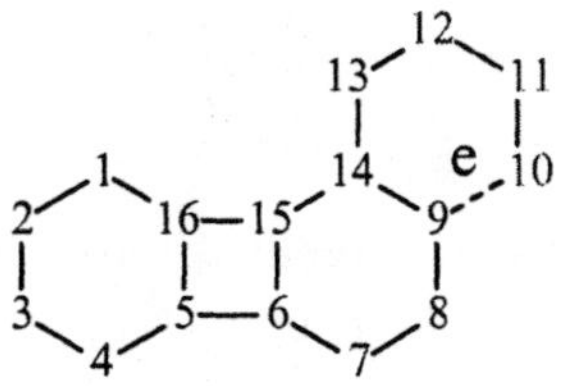

$$G_{5.1} \qquad\qquad\qquad\qquad\qquad G_{5.2}$$

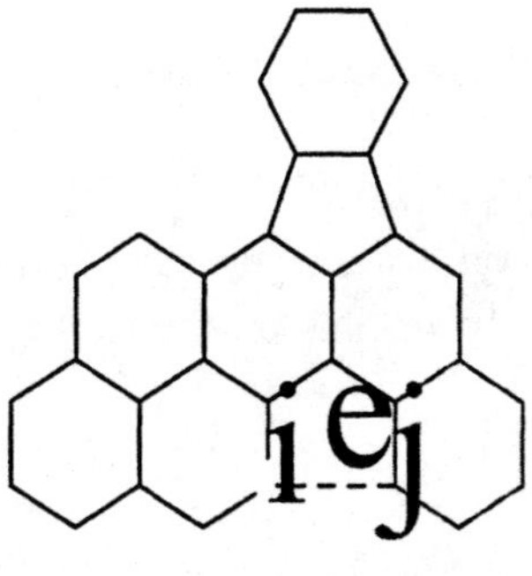

Figure 5.1. Two molecular graphs and an edge $e = (i,j)$ in them; the vertices belonging to the set N_1 are indicated by blue; those belonging to N_2 by red; vertices that do not belong to either N_1 or N_2 are not marked at all; such vertices exist only in non-bipartite graphs (such as $G_{5.2}$) whereas in bipartite graphs (such as $G_{5.1}$) are necessarily absent.

For the molecular graph G_1 we have

$N_{1,e}$ = {1,2,3,4,5,6,7,8,9,13,14,15,16}

$N_{2,e}$ = {10,11,12}

Consequently, $n_{1,e}(G_1)$=13 and $n_{2,e}(G_1)$=3. Note that

$$N_{1,e}(G_1) \cup N_{2,e}(G_1) = V(G_1)$$

and

$$n_{1,e}(G_1) + n_{2,e}(G_1) = |V(G_1)|$$

The latter two relations are properties of all molecular graph which do not possess odd-membered cycles (molecular graphs of alternant hydrocarbons).

For the molecular graph G_2 we have $n_{1,e}$ = 13 and $n_{2,e}$ = 7, see Figure 5.1. Note that G_2 possesses 10 vertices that do not belong to either $N_{1,e}(G_2)$ or $N_{2,e}(G_2)$. Such vertices are encountered in the case when the edge e belongs to an odd-membered cycle, hence in the case of graphs representing non-alternant molecules.

Note that

$$n_{1,e}(G_2) + n_{2,e}(G_2) < |V(G_2)|$$

We are now prepared to formulate our main:

Definition 5.2. The Szeged index $Sz(G)$ of the molecular graph G is equal to

$$Sz = Sz(G) = \sum_{e \in E(G)} n_{1,e}(G) \cdot n_{2,e}(G) \tag{5.6}$$

where the summation goes over all edges of G.

Examples

We now illustrate the calculation of the Szeged index, directly from its definition (5.6). We first do this on the example of 1,1,-dimethyl cyclopentane ($G_{5.3}$), depicted in Figure 5.2.

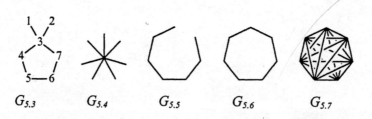

$G_{5.3}$ $G_{5.4}$ $G_{5.5}$ $G_{5.6}$ $G_{5.7}$

Figure 5.2. The molecular graph of 1,1-dimethyl,cyclopentane $G_{5.3}$ and representatives of the special graphs: *star S_N, path P_N, cycle C_N* and *complete graph K_N;* here $G_{5.4} = S_8$, $G_{5.5} = P_7$, $G_{5.6} = C_7$ and $G_{5.7} = K_7$.

First of all, notice that in $G_{5.3}$ the edges e_1 = (1,3) and e_2 = (2,3) are symmetry-equivalent. Therefore the values of the products $n_{1,e1}(G_{5.3})n_{2,e1}(G_{5.3})$ and $n_{1,e2}(G_{5.3})n_{2,e2}(G_{5.3})$ will necessarily be equal. The same is true for the edge-pairs (3,4) and

(3,7), as well as (4,5) and (6,7). Therefore, in actual calculations of Sz we need to determine the value of $n_{1,e}(G_{5.3})n_{2,e}(G_{5.3})$ only for four edges.

Nevertheless, here we examine all the seven edges of $G_{5.3}$:

Edge	N_1	N_2	n_1	n_2
(1,3)	{1}	{2,3,4,5,6,7}	1	6
(2,3)	{2}	{1,3,4,5,6,7}	1	6
(3,4)	{1,2,3,7}	{4,5}	4	2
(3,7)	{1,2,3,4}	{6,7}	4	2
(4,5)	{1,2,3,4}	{5,6}	4	2
(5,6)	{5}	{6}	1	1
(6,7)	{5,6}	{1,2,3,7}	2	4

Then application of formula (5.6) yields:

$$Sz(G_{5.3})=[1x6] + [1x6] + [4x2] + [4x2] + [4x2] + [1x1] + [2x4] = 45$$

As more advanced examples we now deduce the general formulas for the Szeged index of some special graphs, the most frequently occurring ones in graph theory; their structure is shown in Figure 5.2. In what follows N denotes the number of vertices.

The *star* S_N has N-1 edges that all are symmetry-equivalent. For each edge of S_N we have $n_1 = 1$ and $n_2 = N$-1. Therefore,

$$Sz(S_N) =(N\text{-}1)x[(N\text{-}1)x1] = (N\text{-}1)^2 \tag{5.7}$$

The *path* P_N has N-1 edges. On one side of the ith edges there are i vertices, on the other side there are N-i vertices, i=1, 2, ..., N-1. Therefore,

$$Sz(P_N) = \sum_{i=1}^{N-1} i \cdot (N-i) = \frac{N \cdot (N^2 -1)}{6} = \binom{N+1}{3} \tag{5.8}$$

The *cycle* C_N has N edges that all are symmetry-equivalent. If N is even, then for each edge of C_N we have $n_1 = n_2 = N/2$; if N is odd, then $n_1 = n_2 = (N-1)/2$.

Therefore, for even values of N,

$$Sz(C_N) = N \cdot \left[\frac{N}{2} \cdot \frac{N}{2}\right] = \frac{N^3}{4} \tag{5.9}$$

whereas for odd values of N,

$$Sz(C_N) = N \cdot \left[\frac{N-1}{2} \cdot \frac{N-1}{2}\right] = \frac{N(N-1)^2}{4} \tag{5.10}$$

The *complete graph* K_N has $N(N-1)/2$ edges that all are symmetry-equivalent. If $e = (x,y)$ is an edge of K_N then $N_1 = \{x\}$ and $N_2 = \{y\}$, implying $n_1 = n_2 = 1$. This is because every other vertex of K_N is on equal distance to both x and y, this distance, of course, being unity. Consequently,

$$Sz(K_N) = \frac{N(N-1)}{2} \cdot [1 \cdot 1] = \frac{N(N-1)}{2} = \binom{N}{2} \tag{5.11}$$

* * * *

Because the Szeged index is intimately related to the Wiener index, we recall here that the Wiener index $W(G)$ equals the sum of all topological distances in the graph (see eq 4.23)

At the first glance Definitions 1 and 2 look quite dissimilar. However, the following result is an immediate consequences of eqs (5.1) and (5.6).

Theorem 5.1. If G is a tree (i.e. a connected acyclic graph), then

$Sz(G) = W(G)$

(5.12)

Hence, in the case of molecular graphs of alkanes there is no difference between the Wiener and Szeged index and, as far the Szeged index is concerned, there is no need for any additional research. Therefore, in what follows we will be almost exclusively interested in the properties of the Szeged index of cycle-containing (molecular) graphs.

5.3. FURTHER RELATIONS BETWEEN SZEGED AND WIENER INDICES

Examples show[1] that in addition to trees there exist other graphs whose Szeged and Wiener indices coincide. [The simplest such example is the complete graph. Because the distance between any two vertices in K_N is unity, $W(K_N)$ = number of edges of K_N = $N(N-1)/2$. Above we have shown that $Sz(K_N)$ is also equal to $N(N-1)/2$.]

The problem of relation between Sz and W is basically resolved by the following result.[2,3]

Let K be the set of all connected graphs, all blocks of which are complete graphs. Recall that trees and complete graphs belong to K.

Theorem 5.2.

(a) $Sz(G) = W(G)$ if and only if $G \in K$.

(b) If G is connected, but $G \notin K$, then $Sz(G) > W(G)$.

In Figure 5.3 are depicted examples of cycle-containing graphs for which $Sz = W$, that is graphs belonging to the class K.

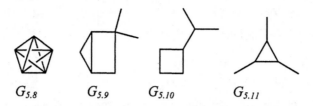

$$G_{5.8} \qquad G_{5.9} \qquad G_{5.10} \qquad G_{5.11}$$

Figure 5.3. Examples of connected cycle-containing graphs, all blocks of which are complete graphs; for these the Szeged and Wiener indices coincide; only very few of these graphs (e.g. $G_{5.10}$ and $G_{5.11}$) are molecular graphs.

Because of Theorem 5.1 the trees with minimal and maximal Szeged indices are precisely those which have minimal and maximal Wiener indices. These latter trees are known for a long time[46,47] and we state the respective result only for the sake of completeness. As before, let S_N and P_N be the N-vertex star and path, cf. Figure 5.2.

Theorem 5.3. If T_N is an N-vertex tree, other than S_N and P_N, then

$$Sz(S_N) < Sz(T_N) < Sz(P_N) \tag{5.13}$$

The connected unicyclic graphs with minimal and maximal SZ have been determined already in the first study of this quantity.[1] Let S_N+e be the graph obtained by adding one more edge to S_N. Let C_N be the N-vertex cycle and let C_N^1 be the graph obtained by attaching a new vertex of degree one to some vertex of C_N.

Theorem 5.4.
(a)   Let N be an odd number, greater than one. If U_N is an N-vertex connected unicyclic graph, other than S_N+e or C_{N-1}^1, then

$$Sz(S_N+e) < Sz(U_N) < Sz(C_{N-1}^1) \tag{5.14}$$

(b)   Let N be an even number, greater than two. If U_N is an N-vertex connected unicyclic graph, other than S_N+e or C_N, then

$$Sz(S_N+e) < Sz(U_N) < Sz(C_N) \tag{5.15}$$

Analytical formulas for Sz of bicyclic graphs have been derived in terms of the length of the cycles and the length of their common part.[31] Formulas for the connected bicyclic graphs with minimal and maximal Sz have been determined in.[37] Let S_N+e+f be the graph obtained by adding two nonadjacent edges to the star S_N. Let X_N and X'_N be the N-vertex bicyclic graphs, the structure of which is shown in Figure 5.4.

Theorem 5.5.

(a)　　Let N be an odd number, greater than five. If B_N is an N-vertex connected bicyclic graph, other than S_N+e+f or X_N, then

$$Sz(S_N+e+f) < Sz(B_N) < Sz(X_N) \tag{5.16}$$

(b)　　Let N be an even number, greater than six. If B_N is an N-vertex connected bicyclic graph, other than S_N+e+f or X'_N, then

$$Sz(S_N+e+f) < Sz(B_N) < Sz(X'_N) \tag{5.17}$$

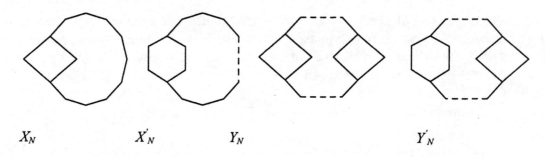

X_N　　　　　　　X'_N　　　　Y_N　　　　　　　　　Y'_N

Figure 5.4. Bicyclic and tricyclic graphs with maximal Szeged index.

In the work[37] also formulas for N-vertex tricyclic graphs with minimal and maximal Sz value have been determined. The structure of tricyclic graphs with maximal Sz value is shown in Figure 5.4: $Sz(Y_N)$ is maximal in the case of even N, $N \geq 12$, and $Sz(Y'_N)$ is maximal if N is odd, $N \geq 13$.

It should be noted that the extremal graphs depicted in Figure 5.4 have a completely different structure than the bicyclic graphs with maximal Wiener index.

It is easy to show[1] that among connected N-vertex graphs and N-vertex bipartite graphs the complete graph K_N and the star S_N, respectively, have minimal Sz value. The analogous result holds also for the Wiener number. The N-vertex graph with maximal Sz value is, however, completely different than the N-vertex graph with maximal W value. (Recall that this latter graph is the path, P_N.[46]) The following result was first conjectured by Klavžar et al.[9] and then proven in a rigorous mathematical manner by Dobrynin.[32]

Theorem 5.6.

Among connected N-vertex graphs the complete bipartite graph[48] $K_{N/2,N/2}$ (if N is even) or $K_{(N-1)/2,(N+1)/2}$ (if N is odd) has maximal Sz value.

We mention in passing that $Sz(K_{a,b}) = a^2b^2$.

* * * * *

A number of profound and somewhat unexpected analogies between the Szeged and Wiener indices have been found. Of them we mention here only two. For those who are not interested in mathematical details (for which nobody can claim to be of great chemical significance), it will suffice to compare the forms of the relations (5.1) and (5.6).

Let $C(h,k)$ be a class of connected bipartite graphs consisting of cycles of equal size, characterized as follows. Let k and h be positive integers. If $h = 1$, then the class $C(1,k)$ consists of one element only, namely the cycle C_{2k+2}. If $h > 1$, then every element of $C(h,k)$ is a graph obtained by joining the endpoints of a path with $2k$ vertices to a pair of adjacent vertices of some graph from $C(h-1,k)$.

It has been shown previously[49] that if G' and G'' are graphs from the class $C(h,k)$, then

$$W(G') \equiv W(G'') \pmod{2k^2} \tag{5.18}$$

To clarify: relation (5.18) means that the remainder after division of $W(G')$ by $2k^2$ is equal to the remainder after division of $W(G'')$ by $2k^2$. In other words, the difference $W(G') - W(G'')$ is divisible by $2k^2$.

A fully analogous result has been established for the Szeged index:[4]

Theorem 5.7. Let G' and G'' be arbitrary graphs from the class C(h,k). Then

$$Sz(G') \equiv Sz(G'') \pmod{2k^2} \tag{5.19}$$

Eventually, several generalizations and extensions of Theorem 5.7 have been put forward.[11,20,28] In what follows we will pay particular attention to a special case of this theorem important in chemical applications.[5] Namely in the case $k = 2$ some, but not all, elements of the class $C(h,k)$ are molecular graphs of catacondensed benzenoid hydrocarbons[50] with h hexagons. In fact, molecular graphs of all h-cyclic catacondensed benzenoid hydrocarbons belong to the class $C(h,k)$. As a curiosity we mention that the below result was independently and practically simultaneously, but prior to the discovery of Theorem 5.7, obtained in Indore (India), Novosibirsk (Russia) and Szeged (Hungary) and was then published jointly in a single paper.[5]

Theorem 5.8. Let H' and H'' be the molecular graphs of two catacondensed benzenoid hydrocarbons with h hexagons. Then

$$Sz(H') \equiv Sz(H'') \pmod 8 \tag{5.20}$$

or, what is the same, the difference $Sz(H') - Sz(H'')$ is divisible by 8.

In Figure 5.5 are depicted three catacondensed benzenoid hydrocarbons with 4 hexagons. Their Szeged indices differ considerably: $Sz(G_{5.12}) = 1269$, $Sz(G_{5.13}) = 1301$,

$Sz(G_{5.14})) = 1381$, but in accordance with Theorem 8, the remainder after division of both 1269, 1301 and 1381 with 8 is 5.

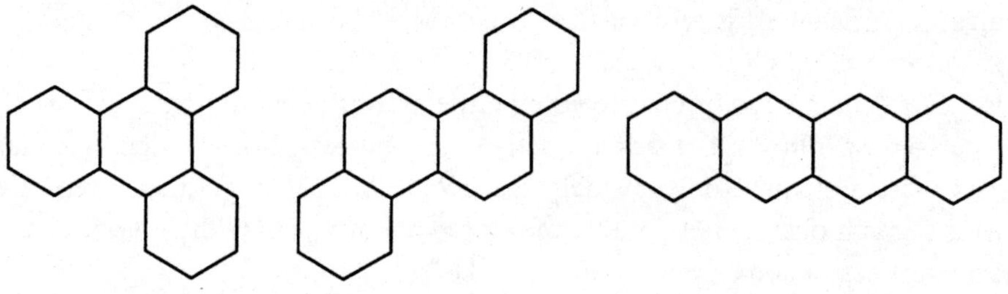

$$G_{5.12} \qquad\qquad G_{5.13} \qquad\qquad G_{5.14}$$

Figure 5.5. Isomeric catacondensed benzenoid systems with $h = 4$;
$Sz(G_{5.12}) = 1269$, $Sz(G_{5.13}) = 1301$, $Sz(G_{5.14}) = 1381$.

Needless to say that a fully analogous modulo 8 regularity holds also for the Wiener indices of catacondensed benzenoids.[51] The chemical meaning of such congruencies is obscure, but these certainly reflect some peculiar intrinsic symmetry in the distance-based properties of the underlying benzenoid molecules. An extension/generalization of Theorem 5.8 (and of its Wiener-number-counterpart) to pericondensed benzenoids[50] has never been accomplished.

* * * * *

Let G and H be graphs with $|V(G)|$ and $|V(H)|$ vertices, respectively. Let $G{\times}H$ denote the Cartesian product of G and H. It is known that[52,53]

$$W(G{\times}H) = |V(G)|^2\, W(H) + |V(H)|^2\, W(G)$$
$$(5.21)$$

For the Szeged index the analogous result reads:[9]

Theorem 5.9.

$$Sz(G{\times}H) = |V(G)|^3\, Sz(H) + |V(H)|^3\, Sz(G) \qquad\qquad (5.22)$$

Because the Cartesian product of graphs plays hardly any role in chemical graph theory, we skip here its definition (which, however, can be found elsewhere[9,52,53]) and only call the readers' attention to the intriguing parallelism between eqs (5.21) and (5.22).

In our opinion such close analogies between W and Sz in polycyclic graphs must have a deeper-lying cause. The discovery of this unifying principle remains a task for the future.

* * * * *

Concluding this section, in which the mathematical results pointing at concealed connections between Sz and W were outlined, we wish to note the following. When investigating the Szeged index, the strategy often is to try to prove properties of Sz analogous to known properties of W. Sometimes, however, the opposite direction of reasoning happened to be useful. For instance, it was noticed[11,28] that some specific subgraphs of benzenoid graphs - those isomorphic to the linear polyacenes - play an important role in certain considerations of Sz.

Bearing this in mind, new results for the Wiener index of benzenoid hydrocarbons were recently obtained, in which W is expressed in terms of such subgraphs.[54-56]

5.4. METHODS FOR THE CALCULATION OF THE SZEGED INDEX

In the general case the Szeged index is computed by using its definition, eq 5.6. This means that for each edge e the sets $N_{1,e}$ and $N_{2,e}$ have to be found (eqs 5.2 and 5.3) and then the numbers $n_{1,e}$ and $n_{2,e}$ (eqs 5.4 and 5.5). For this we need to know the distance between any two vertices in the underlying molecular graph, cf. (5.2) and (5.3), that is the elements of the distance matrix (see Chap. 2).

In summary, the calculation of the Szeged index is done by the following steps:

1. Determine the distance matrix of the molecular graph considered, i.e. find the distances between
all pairs of its vertices.

2. Choose an edge e.

3. Construct the sets $N_{1,e}$ and $N_{2,e}$ using their definition, eqs 5.2 and 5.3.

4. Count the elements of $N_{1,e}$ and $N_{2,e}$ and thus determine $n_{1,e}$ and $n_{2,e}$, eqs 5.4 and 5.5.

5. Repeat Steps 3 & 4 consecutively for all edges of the molecular graph considered.

6. Apply eq 5.6 and calculate Sz.

The distance matrix, required in Step 1, can be calculated by one of the several techniques developed in the theory of graphs and networks (see, for instance Chapters 11 and 12 in the book[57]). The algorithm for the calculation of Sz can always be designed in such a way that Step 3 is not explicitly performed; yet the elements of the sets N_1 and N_2 must somehow be recognized before they are counted in Step 4.

A general algorithm for the calculation of the Szeged index was put forward by Žerovnik.[10] Chepoi and Klavžar showed[17] that in the case of benzenoid hydrocarbons Sz can be obtained in linear time (with regard to the number of vertices).

5.4.1. The Gutman-Klavžar Algorithm

A special algorithm for the calculation of the Szeged indices of the so-called Hamming graphs[58] was designed by Gutman and Klavžar.[7] For chemical applications, the fact that the molecular graphs of benzenoid hydrocarbons are Hamming graphs[58] is particularly important. Thus we have a special, quite convenient and efficient, method for computing the Sz value of benzenoid systems.[7] The Gutman-Klavžar algorithm was eventually quite successfully applied in the theory of the Szeged index.[7,8,13,30,34]

In order to formulate the Gutman-Klavžar algorithm (stated below as Theorem 5.10), we need some preparation.

Benzenoid systems are plane graphs representing benzenoid hydrocarbons.[50] They are always considered as being embedded into the hexagonal (graphite) lattice. All their hexagons are regular, mutually congruent. A benzenoid system is always drawn so that some of its edges are vertical. Denote by B a benzenoid system and by N the number of its vertices. An elementary cut or elementary edge-cut of B is a straight line segment, passing through the centers of some edges of B, being orthogonal to these edges, and intersecting the perimeter of B exactly two times, so that at least one hexagon lies between these two intersection points.

At this point some examples are purposeful. In Figure 5.6 are shown three elementary cuts of benzo[c]phenanthrene $(G_{5.15})$ as well as all elementary cuts of anthracene $(G_{5.16})$.

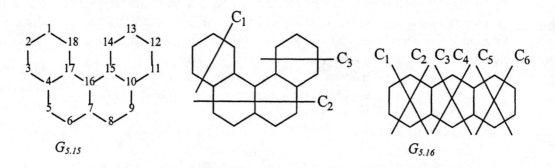

Figure 5.6. Three elementary edge-cuts of *benzo[c] phenanthrene* $(G_{5.15})$ and all elementary edge-cuts of *anthracene* $(G_{5.16})$.

Let B be a benzenoid system, $V(B)$ its vertex set and C one of its elementary cuts. Then C partitions the vertices of B into two non-empty classes $N_1(C;B)$ and $N_2(C;B)$, such that

$$N_1(C;B) \cap N_2(C;B) = \varnothing \qquad\qquad (5.23)$$
$$N_1(C;B) \cup N_2(C;B) = V(B) \qquad\qquad (5.24)$$

The elements of $N_1(C;B)$ are the vertices of B lying on one side of C, the elements of $N_2(C;B)$ are the vertices of B lying on the other side of C. In what follows it makes no difference which side of C corresponds to $N_1(C;B)$ and which to $N_2(C;B)$.

The number of elements of $N_1(C;B)$ and $N_2(C;B)$ are denoted by $n_1(C;B)$ and $n_2(C;B)$, respectively. The number of edges of B intersected by C is denoted by $r(C;B)$.

For instance, in the case of the benzenoid system $G_{5.15}$ and its elementary cuts C_1, C_2 and C_3, depicted in Figure 5.6, we have:

$$N_1(C_1;G_{5.15}) = \{1,2,3\}$$

$$N_2(C_1;G_{5.15}) = \{4,5,6,7,8,9,10,11,12,13,14,15,16,17,18\}$$
$$N_1(C_2;G_{5.15}) = \{1,2,3,4,10,11,12,13,14,15,16,17,18\}$$
$$N_2(C_2;G_{5.15}) = \{4,5,6,7,8,9\}$$
$$N_1(C_3;G_{5.15}) = \{12,13,14\}$$
$$N_2(C_3;G_{5.15}) = \{1,2,3,4,5,6,7,8,9,10,11,14,15,16,17,18\}$$

and further

$n_1(C_1;G_{5.15}) = 3$	$n_2(C_1;G_{5.15}) = 15$	$r(C_1;G_{5.15}) = 2$
$n_1(C_2;G_{5.15}) = 13$	$n_2(C_2;G_{5.15}) = 5$	$r(C_2;G_{5.15}) = 3$
$n_1(C_3;G_{5.15}) = 3$	$n_2(C_3;G_{5.15}) = 15$	$r(C_3;G_{5.15}) = 2$

We are now prepared to describe the Gutman-Klavžar algorithm.[7]

Theorem 5.10. The Szeged index of a benzenoid system B can be calculated by means of the formula

$$Sz(B) = \sum_C r(C;B)n_1(C;B)n_2(C;B) \qquad (5.25)$$

in which the summation embraces all elementary cuts of B.

The advantage of eq 5.25 over eq 5.6 is that there are much fewer elementary cuts than edges, especially in the case of large benzenoid systems. Consequently, the right-hand side of (5.25) contains much fewer summands than the right-hand side of (5.6). In practical applications of formula (5.25) it is not necessary to calculate both $n_1(C;B)$ and $n_2(C;B)$.

Namely the sum $n_1(C;B) + n_2(C;B)$ is independent of the elementary cut C and is equal to the number N of vertices of B. Hence, if we know $n_1(C;B)$, then we calculate $n_2(C;B)$ as $N - n_1(C;B)$; in this case it is reasonable that $n_1(C;B)$ is chosen to be the smallest among $n_1(C;B)$, $n_2(C;B)$.

We illustrate the Gutman-Klavžar algorithm on the example of the anthracene graph $G_{5.16}$, see Figure 5.6. The anthracene graph has 14 vertices, 16 edges, but only seven elementary cuts. By direct counting we get:

$n_1(C_1;G_{16}) = 3$	$n_2(C_1;G_{16}) = 14-3 = 11$	$r(C_1;G_{16}) = 2$
$n_1(C_2;G_{16}) = 3$	$n_2(C_2;G_{16}) = 14-3 = 11$	$r(C_2;G_{16}) = 2$
$n_1(C_3;G_{16}) = 7$	$n_2(C_3;G_{16}) = 14-7 = 7$	$r(C_3;G_{16}) = 2$
$n_1(C_4;G_{16}) = 7$	$n_2(C_4;G_{16}) = 14-7 = 7$	$r(C_4;G_{16}) = 2$
$n_1(C_5;G_{16}) = 3$	$n_2(C_5;G_{16}) = 14-3 = 11$	$r(C_5;G_{16}) = 2$
$n_1(C_6;G_{16}) = 3$	$n_2(C_6;G_{16}) = 14-3 = 11$	$r(C_6;G_{16}) = 2$
$n_1(C_7;G_{16}) = 7$	$n_2(C_7;G_{16}) = 17-7 = 7$	$r(C_7;G_{16}) = 7$

Direct application of eq 5.25 yields then

$$Sz(G_{5.16})=[2x3x11]+[2x3x11]+[2x7x7]+[2x7x7]+[2x3x11]+[2x3x11]+[4x7x7]=656$$

The calculation in the above example was performed without using the symmetry of the molecule considered. This symmetry causes that numerous summands are mutually equal. Needless to say that by taking into account molecular symmetry the algorithm additionally gains on efficiency and simplicity.

By means of the Gutman-Klavžar algorithm general expressions for the Szeged index of a large number of homologous series of benzenoid hydrocarbons could be determined.[7-9,13] Four such expressions, pertaining to the benzenoid systems depicted in Figure 5.7 are given below:

$$Sz(L_k) = \frac{1}{3}(44k^3+72\,k^2+43\,k+3) \tag{5.26}$$

$$Sz(H_k) = \frac{3}{2}(36\,k^6- k^4+ k^2) \tag{5.27}$$

$$Sz(Q_k) = \frac{1}{6}(12\,k^6+72\,k^5+137\,k^4+92\,k^3+13\,k^2-2\,k) \tag{5.28}$$

$$Sz(T_k) = \frac{1}{4}(k^6+12\,k^5+49\,k^4+84\,k^3+58\,k^2+12\,k) \tag{5.29}$$

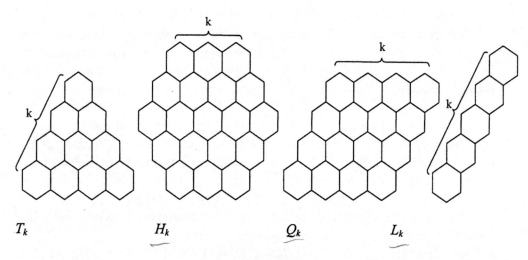

T_k H_k Q_k L_k

Figure 5.7. Some highly symmetric homologous series of benzenoid systems, $k = 1,2,3,..., .$

Elementary cuts in benzenoid systems have found several other applications in the theoretical chemistry of benzenoid hydrocarbons: in studies concerned with Kekule structures,[59] Wiener indices[60-64], hyper-Wiener indices[64,65] and elsewhere.[66]

* * * * *

An attempt to use the symmetry of a (molecular) graph in order to facilitate the calculation of its Sz value were communicated by Žerovnik.[36] Expressions for the Szeged index of some highly symmetric graphs were earlier deduced by Klavžar et al.[9] However, no complete and practically applicable approach, which would enable the systematic exploitation of molecular symmetry (or more precisely: of the automorphism group of the underlying molecular graph) in the theory of the Szeged index is available at the present moment.

5.5 EXTENSIONS: SZEGED MATRICES, HYPER-SZEGED AND HARARY-SZEGED INDICES

Diudea was the first to recognize[16] that the right-hand side of eqs 5.2 and 5.3 are well-defined also in the case when the vertices i and j are not adjacent. The same applies to n_1 and n_2 eqs 5.4 and 5.5. Consequently, the quantity

$$n_{ij} = n_{ij}(G) = |\{v|v \text{ in } V(G) , d_{v,i} < d_{v,j}\}| ; \ (i,j) \in P(G) \qquad (5.30)$$

can be viewed as the (ij)-entry of some square matrix of order N. In addition, $n_{ij} = 0$ whenever $i = j$.

In words: n_{ij} is the number of vertices of the graph G lying closer to vertex i than to vertex j.

Three so-called *Szeged matrices* have been defined (see Sect. 2.8):[16,21,27,38]

the *unsymmetric Szeged matrix*

$$[\mathbf{USZ}]_{ij} = n_{ij} \qquad (5.31)$$

the *full symmetric Szeged matrix*

$$[\mathbf{SZ}_p] = n_{ij}\, n_{ji} \qquad (5.32)$$

and *the sparse symmetric Szeged matrix*

$$[\mathbf{SZ}_e]_{ij} = [\mathbf{A}]_{ij}\, n_{ij}\, n_{ji} \qquad (5.33)$$

where $[\mathbf{A}]_{ij}$ is the (ij)-element of the adjacency matrix (see Chap. 2). The corresponding *detour-Szeged* matrices were also defined (see Sect. 2.8).

Evidently, the sum of the elements of the upper (or lower) triangle of $\mathbf{SZ}_e$ is just the *Szeged index Sz*. The analogous sum of the elements of $\mathbf{SZ}_p$ has been named[21,26,38] the *hyper-Szeged* index, symbolized in this book by *IP(SZD)*. A few more graph invariants based on or related to the Szeged matrices have been put forward.[15,35,38]

The theory of the Szeged matrices, the hyper-Szeged and Harary-Szeged indices is presently at a very early stage of development. Table 5.1 lists some formulas[15] for calculating these indices in paths, cycles and stars.

Table 5.1. Formulas for Hyper-Szeged $IP(SZD)$ and Harary-Szeged $IP(RSZD)$ Indices in Paths, Cycles and Stars

	Index	Formula / Examples
Paths		
1	$IP(SZD)$	$\dfrac{1}{48}(5N^4 - 10N^3 + 16N^2 - 8N - 6Nz + 3z)$ $N = 11;\ 1285 \quad N = 12;\ 1846$
Cycles		
2	$IP(SZD)$	$\dfrac{1}{8}N(N-1)^{(2z+1)}(N^2 - 2N + 4)^{(1-z)}$ $N = 11;\ 1375 \quad N = 12;\ 2046$
3	$IP(RSZD)$	$\dfrac{2(N-1+z)}{(N-2+z)}$ $N = 11;\ 2.2 \quad N = 12;\ 2.2$
Stars		
4	$IP(SZD)$	$\dfrac{1}{2}(N-1)(3N-4)$ $N = 11;\ 145$
5	$IP(RSZD)$	$\dfrac{1}{2}(N^2 - 3N + 4)$ $N = 11;\ 46$

5.6. CHEMICAL APPLICATIONS OF THE SZEGED INDEX

When speaking of the chemical applications of the Szeged index one has first to recall that the Wiener index has found numerous such applications, outlined elsewhere in this book and in a number of reviews.[67] Namely, the Wiener index is correlated with a large number of physico-chemical properties of organic molecules. This is especially the case for alkanes, whose molecular graphs are trees. Since in the case of trees the Wiener and the Szeged index coincide, all studies on the possible chemical applications of Sz had to be done on cyclic molecules.

Unicyclic molecules are the obvious first target for such studies. For them, it has been shown that Sz and W depend on molecular structure in a remarkably similar manner.[12,18] Systematic numerical testing[14,24,25] revealed that the correlations between Sz and various physico-chemical properties of monocycloalkanes are of comparable, sometimes slightly inferior, quality as the analogous correlations with W.

The Sz and W values of benzenoid molecules were also found to be well correlated,[8,13,19] yet for very large benzenoids this correlation is curvilinear.[13] This means that also for this class of polycyclic molecules the structural information contained in Sz is quite similar to that contained in W.

In conclusion: No convincing example of a chemical application of the Szeged index has been discovered so far, in which the Szeged index would yield reasonably better results than the Wiener index. In view of the relatively small number of classes of molecules examined (only monocycloalkanes and benzenoid hydrocarbons), this pessimistic conclusion needs not be taken as something absolute. On the other hand, bearing in mind the numerous mathematical relations between Sz and W (outlined in due detail above), finding cases where the Szeged index is significantly more suitable for modeling physico-chemical or pharmacological properties of organic compounds than the Wiener index should be considered as a kind of surprise.

REFERENCES AND NOTES

1. Gutman, I. A Formula for the Wiener Number of Trees and Its Extension to Graphs Containing Cycles. *Graph Theory Notes of New York,* **1994**, *27*, 9-15.

2. Dobrynin, A.; Gutman, I. On a Graph Invariant Related to the Sum of all Distances in a Graph. *Publ. Inst. Math. (Beograd)*, **1994**, *56*, 18-22.

3. Dobrynin, A.; Gutman, I. Solving a Problem Connected with Distances in Graphs. *Graph Theory Notes of New York,* **1995**, *28*, 21-23.

4. Dobrynin, A. A.; Gutman, I.; Domotor, G. A Wiener-Type Graph Invariant for Some Bipartite Graphs. *Appl. Math. Lett.* **1995**, *8*, 57-62.

5. Khadikar, P. V.; Deshpande, N. V.; Kale, P. P.; Dobrynin, A.; Gutman, I.; Domotor, G. The Szeged Index and an Analogy with the Wiener Index. *J. Chem. Inf. Comput. Sci.* **1995**, *35*, 547-550.

6. Sz should be pronounced as S in English. In Hungarian orthography Sz always corresponds to S in English. For instance, the Hungarian spelling of ``minister'', ``gangster'', ``superstar'', ``tennis'' and ``salami'' is ``miniszter'', ``gengszter'', ``szupersztar'', ``tenisz'' and ``szalami'', whereas the pronunciation (as well as meaning) are essentially the same as in English.

7. 7. Gutman, I.; Klavzar, S. An Algorithm for the Calculation of the Szeged Index of Benzenoid Hydrocarbons. *J. Chem. Inf. Comput. Sci.* **1995**, *35*, 1011-1014.

8. Gutman, I.; Khadikar, P. V.; Rajput, P. V.; Karmarkar, S. The Szeged Index of Polyacenes. *J. Serb. Chem. Soc.* **1995**, *60*, 759-764.

9. Klavžar, S.; Rajapakse, A.; Gutman, I. The Szeged and the Wiener Index of Graphs.

10. *Appl. Math. Lett.* **1996**, *9*, 45-49.

11. Žerovnik, J. Computing the Szeged Index. *Croat. Chem. Acta,* **1996**, *69*, 837-843.

12. Dobrynin, A. A.; Gutman, I. On the Szeged Index of Unbranched Catacondensed Benzenoid Molecules. *Croat. Chem. Acta,* **1996**, *69*, 845-856.

13. Gutman, I. Two Distance--Based Graph Invariants and Their Relation in the Case of Unicyclic Graphs. *Bull. Acad. Serbe Sci. Arts (Cl. Math. Natur.)*, **1996**, *111*, 19-29.

14. Gutman, I.;Dömötör, G.; Lam, P. C. B.; Shiu, W. C.; Popovic, L. Szeged Indices of Benzenoid Hydrocarbons. *Polyc. Arom. Comp.* **1996**, *8*, 259-270.

15. 14 Khadikar, P. V.; Karmarkar, S.; Joshi, S.; Gutman, I. Estimation of the Protonation Constants of Salicylhydroxamic Acids by Means of the Wiener Topological Index. *J. Serb. Chem. Soc.* **1996**, *61*, 89-95.

16. Diudea, M. V. Indices of Reciprocal Property or Harary Indices. *J. Chem. Inf. Comput. Sci.* **1997**, *37*, 292-299.

17. Diudea, M. V. Cluj Matrix Invariants. *J. Chem. Inf. Comput. Sci.* **1997**, *37*, 300-305.

18. Chepoi, V.; Klavžar, S. The Wiener Index and the Szeged Index of Benzenoid Systems in Linear Time. *J. Chem. Inf. Comput. Sci.* **1997**, *37*, 752-755.

19. Gutman, I.; Popović, L.; Khadikar, P. V.; Karmarkar, S.; Joshi, S.; Mandloi, M. Relations between Wiener and Szeged Indices of Monocyclic Molecules. *Commun. Math. Chem. (MATCH)*, **1997**, *35*, 91-103.

20. Gutman, I.; Khadikar, P. V.; Khaddar, T. Wiener and Szeged Indices of Benzenoid
21. Hydrocarbons Containing a Linear Polyacene Fragment. *Commun. Math. Chem.*
22. *(MATCH)*, **1997**, *35*, 105-116.

23. Dobrynin, A. A.; Gutman, I. Szeged Index of some Polycyclic Bipartite Graphs with Circuits of Different Size. *Commun. Math. Chem. (MATCH)*, **1997**, *35*, 117-128.

24. Diudea, M. V.; Minailiuc, O. M.; Katona, G.; Gutman, I. Szeged Matrices and Related Numbers. *Commun. Math. Chem. (MATCH)*, **1997**, *35*, 129-143.

25. Dobrynin, A. A. The Szeged Index for Complements of Hexagonal Chains.
26. Commun. Math. Chem. (MATCH), **1997**, 35, 227-242.

27. Gutman, I. Wiener Numbers of Benzenoid Isomers Containing a Linear Polyacene
28. Fragment. *Indian J. Chem.* **1997**, *36*, 644-648.

29. Karmarkar, S.; Karmarkar, S.; Joshi, S.; Das, A.; Khadikar, P. V. Novel Application of Wiener vis-a-vis Szeged Indices in Predicting Polychlorinated Biphenyls in the Environment. *J. Serb. Chem. Soc.* **1997**, *62*, 227-234.

30. Das, A.; Dömötör, G.; Gutman, I.; Joshi, S.; Karmarkar, S.; Khaddar, D.; Khaddar, T,; Khadikar, P. V.; Popović, L,; Sapre, N. S.; Sapre, N.; Shirhatti, A. A Comparative Study of the Wiener, Schultz and Szeged Indices of Cycloalkanes. *J. Serb. Chem. Soc.* **1997**, *62*, 235-239.

31. Diudea, M. V.; Parv, B.; Topan, M. I. Derived Szeged and Cluj Indices.
32. J. Serb. Chem. Soc. **1997**, 62, 267-276.

33. Kiss, A. A.; Katona, G.; Diudea, M. V. Szeged and Cluj Matrices within the Matrix Operator W(M1,M2,M3). *Coll. Sci. Papers Fac. Sci. Kragujevac,* **1997**, *19*, 95-107.

34. Dobrynin, A. A.; Gutman, I. Congruence Relations for the Szeged Index of Hexagonal Chains. *Publ. Elektrotehn. Fak. (Beograd), Ser. Mat.* **1997**, *8*, 106-113.

35. Marjanović, M.; Gutman, I. A Distance-Based Graph Invariant Related to the Wiener and Szeged Numbers. *Bull. Acad. Serbe Sci. Arts (Cl. Math. Natur.),* **1997**, *114*, 99-108.

36. Gutman, I. Comparing Wiener Numbers of Isomeric Benzenoid Hydrocarbons.
37. ACH - Models in Chemistry, **1997**, 134, 477-486.

38. Lukovits, I. Szeged Index: Formulas for Fused Bicyclic Graphs. Croat. Chem. Acta
39. **1997**, *70*, 863-871.
40. Dobrynin, A. A. Graphs Having the Maximal Value of the Szeged Index. *Croat. Chem. Acta,* **1997**, *70*, 819-825.
41. Gutman, I.; Popović, L.; Pavlović, L. Elementary Edge-Cuts in the Theory of Benzenoid Hydrocarbons - An Application. *Commun. Math. Chem. (MATCH)*, **1997**, *36*, 217-229.
42. Gutman, I.; Klavžar, S. Relations between Wiener Numbers of Benzenoid Hydrocarbons and Phenylenes. *ACH - Models in Chemistry*, **1998**, *135*, 45-55.
43. Minaliuc, O. M.; Katona, G.; Diudea, M. V.; Strunje, M.; Graovac, A.; Gutman, I.
44. Szeged Fragmental Indices. *Croat. Chem. Acta*, **1998**, *71*, 473-488.
45. Žerovnik, J. Szeged Index of Symmetric Graphs. *J. Chem. Inf. Comput. Sci.* **1999**, *39*, 77-80.
46. Simić, S.; Gutman, I.; Baltić, V. Some Graphs with Extremal Szeged Index. *Math. Slovaca* (in press).
47. Diudea, M. V.; Gutman, I. Wiener-Type Topological Indices. *Croat. Chem. Acta*,
48. **1998**, *71*, 21-51.
49. Gutman, I.; Dobrynin, A. A. The Szeged Index - A Success Story. *Graph Theory Notes New York*, **1998**, *34*, 37-44.
50. Wiener, H. Structural Determination of Paraffin Boiling Points. *J. Am. Chem. Soc.*
51. **1947**, *69*,17-20.
52. Lukovits, I. Wiener Indices and Partition Coefficients of Unsaturated Hydrocarbons.
53. *Quant. Struct. Act. Relat.* **1990**, *9*, 227-231.
    42. Lukovits, I. Correlation between the Components of the Wiener Index and Partition
54. Coefficients of Hydrocarbons. *Int. J. Quantum Chem. Quantum Biol. Symp.* **1992**, *19*, 217-223. 43. Juvan, M.; Mohar, B. Bond Contributions to the Wiener Index. *J. Chem. Inf. Comput. Sci.* **1995**, *35*, 217-219.
55. 44. Lukovits, I.; Gutman, I. Edge-Decomposition of the Wiener Number. *Commun. Math. Chem. (MATCH)*, **1994**, *31*, 133-144.
56. 45. Lukovits, I. An Algorithm for Computation of Bond Contributions of the Wiener Index. *Croat. Chem. Acta*, **1995**, *68*, 99-103.
57. 46. Entringer, R. C.; Jackson, D. E.; Snyder, D. A. Distance in Graphs. *Czech. Math. J.*
58. **1976**, *26*, 283-296.
59. 47. Gutman, I. A Property of the Wiener Number and Its Modifications. *Indian J. Chem.*
60. **1997**, *36A*, 128-132.
    48. The complete bipartite graph K_{ab} is a graph with a+b vertices. The vertices of K_{ab} can
61. be partitioned into two classes, possessing *a* and *b* elements. No two vertices belonging to the same class are adjacent; any two vertices belonging to different

classes are adjacent. Note that $K_{1,1} \equiv P_2$, $K_{1,2} \equiv P_3$, $K_{2,2} \equiv C_4$ where P_N and C_N stand for the N-vertex path and cycle, cf. Figure 5.2.

62. Gutman, I. On Distances in Some Bipartite Graphs. *Publ. Inst. Math.* (*Beograd*), **1988**, *43*, 3-8. 50. Gutman, I.; Cyvin, S. J. *Introduction to the Theory of Benzenoid Hydrocarbons*; Springer-Verlag, Berlin,1989.

63. Gutman, I. Wiener Numbers of Benzenoid Hydrocarbons: Two Theorems.

64. *Chem. Phys. Lett.* **1987**, *136*, 134-136.

65. Graovac, A.; Pisanski, T. On the Wiener Index of a Graph. *J. Math. Chem.* **1991**, *8*, 53-62.

66. Yeh, Y. N.; Gutman, I. On the Sum of All Distances in Composite Graphs. *Discr.*

67. *Math.* **1994**, *135*, 359-365.

68. Dobrynin, A. A. A New Formula for the Calculation of the Wiener Index of

69. Hexagonal Chains. *Commun. Math. Chem.* (*MATCH*), **1997**, *35*, 75-90.

70. Dobrynin, A. A. Congruence Relations for the Wiener Index of Hexagonal Chains. *J. Chem. Inf. Comput. Sci.* **1997**, *37*, 1109-1110.

71. Dobrynin, A. A. New Congruence Relations for the Wiener Index of Cata-Condensed

72. Benzenoid Graphs. *J. Chem. Inf. Comput. Sci.* **1998**, *38*, 405-409.

73. Buckley, F.; Harary, F. *Distance in Graphs*; Addison-Wesley, Redwood, 1990.

74. Klavžar, S.; Gutman, I.; Mohar, B. Labeling of Benzenoid Systems which Reflects the Vertex-Distance Relations. *J. Chem. Inf. Comput. Sci.* **1995**, *35*, 590-593.

75. Sachs, H. Perfect Matchings in Hexagonal Systems. *Combinatorica*, **1984**, *4*, 89-99.

76. Gutman, I.; Klavzar, S. A Method for Calculating Wiener Numbers of Benzenoid

77. Hydrocarbons. *ACH- Models in Chemistry*, **1996**, *133*, 389-399. Klavžar, S.; Gutman, I. Wiener Number of Vertex-Weighted Graphs and a Chemical

78. Application. *Discr. Appl. Math.* **1997**, *80*, 73-81.

79. Klavžar, S.; Gutman, I.; Rajapakse, A. Wiener Numbers of Pericondensed Benzenoid

80. Hydrocarbons. *Croat. Chem. Acta*, **1997**, *70*, 979-999.

81. Gutman, I.; Jovasević, V. Wiener Indices of Benzenoid Hydrocarbons Containing two Linear Polyacene Fragments. *J. Serb. Chem. Soc.* **1998**, *63*, 31-40.

82. Klavžar, S. Applications of Isometric Embeddings to Chemical Graphs. *DIMACS Ser. Discr. Math. Theor. Comput. Sci.* (in press).

83. Klavžar, S.; Zigert, P.; Gutman, I. An Algorithm for the Calculation of the Hyper-Wiener Index of Benzenoid Hydrocarbons. 65: (in press).

84. Gutman, I.; Cyvin, S. J. Elementary Edge-Cuts in the Theory of Benzenoid Hydrocarbons. *Commun. Math. Chem.* (*MATCH*), **1997**, *36*, 177-184.

85. Numerous review articles exist on the Wiener index and its chemical applications. Two recent ones are: Gutman, I.; Yeh, Y. N.; Lee, S. L.; Luo, Y. L. Some Recent Results in the Theory of the Wiener Number. *Indian J. Chem.* **1993**, *32A*, 651-661 and Gutman, I.; Potgieter, J.H. Wiener Index and Intermolecular Forces. *J.*

Serb.Chem. Soc. **1997**, *62*, 185-192, where references to earlier works on W can be found.

Chapter 6

CLUJ INDICES

In cycle-containing graphs, the Wiener matrices are not defined. (See Chap. Topological Matrices). Wiener indices are herein calculated by means of the distance-type matrices[1,2] but their meaning is somewhat changed (see Sect. 4.2).

An attempt of Lukovitz to extend the Randić's definition of hyper-Wiener index to simple cycles resulted in a quite strange version of this index.[3]

Cluj matrices try to fill the hall of the Wiener matrices, in the same manner as these matrices do in acyclic graphs (i.e. counting the external paths with respect to the path (i,j)).

6.1. CLUJ INDICES, *CJ* AND *CF*

Among several conceivable versions of the Cluj indices,[4] two variants are here discussed: (1) *at least one path external to the path* (i,j), (see Sect. 2.11.1) leading to *CJ* indices and (2) *all paths external to the path* (i,j), (see Sect. 2.11.2) that provides the *CF* indices. As shown in Sect. 2.11.2, the entries in a **CF** matrix are true fragments (i.e. connected subgraphs).

The Cluj indices are calculated[5-10.] as half-sum of the entries in a Cluj symmetric matrix, **M**, (**M = CJD, CJΔ, CFD, CFΔ** - see Sect. 2.11.1)

$$IE(M) = (1/2)\sum_i \sum_j [\mathbf{M}]_{ij}[\mathbf{A}]_{ij} \tag{6.1}$$

$$IP(M) = (1/2)\sum_i \sum_j [\mathbf{M}]_{ij} \tag{6.2}$$

or from an unsymmetric Cluj matrix, by

$$IE2(UM) = (1/2)\sum_i \sum_j [\mathbf{UM}]_{ij}[\mathbf{UM}]_{ji}[\mathbf{A}]_{ij} \tag{6.3}$$

$$IP2(UM) = (1/2)\sum_i \sum_j [\mathbf{UM}]_{ij}[\mathbf{UM}]_{ji} \tag{6.4}$$

The number defined on edge, IE, is an *index* while the number defined on path, IP is a *hyper-index*. Note that the operators IE and IP, as well as $IE2$ and $IP2$ may be applied to both symmetric and unsymmetric matrices. When the last two operators are calculated on a symmetric matrix, the terms of sum represent squared entries in that matrix. This is the reason for the number 2 in the symbol of these operators. It is obvious that $IE(M) = IE2(UM)$ and $IP(M) = IP2(UM)$ with the condition $\mathbf{M} = (\mathbf{UM})(\mathbf{UM})^{\mathrm{T}}$ where $(\mathbf{UM})^{\mathrm{T}}$ is the transpose of the unsymmetric matrix $\mathbf{UM}$. Only in trees, and only for Cluj distance indices, $IE2(UM) = IP(UM)$. The edge defined indices are identical for the two versions of Cluj indices in all graphs: $IE(CJD) = IE(CFD)$; $IE(CJ\Delta) = IE(CF\Delta)$. Values of the above discussed indices for a set of 45 cycloalkanes[5] are listed in Table 6.1.

The boiling point of the set of cycloalkanes included in Table 6.1 correlates r=0.991; s=5.93; F = 2333.7 with $\ln IP(CJD)$[5] and r=0.989; s=6.60; F = 1876.22 with $\ln IP(CFD)$.

A systematic search has been undertaken, including the calculation of the sensitivity, S (see Chap. Topological Indices) of these indices on the set of all cycloalkane isomers having ten vertices/atoms and three to ten membered cycles (376 structures).[9] That study indicated that the sensitivity of $IP2(UCJD)$ to distinguish among the above mentioned isomers is about 0,525. This value is superior to the sensitivity of the Wiener W (0,216) and hyper-Wiener WW (0,408) indices.

The cycloalkane isomers were generated by the program FRAGGEN, written in Turbo Pascal at the TOPO Group Cluj.

6.2. CLUJ INDICES OF PARTICULAR GRAPHS

6.2.1 Cluj Indices of Path Graphs

Cluj indices were designed to reproduce the Wiener indices in path graphs and to extend the Wiener definition (see eqs 4.19, 4.20) to cycle-containing graphs. It is immediate that a relation of the type (4.29) also holds for the Cluj indices

$$IP(CJD) = IE(CJD) + I_\Delta (CJD) \tag{6.5}$$

For path graphs, by replacing the formulas of Cluj indices existing in Table 6.2, entries 1 and 3, one obtains

$$I_\Delta (CJD)(P_N) = N(N-1)(N-2)(N+1)/24 \tag{6.6}$$

Table 6.1. Boiling Points and Cluj-Type Indices for Some Cycloalkanes.

No	Graph*	BP	IP(CJD)	IP(CJ△)	IP(CFD)	IP(CF△)	IE(CJD)	IE(CJ△)
1	C4	13.1	18	6	18	6	16	4
2	11MC3	21	24	20	24	20	15	15
3	EC3	35.9	32	26	32	26	17	17
4	MC4	40.5	37	17	39	17	28	10
5	C5	49.3	40	10	40	10	20	5
6	112MC3	56.5	49	39	49	39	26	26
7	123MC3	66	54	42	54	42	27	27
8	EC4	70.7	73	41	77	41	45	21
9	MC5	71.8	71	25	75	25	33	12
10	C6	80.7	90	24	90	24	54	6
11	PC4	110	132	84	138	84	68	38
12	11MC5	88.9	105	43	113	43	48	21
13	12MC5	91.9**	109	48	121	48	49	22
14	13MC5	91.7**	119	46	127	46	51	21
15	MC6	100.9	142	48	149	49	78	14
16	C7	117	154	42	154	42	63	7
17	112MC5	114	150	74	170	74	67	34
18	113MC5	105	170	70	182	70	71	32
19	123MC5	115	164	77	184	77	70	34
20	1M2EC5	124	178	93	198	93	72	39
21	1M3EC5	121	199	88	211	88	76	37
22	PC5	131	215	113	227	113	78	45
23	IPC5	126.4	186	92	198	92	73	40
24	11MC6	119.5	197	75	211	77	104	24
25	12MC6	123.4**	202	81	222	85	106	25
26	13MC6	124.5**	211	80	227	82	108	24
27	14MC6	120	220	80	234	82	110	24
28	EC6	131.8	226	94	242	96	109	29
29	MC7	134	225	71	235	73	88	16
30	C8	146	288	64	288	64	128	8
31	1123MC5	132.7	222	109	250	109	93	48
32	113MC6	136.6	285	117	310	120	140	36
33	124MC6	136	296	120	324	124	144	37
34	135MC6	138.5	291	114	318	117	144	36
35	1M2EC6	151	300	142	336	149	142	44
36	1M3EC6	149	322	141	349	144	146	42
37	PC6	154	352	170	377	173	148	52
38	IPC6	146	313	143	338	146	142	46
39	EC7	163.5	337	127	361	131	121	33
40	C9	170	450	90	450	90	144	9
41	1M2IPC6	171	401	206	453	216	180	65
42	1M3IPC6	167.5	436	205	474	209	186	62
43	13EC6	170.5	467	221	507	225	192	64
44	PC7	183.5	503	219	541	225	163	59
45	C10	201	705	145	705	145	250	10

M = methyl; E = ethyl; P = propyl; IP = isopropyl; Cn = n-membered cycle

** values for the *trans*-isomer

Table 6.2 include formulas[7,9,11] for the Cluj indices derived from the basic and reciprocal Cluj matrices (see Sect. 6.3). These formulas were derived by analyzing the corresponding Cluj matrices and transforming, where possible, the sums in simple formulas.

Note that the two variants, *CJ* and *CF*, are both calculable by the same formulas in the simple graphs: *paths, cycles* and *stars*. As the reader can see, in Table 6.2 the Cluj indices are symbolized by *CJ*.

6.2.2 Cluj Indices of Simple Cycles

In simple cycles, the edge-defined Cluj indices are very simple, as shown in Table 6.2, entries 5 and 6. The path-defined Cluj indices (entries 7 and 8) show a *N mod* 4 dependency[7,9] and the formulas are still simple.

Note that only the *CJD* indices depend on the parity of the cycle (by $z = N \bmod 2$).

The composition formula (6.5) also holds in cycle-containing graphs. From the formulas in entries 5, 7 and 6, 8 (Table 6.2) it results in

$$I_\Delta(CJD)(C_N) = \frac{N}{24}\left(\begin{array}{c} 72z^2N - 9y^2N - 16z^3 - 3y^2 - 12yz^2 + 12yzN + 12y^2z - 27N^2 - 2y^3 + \\ 7N^3 + 8N + 8y - 60zN - 12z^2 + 18yN - 12yz + 16z \end{array}\right) \tag{6.7}$$

$$I_\Delta(CJ\Delta)(C_N) = \frac{N}{6}(4k^3 + 3yk^2 + 6k^2 + 6yk + 2k + 3y - 6) \tag{6.8}$$

The quantities *k, y, z* and *N* have the same meaning as in Table 6.2. It is obvious that $I_\Delta(M)$ is the part of the Cluj index defined on paths larger than 1.

6.2.3. Cluj Indices of Stars

In simple stars, the Cluj indices are functions of the number of edges, *N-1*. They can be calculated according to the formulas[9] given in the entries 13 and 15 of the Table 6.2.

6.2.4 Graphs with Minimal *IP(CJΔ)* Value

As in the case of the Wiener index *W*, the *detour-Cluj index IP(CJΔ)* shows its minimal value[7] in case of the *complete graphs*, K_N (see the graph $G_{6.1}$, Figure 6.1)

$$IP(CJ\Delta)(K_N) = N(N-1)/2$$
$$(6.9)$$

Table 6.2. Formulas for Cluj Indices of Paths, Cycles and Stars

Index	Sums	Final Relations	Examples
Paths:	$I(CJ) = I(CF);\quad I(RCJ) = I(RCF).$		
1 $IE(CJD)$	$\displaystyle\sum_{i=1}^{N-1}(N-i)i$	$\dfrac{1}{6}N(N-1)(N+1)$	$N = 11;\ 220$
2 $IE(RCJD)$	$\displaystyle\sum_{i=1}^{N-1}(N-i)^{-1}i^{-1}$		$N = 11;\ 0.533$
3 $IP(CJD)$	$\displaystyle\sum_{i=1}^{N-1}\sum_{j=1}^{N-i}ij$	$\dfrac{1}{24}N(N-1)(N+2)(N+1)$	$N = 11;\ 715$
4 $IP(RCJD)$	$\displaystyle\sum_{i=1}^{N-1}\sum_{j=1}^{N-i}i^{-1}j^{-1}$		$N = 11;\ 7.562$
Cycles:	$I(CJ)=I(CF);\quad I(RCJ)=I(RCF).$		
5 $IE(CJD)$	$N(N-z)^2/4$		$N = 11;\ 275$
6 $IE(CJ\varDelta)$	N		$N = 11;\ 11$

7 $IP(CJD)$

$$(1-z)N\left(\frac{N-z}{2}\right)^2 + 2N\sum_{i=1-z}^{\left(\frac{N-z}{4}-\left(1-\frac{y-z}{4}\right)\right)}\left(\frac{N-z}{2}-i\right)^2 + (1-y)\frac{N}{2}\left(\frac{N-z}{2}-\left(\frac{N-z}{4}-\frac{y-z}{4}\right)\right)^2$$

$$N(72z^2N-9y^2N-16z^3-3y^2-12yz^2+12yzN+12y^2z-3N^2-2y^3+7N^3+8N+8y$$
$$-108zN+12z^2+18yN-12yz+16z)/96\ ;\ y=N\bmod4;\ z=N\bmod2$$

	$N = 9;\ 450$
	$N = 10;\ 705$
	$N = 11;\ 1001$
	$N = 12;\ 1470$

8 $IP(CJ\varDelta)$ $(k+1)N(4k^2+3yk+2k+3y)/6$
$k=[(N-1)/4];\ y=(N-1)\bmod 4$

$N = 9;\ 90$
$N = 10;\ 145$
$N = 11;\ 209$
$N = 12;\ 282$

| 9 $IE(RCJD)$ | $4N/(N-z)^2$ | | $N = 11;\ 0.44$ |

10 $IP(RCJD)$

$$(1-z)N\left(\frac{N-z}{2}\right)^{-2} + 2N\sum_{i=1-z}^{\left(\frac{N-z}{4}-\left(1-\frac{y-z}{4}\right)\right)}\left(\frac{N-z}{2}-i\right)^{-2} + (1-y)\frac{N}{2}\left(\frac{N-z}{2}-\left(\frac{N-z}{4}-\frac{y-z}{4}\right)\right)^{-2}$$

$$\frac{4(1-z)N}{(N-z)^2} + 2N\left(\Psi\left(1,1-\frac{z}{2}-\frac{N}{2}\right)-\Psi\left(1,\frac{y}{4}-\frac{N}{4}\right)\right) + \frac{8(1-y)N}{(N-2z+y)^2}$$

$N = 9;\ 3.125$
$N = 10;\ 3.317$
$N = 11;\ 3.477$
$N = 12;\ 3.460$

11 $IE(RCJ\varDelta)$	N		$N = 11;\ 11$
12 $IP(RCJ\varDelta)$	$\displaystyle 2N\sum_{i=1}^{k}i^{-2} + Ny(k+1)^{-2}/2$	$2N(-\Psi(1,k+1)+\pi^2/6)+Ny/2(k+1)^2$ $k=[(N-1)/4];\ y=(N-1)\bmod 4$	$N = 9;\ 22.500$ $N = 10;\ 25.556$ $N = 11;\ 28.722$ $N = 12;\ 32.000$
Stars:	$I(CJ) = I(CF);\quad I(RCJ) = I(RCF).$		
13 $IE(CJD)$	$(N-1)(1\mathrm{x}(N-1))$		$N = 11;\ 100$
14 $IE(RCJD)$	$(N-1)(1\mathrm{x}(N-1)^{-1})$		$N = 11;\ 1$
15 $IP(CJD)$	$(N-1)(1\mathrm{x}(N-1)) + \displaystyle\sum_{i=1}^{N-2}i\mathrm{x}1\mathrm{x}1$		$N = 11;\ 145$
16 $IP(RCJD)$	$(N-1)(1\mathrm{x}(N-1)^{-1}) + \displaystyle\sum_{i=1}^{N-2}i\mathrm{x}1\mathrm{x}1$		$N = 11;\ 46$

$y = N \bmod 4;\ z = N \bmod 2;\ \gamma(x) = \mathrm{int}(\exp(-t)t^\wedge(x-1),\ t = 0,..\text{infinity};$

$\psi(x) = \text{diff}(\ln(\gamma(x),x)); \quad \psi(n,x) = \text{diff}(\psi(x), x\$n); \quad \psi(0,x) = \psi(x);$

The minimal value, given by eq 6.9 is also obtained for *star-triangulanes*, $G_{6.2}$, *strips* with *odd girth g*, $G_{6.3}$, *Möbius strips* (irrespective of *g*), $G_{6.4}$, and *dipyramids* (of any *g*), $G_{6.5}$. Note that the strips with *even g*, $G_{6.6}$, do not have a minimal value for $IP(CJ\Delta)$.

The graphs showing a minimal value $IP(CJ\Delta)$ are full Hamiltonian detour graphs,[7] *FHΔ* (see Sect. 2.11.2) and, consequently, the nondiagonal entries in the **CJΔ** matrix are unity. In such graphs, $I(CJ\Delta) = I(CF\Delta)$ and the edge-defined index $IE(CJ\Delta)$ equals the number of their edges. Formulas[7] for calculating the detour-Cluj indices in the above discussed graphs are given in Table 6.3. The formulas are given as functions of *g*. Also included in Table 6.3. are the classical detour indices, *w* and *ww*, as functions of $IP(CJ\Delta)$.

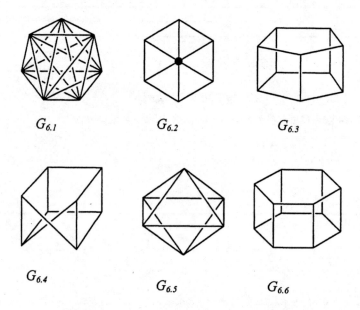

$G_{6.1}$ $G_{6.2}$ $G_{6.3}$

$G_{6.4}$

$G_{6.5}$ $G_{6.6}$

Figure 6.1. Graphs with minimal $IP(CJ\Delta)$ value

6.3. DISTANCE EXTENDED CLUJ-TYPE INDICES

The Tratch's and cowork.[12] *extended distance matrix*, **E**, can be interpreted as a *distance extended Wiener* matrix $\mathbf{D\_W}_p$, at least in acyclic structures (see Sect. 2.12). The $\mathbf{D\_W}_p$ matrix is just the Hadamard product[13] of the $\mathbf{D}_e$ and $\mathbf{W}_p$ matrices. The half sum of its entries gives a *distance extended Wiener* index.

Similarly, Diudea[6] performed the Hadamard product

$$\mathbf{D}_e \bullet \text{UCJD} = \text{D\_UCJD} \tag{6.10}$$

Table 6.3. Formulas for Calculating Detour-Cluj and Detour Indices of Some Particular Graphs

Graphs	N	Q	$IP(CJ\Delta)$ *	$IE(CJ\Delta)$	ww	w
Complete graphs	g	$g(g-1)/2$	$g(g-1)/2$	$g(g-1)/2$	$(IP(CJ\Delta))^2$	$(g-1)IP(CJ\Delta)$
Triangulanes	$g+1$	$2g$	$g(g+1)/2$	$2g$	$(IP(CJ\Delta))^2$	$gIP(CJ\Delta)$
Strips (g-even)	$2g$	$3g$	$g(5g-4)$	$3g$	$g(2g-1)(2g^2-2g+1)$	$g(4g^2-5g+2)$
Strips (g-odd) & Möbius Strips	$2g$	$3g$	$g(2g-1)$	$3g$	$(IP(CJ\Delta))^2$	$(2g-1)IP(CJ\Delta)$
Dipyramids	$g+2$	$3g$	$(g+2)(g+1)/2$	$3g$	$(IP(CJ\Delta))^2$	$(g+1)IP(CJ\Delta)$

* $IP(CJ\Delta) = IP(CF\Delta)$; $IE(CJ\Delta) = IE(CF\Delta)$

and the resulting *distance extended Cluj* matrix **D_UCJD** offered, in trees T, a new definition of the hyper-Wiener index WW (see eq 2.47 and also Chap. Topological Indices)

$$IP(D\_UCJD)(T) = WW \tag{6.11}$$

Various other combinations: **D_M** or **Δ_M**, **M** being a symmetric or unsymmetric Cluj matrix, were performed on trees or on cycle-containing graphs, by means of the CLUJ software program.

Since the Cluj matrices are, in general, unsymmetric, an index of the form $IP2(UM)$ (eq 6.4) on the matrix **D_UCJD** would involve squared distances and promise a better selectivity

$$IP2(D\_UCJD) = \sum_{(i,j)} ([\mathbf{D}_e]_{ij})^2 [\mathbf{UCJD}]_{ij} \tag{6.12}$$

Indeed, among the 2562 structures of the set of all unbranched cata-condensed benzenoid graphs with three to ten rings[14] $IP2(D\_UCJD)$ showed no degeneracy.

An extension of eq 6.10 to a $3D$-$\mathbf{D}_e$ matrix[6] (e.g. by using the geometric matrix, **G**) allows the construction of various *3D-distance extended matrices*, such as **G_UCJD** (see Figure 2.19). They can offer *3D- sensitive* indices (see Chap. Fragmental Property Indices).

6.4 INDICES DEFINED ON RECIPROCAL CLUJ MATRICES

The half sum of entries in reciprocal matrix **RM** of a square matrix **M** is an index, referred to as a *Harary-type* index (see Sect. 4.3). The symbol used in ref[9] is H_M, where M recall the *info* matrix (i.e. the square matrix whose reciprocal entries give **RM**).

Because of many Cluj matrices (symmetric or not, edge-defined or path-defined), we adopted here symbols of the type (6.1)-(6.4). Thus, $IE(RCJD)$ denote an edge-defined index (i.e. an *index*) on the symmetric reciprocal Cluj-Distance matrix while $IP2(URCJ\Delta)$ is a path-defined index (i.e. a *hyper index*) on the unsymmetric reciprocal Cluj-Detour matrix.

Table 6.2 includes formulas[9,11] for the indices defined on reciprocal Cluj matrices.

These formulas were derived by analyzing the corresponding Cluj matrices and transforming the sums in simple formulas. Some formulas for these Harary-type indices in cycles (entries 10 and 12) involve the well-known gamma, digamma and polygamma function, given at the end of Table 6.2. Numerical values of these formulas are also given. In path graphs, simple formulas could, however, not be derived.

A systematic testing has been undertaken of the sensitivity, S (see Chap. Topological Indices) of the indices defined on reciprocal Cluj matrices within the set of all cycloalkane isomers having ten vertices/atoms and three to ten membered cycles (376 structures).[9] The study indicated that the sensitivity of $IP(RCJD)$ to distinguish among these isomers is about unity.[9,14] Its discriminating ability is superior to that of the Wiener W (0,216) and hyper-Wiener WW (0,408) indices.

In the class of all unbranched cata-condensed benzenoid graphs with three to ten rings[14] (2562 structures), $IP(RCJD)$ showed $S = 0.988$.

In addition to an increased sensitivity, the *Harary-Cluj* indices showed good *correlating ability*. The boiling point of a set of 30 cycloalkanes[9] correlated 0.978 with the classical Harary index, H_{De} and the Harary-Cluj index $IP(RCJD)$. The viscosity of a set of 25 cycloalkanes[14] with a saturated aliphatic side chain correlated 0.974 with $IP(RCJD)$ and 0.996 when combined with the Wiener index, as $\ln W$, and the $IP(RCJ\Delta)$ index.

6.5. INDICES DEFINED ON SCHULTZ-CLUJ MATRICES

The Schultz matrices, $\mathbf{SCH}(G)$ (see Sect. 2.15) are related to the molecular topological index, MTI, or the Schultz index,[15] (see Chap. Topological Indices). In the extension of Diudea and Randić[16] the Schultz matrix is defined as

$$\mathbf{SCH}_{(M_1,A,M_3)} = \mathbf{M}_1(\mathbf{A} + \mathbf{M}_3) = \mathbf{M}_1\mathbf{A} + \mathbf{M}_1\mathbf{M}_3 \tag{6.13}$$

and a *composite* index (edge-defined or path-defined) can be calculated by

$$I(SCH_{(M1,A,M3)}) = \sum_i \sum_j [\mathbf{SCH}_{(M1,A,M3)}]_{ij} = \mathbf{u}\mathbf{SCH}_{(M1,A,M3)}\mathbf{u}^T \tag{6.14}$$

A Schultz-extended number is *walk matrix* calculable as[16,17]

$$I(SCH_{(M1,A,M3)}) = \mathbf{u}\mathbf{W}_{(M_1,1,(A+M_3))}\mathbf{u}^T = \mathbf{u}\mathbf{W}_{(M_1,1,A)}\mathbf{u}^T + \mathbf{u}\mathbf{W}_{(M_1,1,M_3)}\mathbf{u}^T \tag{6.15}$$

When one of the $\mathbf{M_1}$ or $\mathbf{M_2}$ matrices is unsymmetric, the resulting Schultz matrix will also be unsymmetric. In such a case an index of the form $I2(UM)$ can be derived

$$I2(USCH_{(M1,A,M3)}) = \sum_p [\mathbf{USCH}_{(M1,A,M3)}]_{ij}[\mathbf{USCH}_{(M1,A,M3)}]_{ji} \qquad (6.16)$$

Of course, the relation $I(M) = I2(UM)$ (see Sect. 6.1) is preserved, with the condition $\mathbf{M} = (\mathbf{UM})(\mathbf{UM})^{\mathbf{T}}$. Within this book, $\mathbf{M_1}$ is a symmetric matrix (e.g., $\mathbf{A}$, $\mathbf{D_e}$, $\mathbf{\Delta_e}$) and $\mathbf{M_3}$ is an unsymmetric Cluj matrix (e.g., $\mathbf{CJD}$, $\mathbf{CF\Delta}$).

If we write now a Schultz-type index as

$$I(SCH_{(M_1,A,M_3)}) = \mathbf{u}\mathbf{W}_{(\mathbf{M_1},\mathbf{1},\mathbf{A})}\mathbf{u}^{\mathbf{T}} + \mathbf{u}\mathbf{W}_{(\mathbf{M_1},\mathbf{1},\mathbf{M_3})}\mathbf{u}^{\mathbf{T}} = I_{M_1,A} + I_{M_1,M_3} \qquad (6.17)$$

the quantities $I_{M1,A}$ and $I_{M1,M3}$ can also be viewed as composite indices.[16,18] Table 6.4 lists values of $I_{M1,A}$ along with the hyper-Cluj $IP2(UCJD)$ and Schultz-Cluj $IP2(USCH_{(De,A,UCJD)})$ index values for the octane isomers. It can be seen that $I_{UCJD,A}$ is the arithmetic mean of $I_{De,A}$ and $I_{We,A}$ indices as a consequence of the relations

$$\Sigma_i\Sigma_j [((\mathbf{UCJD})\mathbf{A} + \mathbf{A}(\mathbf{UCJD}))/2]_{ij} = \Sigma_i\Sigma_j [(\mathbf{D_e}\mathbf{A} + \mathbf{A}\mathbf{W_e})/2]_{ij} = (I_{De,A} + I_{We,A})/2 =$$
$$I_{UCJD,A} \qquad (6.18)$$

(see also Sect. 2.15 and 4.4). They were tested for correlating ability.[18]

Table 6.4. $IP2(UCJD)$, $I_{M1,A}$ and $IP2(USCH_{(De,A,M3)})$ Indices of Octane Isomers

GRAPH*	IP2(UCJD)	$I_{De,A}$	$I_{We,A}$	$I_{UCJD,A}$	$M_3 =$ UCJD
P_8	210	280	322	301	105977
$2MP_7$	185	260	324	292	80240
$3MP_7$	170	248	318	283	68553
$4MP_7$	165	244	316	280	65252
$3EP_6$	150	232	306	269	53945
$25M_2P_6$	161	240	326	283	59061
$24M_2P_6$	147	228	320	274	49804
$23M_2P_6$	143	224	318	271	47537
$34M_2P_6$	134	216	314	265	41753
$3E_2MP_5$	129	212	308	260	38668
$22M_2P_6$	149	228	330	279	50940
$33M_2P_6$	131	212	322	267	39884
$234M_3P_5$	122	204	320	262	33326
$3E_3MP_5$	118	200	314	257	32185
$224M_3P_5$	127	208	332	270	35717
$223M_3P_5$	115	196	326	261	29504
$233M_3P_5$	111	192	324	258	27501
$2233M_4P_4$	97	176	338	257	19885

* M = methyl; E = ethyl.

The Schultz-Cluj *composite indices* showed a powerful ability to discriminate isomeric structures. A family of spiro-graphs showing degenerate sequences of terminal paths, *TPS*, all paths sequences, *APS*, distance degree sequences, *DDS*, detour degree sequences, *ΔDS* and cycle sequence, *CyS* and, consequently, degenerate indices based on these quantities was successfully separated by $IP2(USCH_{(De,A,UCFD)})$ and $IP2(USCH_{(De,A,UCJD)})$ indices (see Sect. 8.6.1, Table 8.14).

6.6. CLUJ INDICES OF DENDRIMERS

Dendrimers are hyperbranched macromolecules, with a rigorous structure.[10] The topology of dendrimers is basically that of a tree (dendron in Greek means tree). The number of edges emerging from each branching point is called progressive degree, p.[10,19] It equals the classical degree δ, minus one: $p = \delta - 1$.

A regular dendrimer has all its branching points of the same degree, otherwise it is irregular. In graph theory, dendrimers correspond to the Cayley trees or Bethe lattices.[20,21]

A tree has either a monocenter or a dicenter[22] (i.e. two points joined by an edge). Accordingly, a dendrimer is called *monocentric* ($G_{6.7}$) and *dicentric* ($G_{6.8}$), respectively (Figure 6.2).

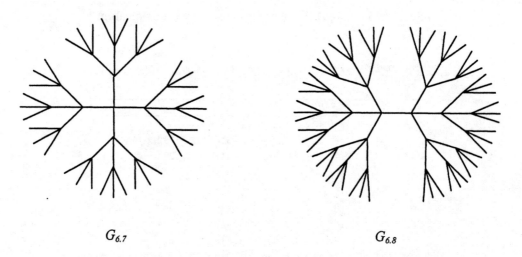

$G_{6.7}$ $G_{6.8}$

Figure 6.2. A *monocentric* ($G_{6.7}$) and a *dicentric* ($G_{6.8}$) regular dendrimer

The numbering of *orbits* (i.e., generations) starts with zero at the core and ends with r, which is the *radius* of the dendrimer (i.e. the number of edges along a radial chain, starting from the core and ending to the periphery).

A *wedge* is a *fragment* of a dendrimer resulting by cutting an edge in a dendrimer.

6.6.1. Enumeration in Regular Dendrimers

A first problem in the topology of dendrimers is the enumeration of its constitutive parts: vertices, edges, and fragments.[65,70] The number of vertices N_i in the ith orbit of a regular dendrimer can be expressed as a function of the progressive degree p and a parameter z: $z = 1$ for a monocentric dendrimer and $z = 0$ for a dicentric one

$$N_i = (2-z)(p+z)p^{(i-1)}; \quad i > 0 \tag{6.19}$$

For the core, the number of vertices is $N_0 = 2\text{-}z$, and the number of external vertices (i.e. the vertices on the rth orbit) can be calculated by

$$\tag{6.20}$$

The total number of vertices N in a dendrimer is obtained by summing the populations on all orbits

$$N = (2-z)+(2-z)(p+z)\sum_{i=1}^{r} p^{(i-1)} = (2-z)+(2-z)(p+z)\left(\frac{p^r-1}{p-1}\right) \tag{6.21}$$

A recursive formula relates the members of a dendrimer family

$$N_{r+1} = pN_r + 2 \tag{6.22}$$

The number of vertices in a wedgeal fragment F_i, starting at the ith orbit and ending at the periphery can be evaluated by

$$F_i = \sum_{s=i}^{r} p^{(r-s)} \tag{6.23}$$

The number of fragments (i.e. wedges) starting at the ith orbit equal the number of vertices lying on that orbit and is calculated by eq 6.19.

6.6.2. Cluj Indices of Regular Dendrimers

In regular dendrimers, Cluj indices are evaluated[10] according to eqs 4.19 and 4.20, by using the fragmental enumeration (see above) . The procedure is illustrated in Figure 6.3. Note that N_i is actually F_i.

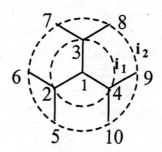

UCJD

	1	2	3	4	5	6	7	8	9	10
1	0	7	7	7	7	7	7	7	7	7
2	3	0	3	3	9	9	3	3	3	3
3	3	3	0	3	3	3	9	9	3	3
4	3	3	3	0	3	3	3	3	9	9
5	1	1	1	1	0	1	1	1	1	1
6	1	1	1	1	1	0	1	1	1	1
7	1	1	1	1	1	1	0	1	1	1
8	1	1	1	1	1	1	1	0	1	1
9	1	1	1	1	1	1	1	1	0	1
10	1	1	1	1	1	1	1	1	1	0

$IP2(UCJD) = A + B + C = 237$

$A = N_1(N_1-1)/2 \ [F_1]^2 + N_2(N_2-1) \ [F_2]^2$

$A = 3(3-1)/2 \ [3]^2 + 6(6-1)/2 \ [1]^2 = 42$

$B = N_1 \ [F_1(N-F_1)] + N_2 \ [F_2(N-F_1)] + N_2[F_2(N-F_2)]$

$B = 3 \ [3\cdot7] + 6 \ [1\cdot7] + 6 \ [1\cdot9] = 159$

$C = N_1(N_2-2)[F_1\cdot F_2]$

$C = 3(6-2) \ [1\cdot3] = 36$

Figure 6.3. Calculation of $IP2(UCJD)$ index of dendrimers

Recall that $IP2(UCJD) = IP(CJD)$, where **CJD** is the symmetric Cluj matrix. Following the above procedure, one obtains:

$$IP(CJD) = \sum_{i=z}^{r} \binom{N_i}{2}[F_i]^{2g} + \sum_{i=1}^{r}\sum_{j=1}^{r} N_j[F_j(N-F_i)]^g + \sum_{i=z}^{r-1}\sum_{j=i+1}^{r} N_i(N_j - p^{(j-i)})[F_i F_j]^g$$

$$(6.24)$$

$$IE(CJD) = (1-z)[F_0]^{2g} + \sum_{i=1}^{r} N_i[F_i(N-F_i)]^g$$

$$(6.25)$$

By virtue of the identity between Wiener and Cluj matrices, in acyclic graphs, the following identities: $IE(CJD) \equiv W_{We}$; $IP(CJD) \equiv WW_{Wp}$ hold. The same is true for the corresponding Harary-type indices: $H_{We} \equiv IE(RCJD)$; $H_{Wp} \equiv IP(RCJD)$.
Expansion of the above symbolic relations lead to the simple formulas:[10]

Monocentric dendrimers:

$$IP(CJD) = \{2p^{2r} (p^2 -1)^2 r^2 + p^{2r} (p^2 -1)(p^2 - 8p -5)r + (p+1)(p^r - 1)[p^r (p^2 + 10p + 3) - 2]\} / 2(p-1)^4 \tag{6.26}$$

$$IE(CJD)= \{[r(p+1)^3 - 2(r+1)(p+1)^2 + (p+1)]p^{2r} + 2(p+1)^2 p^r - (p+1)\}(p-1)^{-3} \tag{6.27}$$

Dicentric dendrimers:

$$IP(CJD) = \{4p^{2r+2} (p -1)^2 r^2 + 4p^{2r+2} (p-4)(p -1)r + p^{2r+2} (p^2 - 3p + 16) - p^{r+1} (p^2 + 10p + 5) + (p+1)\} / (p-1)^4 \tag{6.28}$$

$$IE(CJD) = [4p^{(2r+2)}(p-1)r + (p-1)(p^{(r+1)} -1)^2 - 2p(p^r -1)(3p^{(r+1)} - 1)] (p-1)^{-3} \tag{6.29}$$

Values of Cluj indices in regular dendrimers having p = 2 and 3, up to generation ten are listed in Table 6.5. Values for the corresponding Harary-Cluj indices are presented in Table 6.6.

From Table 6.6 one can see that the $IE(RCJD)$ values decrease as the radius (i.e., generation) of dendrimer increases. For the family of dendrimers having the progressive degree 2, the limit of convergence is 0.6067 while for the family with the progressive degree 3, the limit is 0.7286, irrespective they are mono- or dicentric-dendrimers. The convergence is a characteristic feature of $IE(RCJD)$ index.

6.6.3. Enumeration in Triangulanes and Quatranes

Triangulanes and *quatranes*[10] are the line graphs (see Chap. 1 and Sect. 8.2.1) of the dendrimers with the progressive degree p = 2 and 3 respectively (i.e. the branching usually encountered in organic chemical structures). Their line graphs are lattices of complete graphs (of three and four vertices, respectively) generated around each branching vertex in dendrimer and then transformed into a *dendritic spiro-structure* (see Figure 6.4).

Table 6.5. Cluj Indices $IE(CJD)$ and $IP(CJD)$ in Regular Dendrimers
Having $p = 2$ and 3 and Generation up to 10.

p	r	IE(CJD)		IP(CJD)	
		$z = 0$	$z = 1$	$z = 0$	$z = 1$
2	1	29	9	47	12
	2	285	117	667	237
	3	1981	909	6195	2535
	4	11645	5661	46179	20427
	5	62205	31293	301251	139923
	6	312829	160893	1798531	863523
	7	1510397	788733	10085123	4958787
	8	7084029	3740157	53986819	27022467
	9	32518141	17310717	278891523	141535491
	10	146825213	78661629	1400838147	718754307
3	1	58	16	97	22
	2	1147	400	2842	862
	3	16564	6304	55546	18988
	4	207157	82336	885067	322684
	5	2392942	975280	12486859	4737346
	6	26310703	10897456	162614932	63370330
	7	279816808	117191488	2001654484	795156568
	8	2905693033	1226857792	23632595701	9524050936
	9	29637785506	12591244624	270225628693	110124165742
	10	298120420579	127267866832	3012581235310	1238679833686

Table 6.6. Cluj Indices $IE(RCJD)$ and $IP(RCJD)$ in Regular Dendrimers
Having $p = 2$ and 3 and Generation up to 10.

p	r	IE(RCJD)		IP(RCJD)	
		$z = 0$	$z = 1$	$z = 0$	$z = 1$
2	1	0.91111	1.00000	8.24444	4.00000
	2	0.75700	0.80952	39.48428	21.00000
	3	0.67978	0.70526	171.93340	93.99806
	4	0.64248	0.65479	718.89205	398.36215
	5	0.62434	0.63034	2942.94684	1642.52530
	6	0.61544	0.61840	11913.41433	6674.41687
	7	0.61105	0.61251	47945.57042	26914.25133
	8	0.60886	0.60959	192376.55943	108099.91432
	9	0.60778	0.60814	770707.04503	433297.01294
	10	0.60724	0.60742	3085243.85345	1734996.10484
3	1	0.91964	1.00000	17.41964	7.00000
	2	0.79410	0.82692	179.54978	77.12500
	3	0.75027	0.76122	1696.45922	744.63142
	4	0.73578	0.73939	15533.98437	6873.80590
	5	0.73099	0.73219	140639.34503	62412.87672
	6	0.72941	0.72980	1268302.06937	563405.57044
	7	0.72888	0.72901	11422421.63083	5075774.57299
	8	0.72870	0.72875	102824974.44435	45697411.49634
	9	0.72864	0.72866	925494391.69444	411323103.15047
	10	0.72862	0.72863	8329658484.38390	3702047217.72089

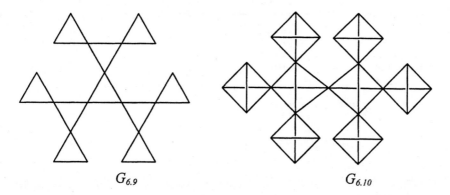

$$G_{6.9} \qquad\qquad\qquad G_{6.10}$$

Figure 6.4. A monocentric triangulane ($G_{6.9}$) and a dicentric quatrane ($G_{6.10}$)

A *spiro-graph* results by fusing a vertex of degree δ_i, belonging to a ring, with a vertex of degree δ_j, of another ring, for giving a *spiro-vertex* of degree $\delta_k = \delta_i + \delta_j$ in the resulted structure (analogue to *spiranes* in organic chemistry).

In the line graphs of a dendrimer, the degree of a vertex $i \in V(L(G))$ is calculated by

$$\delta_i = p_u + p_v; \quad (u,v) \in E(G) \tag{6.30}$$

where p_u and p_v are the progressive degrees of the endpoints of the edge (u,v) and δ_i is just the degree of that edge in G. The branching points of a regular dendrimer have the same progressive degree, so that $\delta_i = 2p$. For the external edges in G it results $\delta_i = p$.

The progressive degree of the complete graph units in the line graphs of dendrimers in discussion is derived from their *reduced graphs* (i.e. the graphs resulted by replacing each unit by a point and then joining those points of which corresponding units have a common spiro-vertex). In triangulanes, $p = 2$ while in quatranes $p = 3$.

The number of vertices on the ith orbit in the reduced graph of the line graph of a regular dendrimer is given by

$$N_i = (2-z)(p+z)p^i \tag{6.31}$$

where $z = 1$ for monocentric and $z = 0$ for dicentric triangulanes and quatranes.

The total number of vertices is obtained by summing the orbital contributions N_i, over all orbits (the core included) of the reduced graph

$$N = (1-z) + (2-z)(p+z)\sum_{i=0}^{r} p^{(r-i)} \tag{6.32}$$

which (after developing the sum) becomes

$$N = (1-z) + (2-z)(p+z)\left(\frac{p^{(r+1)} - 1}{p-1}\right) \tag{6.33}$$

By definition, the number of vertices in $L(G)$ equals the number of edges or the number of vertices less one, in the corresponding dendrimer (having an additional generation, $r+1$, as compared to the reduced graph of $L(G)$)

$$N(L(G)) = E(G) = N(G) - 1 \tag{6.34}$$

A recursive relation among the members of a family of $L(G)$ in regular dendrimers is as follows

$$N_{r+1}(L(G)) = pN_r(L(G)) + (p+1) \tag{6.35}$$

The number of edges in $L(G)$ can be counted by considering the line graph as a collection of spiro-complete graphs built up around each vertex of the corresponding dendrimer, till the $(r-1)$th generation. Since the number of vertices in a complete graph unit, equals $(p+1)$ and its number of edges is combinations of $(p+1)$ choose 2 and keeping in mind eq 6.21, the number of edges in the line graph of a regular dendrimer is

$$E(L(G)) = \binom{p+1}{2} N(G) = \frac{p(p+1)}{2}\left((2-z) + (2-z)(p+z)\frac{p^r - 1}{p-1}\right)$$

$$\tag{6.36}$$

where r in the right hand side of eq 6.36 is the radius of the reduced graph of $L(G)$. Vertices on a wedgeal fragment starting on the ith orbit are counted by

$$F_i = \sum_{s=i}^{r} p^{(r-s)} \tag{6.37}$$

The number of all vertices, N, number of vertices on the last added generation and the number of all edges in triangulanes and quatranes, up to generation ten, are listed in Table 6.7.

6.6.4. Cluj Indices of Triangulanes and Quatranes

Cluj indices are defined in any connected graph, so that it is tempting to calculate them for cycle-containing networks such as triangulanes and quatranes. A procedure similar to that described for dendrimers is illustrated in Figure 6.5 allowed the description[10] of the dendritic line graphs with $p = 2$ and 3, according to the Cluj definitions (see Sect. 2.11). Note that, in dendrimers, as well as in triangulanes and quatranes, both CJ and CF Cluj indices give identical values.

Table 6.7. Enumeration in Triangulanes ($p = 2$) and Quatranes ($p = 3$): Global Vertex Population, N, Periphery Orbital Population, N_r, Number of Edges, E.

p	r	N	N_r	E	N	N_r	E
$z = 1$					$z = 0$		
2	1	9	6	12	13	8	18
	2	21	12	30	29	16	42
	3	45	24	66	61	32	90
	4	93	48	138	125	64	186
	5	189	96	282	253	128	378
	6	381	192	570	509	256	762
	7	765	384	1146	1021	512	1530
	8	1533	768	2298	2045	1024	3066
	9	3069	1536	4602	4093	2048	6138
	10	6141	3072	9210	8189	4096	12282
3	1	16	12	30	25	18	48
	2	52	36	102	79	54	156
	3	160	108	318	241	162	480
	4	484	324	966	727	486	1452
	5	1456	972	2910	2185	1458	4368
	6	4372	2916	8742	6559	4374	13116
	7	13120	8748	26238	19681	13122	39360
	8	39364	26244	78726	59047	39366	118092
	9	118096	78732	236190	177145	118098	354288
	10	354292	236196	708582	531439	354294	1062876

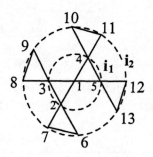

UCJΔ

	1	2	3	4	5	6	7	8	9	10	11	12	13
1	0	7	7	7	7	7	7	7	7	7	7	7	7
2	3	0	3	3	3	11	11	3	3	3	3	3	3
3	3	3	0	3	3	3	3	11	11	3	3	3	3
4	3	3	3	0	3	3	3	3	3	11	11	3	3
5	3	3	3	3	0	3	3	3	3	3	3	11	11
6	1	1	1	1	1	0	1	1	1	1	1	1	1
7	1	1	1	1	1	1	0	1	1	1	1	1	1
8	1	1	1	1	1	1	1	0	1	1	1	1	1
9	1	1	1	1	1	1	1	1	0	1	1	1	1
10	1	1	1	1	1	1	1	1	1	0	1	1	1
11	1	1	1	1	1	1	1	1	1	1	0	1	1
12	1	1	1	1	1	1	1	1	1	1	1	0	1
13	1	1	1	1	1	1	1	1	1	1	1	1	0

$IP2(UCJ\Delta) = A + B + C = 382$
$A = N_0(N_0-1)/2\ [F_0]^2 + N_1(N_1-1)\ [F_1]^2$
$A = 4(4-1)/2\ [3]^2 + 8(8-1)/2\ [1]^2 = 82$
$B = N_0[F_0(N-2F_0)] + N_1[F_1(N-2F_0)] + N_1[F_1(N-2F_1)]$
$B = 4\ [3 \cdot 7] + 8\ [1 \cdot 7] + 8\ [1 \cdot 11] = 228$
$C = N_0(N_1-2)\ [F_0F_1]$
$C = 4(8-2)\ [1 \cdot 3] = 72$

Figure 6.5. Calculation of *IP2(UCJD)* index of triangulanes

$$IP(CJD) = z\binom{N_0}{2}[F_0]^{2g} + \sum_{i=z}^{r} N_i[F_i(N - pF_i)]^g + p\sum_{i=0}^{r-1}\sum_{j=i+1}^{r} N_j[pF_j(N - F_i)]^g$$

$$+ \frac{1}{4-p}\sum_{i=z}^{r} N_i(F_i)^{2g} + \frac{1}{2}\sum_{i=z}^{r} N_i(N_i - p)(pF_i)^{2g} + (1-z)p\sum_{i=1}^{r} N_i(pF_i pF_0)^g +$$

$$+ \sum_{i=1}^{r-1}\sum_{j=i+1}^{r}(N_i - p)N_j[pF_i pF_j]^g$$

$$(6.38)$$

$$IP(CJ\Delta) = \sum_{i=0}^{r}\binom{N_i}{2}[F_i]^{2g} + \sum_{i=z}^{r}\sum_{j=1}^{r} N_j[F_j(N - pF_i)]^g + \sum_{i=0}^{r-1}\sum_{j=i+1}^{r} N_i(N_j - p^{(j-i)})[F_i F_j]^g$$

$$(6.39)$$

$$IE(CJD/\Delta) = z\binom{N_0}{2}[F_0]^{2g} + \frac{1}{4-p}\sum_{i=z}^{r} N_i[F_i]^{2g} + \sum_{i=z}^{r} N_i[F_i(N - pF_i)]^g \quad (6.40)$$

As can be seen, the edge-defined indices are identical both for distance- and detour-Cluj indices. Note the formal similarity between the relations for calculating the Cluj indices in dendrimers and those in their line graphs, particularly for the detour-Cluj indices (eqs 6.24; 6.25 and 6.39; 6.40). In the opposite, the formula for the distance-Cluj index $IP(CJD)$, (6.38) is far more complicated. Expansion of the above symbolic relations did, however, not offer simple formulas.

Values of Cluj indices in triangulanes (p = 2) and quatranes (p = 3) up to generation ten[10] are listed in Tables 6.8. to 6.11.

Table 6.8. Cluj Indices $IE(CJD)$ and $IP(CJD)$ of Triangulanes ($p = 2$) and Quatranes ($p = 3$) with Generation up to 10.

p	r	$IE(CJD)$		$IP(CJD)$	
		$z = 0$	$z = 1$	$z = 0$	$z = 1$
2	1	194	72	946	264
	2	1546	678	12218	4470
	3	9754	4626	108874	45186
	4	54330	26922	793194	353754
	5	280698	142938	5100714	2381706
	6	1380602	715962	30157098	14532330
	7	6562298	3447162	167924266	82787754
	8	30426106	16134906	894215210	448494378
	9	138446842	73950714	4600711210	2338518570
	10	620826618	333499386	23034236970	11833568298
3	1	822	264	8679	2046
	2	13404	4926	243606	74748
	3	177996	69456	4677222	1607484
	4	2128314	857910	73818741	26995314
	5	23922498	9836760	1035358077	393561786
	6	258303288	107632110	13430112780	5240628840
	7	2712012312	1140784032	164846025516	65548466040
	8	27894481878	11816462694	1942134733659	783264588510
	9	282430156494	120294475176	22170706080627	9040351900278
	10	2824297224852	1208172034974	246843377284962	101541238563540

Table 6.9. Cluj Indices $IE(RCJD)$ and $IP(RCJD)$ of Triangulanes ($p = 2$) and Quatranes ($p = 3$) with Generation up to 10.

p	r	$IE(RCJD)$		$IP(RCJD)$	
		$z = 0$	$z = 1$	$z = 0$	$z = 1$
2	1	5.13997	4.19048	13.38442	8.19048
	2	9.23189	7.15947	48.71617	28.15947
	3	17.65167	13.42958	189.58507	107.42764
	4	34.59912	26.11971	753.49117	424.48187
	5	68.54440	51.56906	3011.49124	1694.09436
	6	136.45903	102.50041	12049.87337	6776.91728
	7	272.29997	204.37886	48217.87039	27118.63019
	8	543.98760	408.14347	192920.54704	108508.05779
	9	1087.36569	815.67649	771794.41072	434112.68943
	10	2174.12328	1630.74441	3087417.97674	1736626.84925
3	1	19.30857	13.29808	36.72821	20.29808
	2	55.94973	37.59520	235.49951	114.72020
	3	166.23298	111.08109	1862.69220	855.71251
	4	497.19102	331.70877	16031.17538	7205.51469
	5	1490.09927	993.64410	142129.44430	63406.52082
	6	4468.83511	2979.46685	1272770.90449	566385.03730
	7	13405.04630	8936.94059	11435826.67713	5084711.51358
	8	40213.68106	26809.36364	102865188.12541	45724220.85999
	9	120639.58575	80426.63339	925615031.28020	411403529.78386
	10	361917.29996	241278.44285	8330020401.68386	3702288496.16374

Table 6.10. Cluj Indices $IE(CJ\Delta)$ and $IP(CJ\Delta)$ of Triangulanes ($p = 2$) and Quatranes ($p = 3$) with Generation up to 10.

p	r	$IE(CJ\Delta)$		$IP(CJ\Delta)$	
		$z = 0$	$z = 1$	$z = 0$	$z = 1$
2	1	194	72	382	120
	2	1546	678	4214	1626
	3	9754	4626	34534	14766
	4	54330	26922	239046	108630
	5	280698	142938	1485702	702630
	6	1380602	715962	8574726	4170054
	7	6562298	3447162	46902790	23282310
	8	30426106	16134906	246373382	124224774
	9	138446842	73950714	1254012934	640092678
	10	620826618	333499386	6224179206	3208516614
3	1	822	264	1695	462
	2	13404	4926	38982	12684
	3	177996	69456	677910	240348
	4	2128314	857910	10093917	3762066
	5	23922498	9836760	136304229	52472874
	6	258303288	107632110	1721837676	677965080
	7	2712012312	1140784032	20726902668	8297193144
	8	27894481878	11816462694	240587843187	97532921118
	9	282430156494	120294475176	2714460814731	1111411966854
	10	2824297224852	1208172034974	29937528342642	12356290538148

Table 6.11. Cluj Indices $IE(RCJ\Delta)$ and $IP(RCJ\Delta)$ of Triangulanes ($p = 2$) and Quatranes ($p = 3$) with Generation up to 10.

r	$IE(RCJ\Delta)$		$IP(RCJ\Delta)$	
	$z = 0$	$z = 1$	$z = 0$	$z = 1$
1	5.13997	4.19048	38.72727	20.19048
2	9.23189	7.15947	171.25362	93.29280
3	17.65167	13.42958	718.24957	397.70737
4	34.59912	26.11971	2942.32250	1641.89496
5	68.54440	51.56906	11912.79889	6673.79848
6	136.45903	102.50041	47944.95937	26913.63882
7	272.29997	204.37886	192375.95057	108099.30473
8	543.98760	408.14347	770706.43725	433296.40480
9	1087.36569	815.67649	3085243.24622	1734995.49742
10	2174.12328	1630.74441	12345822.02431	6943615.07018
1	19.30857	13.29808	178.75568	76.29808
2	55.94973	37.59520	1695.70896	743.87020
3	166.23298	111.08109	15533.24859	6873.06652
4	497.19102	331.70877	140638.61404	62412.14453
5	1490.09927	993.64410	1268301.33997	563404.84065
6	4468.83511	2979.46685	11422420.90195	5075773.84398
7	13405.04630	8936.94059	102824973.71565	45697410.76760
8	40213.68106	26809.36364	925494390.96580	411323102.42181
9	120639.58575	80426.63339	8329658483.65528	3702047216.99226
10	361917.29996	241278.44285	74967553341.83228	33318842928.83120

REFERENCES

1.   Hosoya, H. Topological Index. A Newly Proposed Quantity Characterizing the Topological Nature of Structural Isomers of Saturated Hydrocarbons. *Bull. Chem. Soc. Jpn.* **1971**, *44*, 2332-2339.
2.   Diudea, M. V. Walk Numbers $^{e}W_{M}$: Wiener Numbers of Higher Rank, *J. Chem. Inf. Comput. Sci.* **1996**, *36*, 535-540.
3.   Lukovits, I.; Linert, W. A Novel Definition of the Hyper-Wiener Index for Cycles, *J. Chem. Inf. Comput. Sci.* **1994**, *34*, 899-902.
4.   Gutman, I.; Diudea, M. V. Defining Cluj Matrices and Cluj Matrix Invariants,
5.   J. Serb. Chem. Soc. **1998**, 63, 497-504.
6.   Diudea, M. V. Cluj Matrix Invariants
7.   J. Chem. Inf. Comput. Sci. **1997**, 37, 300-305.
8.   Diudea, M. V., Cluj Matrix CJ_u : Source of Various Graph Descriptors,
9.   Commun. Math. Comput. Chem. (MATCH), **1997**, 35, 169-183.
10. Diudea, M. V.; Pârv, B.; Gutman, I. Detour-Cluj Matrix and Derived Invariants,
11. J.Chem.Inf.Comput.Sci. **1997**, 37, 1101-1108.
12. Diudea, M. V.; Pârv, B.; Topan, M. I. Derived Szeged and Cluj Indices,
13. *J. Serb. Chem. Soc.* **1997**, *62*, 267-276.
14. Diudea, M. V. Indices of Reciprocal Property or Harary Indices,
15. J.Chem.Inf.Comput.Sci. **1997**, 37, 292-299.
16. Diudea, M. V.; Katona, G. Molecular Topology of Dendrimers,
17. in *Advances in Dendritic Macromolecules*, Ed. G.A. Newkome, JAI Press Inc., Stamford, Con. 1999, vol.4, pp. 135-201.
18. Diudea, M. V.; Katona, G.; Lukovits, I.; Trinajstić, N. Detour and Cluj-Detour
19. Indices, *Croat. Chem. Acta* **1998**, *71*, 459-471.
20. Tratch, S. S.; Stankevich, M. I.; Zefirov, N. S. Combinatorial Models and Algorithms
21. in Chemistry. The Expanded Wiener Number- a Novel Topological Index,
22. J. Comput. Chem. **1990**, 11, 899-908.
23. Horn, R. A.; Johnson, C. R. *Matrix Analysis*; Cambridge Univ. Press, Cambridge, **1985**.
24. Konstantinova, E. V.; Diudea, M. V. The Wiener Polynomial Derivatives and Other
25. Topological Indices in Chemical Research,
26. Croat. Chem. Acta (submitted).
27. Schultz, H. P. Topological Organic Chemistry. 1. Graph Theory and Topological Indices of   Alkanes. *J. Chem. Inf. Comput. Sci.* **1989**, *29*, 227-228.
28. Diudea, M. V.; Randić, M. Matrix Operator, $W_{(M1,M2,M3)}$ and Schultz-Type Numbers,
29. J. Chem. Inf. Comput. Sci. **1997**, 37, 1095-1100.
30. Diudea, M. V. Valencies of Property, *Croat. Chem. Acta*, (in press)
31. Diudea, M. V.; Gutman, I. Wiener-Type Topological Indices,
32. Croat. Chem. Acta. **1998**, 71, 21-51.

33. Diudea, M. V. Orbital and Wedgeal Subgraph Enumeration in Dendrimers,
34. Commun. Math. Comput. Chem. (MATCH), **1994**, 30, 79-91.
35. Balasubramanian, K. Recent Developments in Tree-Pruning Methods and Polynomials for   Cactus Graphs and Trees, *J. Math. Chem.* **1990**, *4*, 89-102.
36. Cayley, E. On the Mathematical Theory of Isomers, *Philos. Mag.* **1874**, *67*, 444-446.
37. Harary , F. *Graph Theory*, Addison - Wesley, Reading, M.A., 1969.

Chapter 7

FRAGMENTAL PROPERTY INDICES

7.1. INTRODUCTION

In the last decade, structural indices used in QSAR/QSPR (Quantitative Structure-Activity Relationships/ Quantitative Structure-Property Relationships) are rather calculated from steric/geometric and/or electrostatic/partial charges considerations[1,2,3] than from (older) topological basis.[4] In the view of a QSAR analysis, the set of molecules under study is somehow aligned.[5] CoMFA method[6] proposes an algorithm consisting of the following steps: (1) build a set of molecules with known activities and their 3D-structure (eventually obtain the 3D-structure from any specific program, such as: MOPAC, SYBYL,[7,8] HyperChem, [9,10] Alchemy2000,[7] MolConn[9,11]); (2) align the set in the 3D space according to a chosen superimposing method (e.g. those maximizing the steric overlap of some fragments in the molecules,[7,12,13] or those based on a pharmacophore theory[14,15]); (3) construct a grid of points surrounding the superposed molecules (in standard form[6] or in modified form[16]); (4) atomic charges are then calculated for each molecule, at a chosen level of theory, (5) the fields: steric (Lennard-Jones), electrostatic (Coulomb[6]), hydrophobic (e.g. HINT)[17], hydrogen-bond potential[18], molecular orbital field[19,20], or any other user-defined field[21], are further calculated for each molecule by interaction with a probe atom[8,21] at a series of grid points; (6) The resulting descriptors are correlated, by the use of partial least squares (PLS), with the chosen property. A cross-validation procedure will give the measure of the predictive ability of the model. The best results were obtained in a series of congeners but non-congeneric series were also investigated.

CoMFA is a good tool in investigating a variety of biological activities, such as cytotoxicity,[22] enzyme inhibition,[19] binding properties.[23,24,25] Moreover, CoMFA is ultimately used in *drug design,*[14,26] eventually by searching for active substructures[27] in a database. Modifications of the above discussed method were used in 3D-QSAR/QSPR studies.[28,29,30,31,32]

In this chapter a new approach, leading to a family of *fragmental property indices, FPI*, is proposed. These indices are calculated as local descriptors of some fragments of the molecule and, a global index is then obtained by summing the fragmental contributions. This idea is implemented on a set of four models with four default properties, eight descriptors of property, five models of superposition, and four type of summative indices, resulting in 2560 indices for one method of breaking up.

7.2. FORMULAS FOR FRAGMENT CALCULATION

The calculation of *fragmental property indices* starts with a decomposition of molecule into fragments (i.e. spanning subgraphs) corresponding to all pairs of vertices (i,j) in the molecule, i being the reference vertex (see below).

In *Cluj fragmentation criteria*, the *path p* joining the vertices i and j of the pair (i,j) play the central role in selecting the fragments. In cycle-containing graphs, more than one path could join the pair (i,j) thus resulting in more than one fragment referred to i. In such a fragmentation, the most frequently occurring fragments will bring the greatest contribution to a global value of the calculated index for the molecule. In *Szeged fragmentation criteria*, for each pair (i,j) results one fragment.

Before introducing the *fragmental property indices* some formulas for calculating the fragments are needed.

Recall that these fragments are entries in the Cluj and Szeged matrices, respectively (see Chap. 2).

Let $G = (V, E)$ be a graph and $i, j \in V$. Let $p = (i = v_1, v_2, ..., v_{|p|-1}, v_{|p|} = j) \in P(G)_{i,j}$ be a path from i to j in G.

7.2.1 CJ and CF Fragments

Collections of *maximal fragments* of type CJ ($CJDiS_{i,j}^M$ and $CJDeS_{i,j}^M$) and CF ($CFDiS_{i,j}^M$ and $CFDeS_{i,j}^M$) are available by applying the following equations:

$$CXDyS_{i,j}^M = \left\{ CXDy_{i,j,p}^M, \left| CXDy_{i,j,p}^M \right| = \max \left\{ \left| cxdy_{i,j} \right|, \ cxdy_{i,j} \in CXDyS_{i,j} \right\} \right\} \quad (7.1)$$

$$CXDyS_{i,j} = \{ CX_{i,j,p} \in CXS_{i,j} \mid p \in P(G)_{i,j}, p \in Y(G) \} \quad (7.2)$$

with the meaning for X, x, Y, y:

$X; x$	$Y; y$	maximal fragment set
$X = J; x = j$	$Y = D; y = i;$	$CJDiS_{i,j}^M$
$X = J; x = j$	$Y = \Delta; y = e;$	$CJDeS_{i,j}^M$
$X = F; x = f$	$Y = D; y = i;$	$CFDiS_{i,j}^M$
$X = F; x = f$	$Y = \Delta; y = e;$	$CFDeS_{i,j}^M$

where Di is related to the distance d_{ij} while De to the detour δ_{ij} in the graph (see Chap. 1).

More explicitly, the quantities in eq 7.2 (see also (2.37) and (2.42)) are:

$$CJ_{i,j,p} = \{v \mid v \in V(G); d(G)_{i,v} < d(G)_{j,v}; \text{ and } \exists w \in W_{v,i}, w \cap p = \{i\}; p \in D(G) \text{ or } \Delta(G)\}$$

(7.3)

$$CF_{i,j,p} = \{v \mid v \in V(G); d(G_p)_{iv} < d(G_p)_{j,v}; G_p = G - p; p \in D(G) \text{ or } \Delta(G)\}$$

(7.4)

where $d(G_p)_{iv}$ and $d(G_p)_{jv}$ are distances measured in the spanning subgraph $G_p = G - p$ resulted by cutting off the path p except its endpoints. $CJ_{i,j,p}$ and $CF_{i,j,p}$ represent fragments (connected or not) in G, constructed according to eqs (7.3) and (7.4) respectively, with respect to the endpoints i and j of the path p.

7.2.2 Sz Fragments

The Szeged fragments are constructed by the equations:

$$SzDi_{i,j} = \{v \mid v \in V(G); \ d(G)_{v,i} < d(G)_{v,j}\} \tag{7.5}$$

$$SzDe_{i,j} = \{v \mid v \in V(G); \ \delta(G)_{v,i} < \delta(G)_{v,j}\} \tag{7.6}$$

Note that in the definition of the Szeged fragments, the path between the vertex i and vertex j is irrelevant.

7.3. FRAGMENTAL PROPERTY INDICES

7.3.1 Model Parameters

It is well known that the physical laws govern the natural phenomena. Macroscopic interactions are interactions of field-type. This means that the field is produced by a scalar function of potential. Let $f(x, y, z)$ be a scalar function. This function induces a field given in terms of the gradient of f:

$$\vec{\nabla} \cdot f = \left(\frac{\partial}{\partial x}\vec{i} + \frac{\partial}{\partial y}\vec{j} + \frac{\partial}{\partial z}\vec{k}\right) \cdot f(x,y,z) = \frac{\partial f}{\partial x}\vec{i} + \frac{\partial f}{\partial y}\vec{j} + \frac{\partial f}{\partial z}\vec{k}$$

(7.7)

For the potential of type

$f(x, y, z) = pz$ (7.8)

and applying eq 7.7 we obtain the associated field of the form:

$$\vec{\nabla} \cdot f = \frac{\partial f}{\partial x}\vec{i} + \frac{\partial f}{\partial y}\vec{j} + \frac{\partial f}{\partial z}\vec{k} = \frac{\partial(pz)}{\partial x}\vec{i} + \frac{\partial(pz)}{\partial y}\vec{j} + \frac{\partial(pz)}{\partial z}\vec{k} =$$
$$= 0\vec{i} + 0\vec{j} + p\vec{k} = p\vec{k} = \vec{p}$$
(7.9)

This is the case of the well-known uniform gravitational field:

$$\vec{G} = m\vec{g}$$ (7.10)

the potential of which is given by

$$E_p = E_p(z) = mgz$$ (7.11)

where m is the mass of probe and z is the reference coordinate.

Note that eq 7.9 is applicable not only to the Newtonian (gravitational) interactions but also to the Coulombian (electrostatic) interactions. In both cases the relation is valid if the mass M (or the charge Q) that generates the potential f and associated field $\vec{\nabla} \cdot f$ is far enough ($r >> z$) for the approximation

$(r + z)^2/r^2 = (r^2 + 2rz + z^2)/r^2 = 1 + 2z/r + (z/r)^2 \cong 1$

be applied in the equation of field produced by M or Q (see below).

For the potential of type:

$f(x, y, z) = p/z$ (7.12)

eq 7.7 leads to the associated field:

$$\vec{\nabla} \cdot f = \frac{\partial f}{\partial x}\vec{i} + \frac{\partial f}{\partial y}\vec{j} + \frac{\partial f}{\partial z}\vec{k} = \frac{\partial(p/z)}{\partial x}\vec{i} + \frac{\partial(p/z)}{\partial y}\vec{j} + \frac{\partial(p/z)}{\partial z}\vec{k} =$$
$$= 0\vec{i} + 0\vec{j} + \frac{-p}{z^2}\vec{k} = -\frac{p}{z^2}\vec{k} = -\frac{p}{z^3}\vec{z} = -\frac{\vec{p}}{z^2}$$
(7.13)

This is the case of well-known (non-uniform) gravitational field given by:

$$\vec{G} = \vec{G}(m,r) = -k\frac{m}{r^3}\vec{r}$$ (7.14)

and the associated potential of the form:

$$U = U(m,r) = k\frac{m}{r} \tag{7.15}$$

where m is mass of probe and r is the position relative to the location of the point that produces the field.

For the Coulombian field eq 7.13 becomes:

$$\vec{F}_C = \vec{F}_C(r) = -k\frac{q}{r^3}\vec{r} \tag{7.16}$$

and the potential associated to the Coulombian field:

$$U = U(q,r) = k\frac{q}{r} \tag{7.17}$$

For *fragmental property indices* four models of interaction are implemented: two of them *topological* (dense topological and rare topological) and two others *geometric* (dense geometric and rare geometric).

The *models* are related to two types of field interactions: one of *weak dependence on distance* for the potential of the type (7.8) generating a uniform field (7.9) and the second, of *strong dependence on distance* for the potential of the type (7.12) that generate a non-uniform field (7.13).

The variables in the models are *metrics of distance d* (topological d_T and geometrical d_E), *property* Φ (mass M, electronegativity E, cardinality C, partial charge or any other atomic property P), *property descriptor* Ω (p, d, pd, $1/p$, $1/d$, p/d, $p/d2$, $p2/d2$) and method of *superposition* Ψ (S, P, A, G, H).

Given rational numbers $x_1, \ldots, x_n$, the (mathematical) *superposition* is

$$\Psi: \qquad S=\sum_{i=1}^{n} x_i \ ; \quad P=\prod_{i=1}^{n} x_i \ ; \quad A = S/n \ ; \quad G = \left(\mathrm{sgn}(P)\right)^n \cdot \sqrt[n]{abs(P)} \ ;$$

$$H = \left(\sum_{i=1}^{n}\frac{1}{x_i}\right)^{-1} \tag{7.18}$$

The expressions for the *property descriptors* are:

$$\Omega \ : \ p = p; \ \ d = d; \ \ pd = p \cdot d; \ \ 1/p = \frac{1}{p}; \ \ 1/d = \frac{1}{d}; \ \ p/d = \frac{p}{d}; \ \ p/d2 = \frac{p}{d^2};$$

$$p2/d2 = \frac{p^2}{d^2} \tag{7.19}$$

where p is any property ($p \in \Phi$) and d is any metric of distance.

These variables are most frequently used in building of our models by the reasons:

- The expressions of the *property descriptor* Ω simulate the most occurring physical interactions (e.g. p, pd, p/d, $p/d2$, $p2/d2$)[33] and the most usual descriptor in topological and geometric models.[34,35,36] The property descriptor is used either in the calculation of the *vertex descriptor* (when d is the distance from the vertex v to j and p is any atomic property) or in the evaluation of the *fragment descriptor* (when d is the distance between the center of property of the fragment and j, while p is a calculated fragment property).

- The (mathematical) superposition is applied upon a string of vertex descriptors for giving a fragment descriptor. Note that S = *sum operator*; P = *product operator*; A = *arithmetic mean operator*; G = *geometric mean operator*; H = *harmonic sum operator*. The summation is suitable in the case of any additive property (mass, volume, partial charges, electric capacities, etc.)[37]. The multiplication occurs in concurrent phenomena (probabilistically governed)[38,39,40]. The arithmetic mean is useful in evaluating some mean contributions (corresponding to some uniform probabilistic distribution)[41,42]. The geometric mean is used in calculating the group electronegativities[43,44]. Finally, the harmonic sum is present in connection with the elastic forces, electric fields and group mobility in viscous media.[45,46,47]

7.3.2 Description of the Models

Let (i,j) be a pair of vertices and Fr_{ij} any fragment referred to i with respect to j.

Dense Topological Model

Let v be a vertex in the fragment Fr_{ij}. The vertex descriptor applies the property descriptor to the vertex property and topological distance $d_{T\,vj}$. The global property descriptor, resulting by the vertex descriptor superposition, gives the interaction of the whole fragment Fr_{ij} with the point j:

$$PD(Fr_{ij}) = \underset{v \in Fr_{ij}}{\Psi} \left(\Omega\left(d_{Tvj}, p_v \right) \right) \tag{7.20}$$

The j point can be conceived as an *internal probe atom* (see the *CoMFA* approach). However, the chemical identity of j is not considered.

Rare Topological Model

Within this model the global property results by superposing the vertex properties p_v. The vertex descriptor applies the property descriptor to the global property and topological distance $d_{T\,ij}$. The global property descriptor models the interaction of the fragment Fr_{ij} with the point j and the global property being *concentrated* in the vertex i:

$$PD(Fr_{i,j}) = \Omega\left(d_{T\,i,j}, \underset{v\in Fr_{i,j}}{\Psi}(p_v)\right) \tag{7.21}$$

Dense Geometric Model

The global descriptor is the vector sum of the vertex vector descriptors. It applies the property descriptor to the vertex property p_v and the Euclidean distance $d_{E\,v,j}$ in providing a *point of equivalent* (*global*) *property* located at the Euclidean distance $d_{E\,CPj}$ (with $d_{E\,CPj}$ being the *distance of property*). The global property descriptor vector has the orientation of this *distance vector*. The model simulates the interactions in non-uniform fields (gravitational, electrostatic, et al):

$$PD(Fr_{i,j}) = \left\|\sum_{v\in Fr_{i,j}}\vec{\Omega}\left(d_{E\,v,j}, p_v\right)\right\|;\ \vec{\Omega} = \Omega\cdot\frac{\vec{d}_{E\,v,j}}{d_{E\,v,j}};\ P(Fr_{i,j}) = \underset{v\in Fr_{i,j}}{\Psi}(p_v);$$

$$d_{E\,CPj} = \Omega_p^{-1}(DG(Fr_{i,j}), P(Fr_{i,j})), \tag{7.22}$$

or, in words, $d_{E\,CPj}$ is the distance that satisfies: $\Omega(d_{E\,CPj}, P(Fr_{i,j})) = PD(Fr_{i,j})$

Rare Geometric Model

The scalar global descriptor applies the property descriptor to the *center of fragment property* and Euclidean distance between this center and the vertex j.

The model simulates the interactions in uniform fields (uniform gravitational, electrostatic, etc.):

$$PD(Fr_{i,j}) = \Omega\left(d_{E\,CP_i,j}, \underset{v\in Fr_{i,j}}{\Psi}(p_v)\right);$$

$$CP_i(x_{CP_i,j}, y_{CP_i,j}, z_{CP_i,j});\ x_{CP_i,j} = \sum_{v\in Fr_{i,j}}x_v\cdot p_v \bigg/ \sum_{v\in Fr_{i,j}}p_v \tag{7.23}$$

$$y_{CP_i,j} = \sum_{v\in Fr_{i,j}}y_v\cdot p_v \bigg/ \sum_{v\in Fr_{i,j}}p_v;\ z_{CP_i,j} = \sum_{v\in Fr_{i,j}}z_v\cdot p_v \bigg/ \sum_{v\in Fr_{i,j}}p_v$$

7.3.3 Fragmental Property Matrices

The fragmental property matrices are square matrices of the order N (i.e. the number of non-hydrogen atoms in the molecule). The non-diagonal entries in such matrices are fragmental properties, evaluated for the maximal fragments (equations 7.1-7.6) corresponding to a pair of vertices (i,j) by a chosen model.

In Szeged criteria (eqs 7.5 and 7.6), the fragmentation related to the pair of vertices (i,j) results in a unique fragment $Fr_{i,j}$.

In case of Cluj criteria, the fragmentation can supply more than one maximal fragment for the pair (i,j). In such a case, the matrix entry is the arithmetic mean of the individual values.

Thus, if i, j in $V(G)$, $i \neq j$ and $P_{i,j} = \{ p_{i,j}^1, p_{i,j}^2, ..., p_{i,j}^k \}$ paths joining i and j, then cf. CJ or CF definition (eqs 7.1-7.4), the fragments $Fr_{i,j}^1, Fr_{i,j}^2, ..., Fr_{i,j}^k$ are generated. Let m be the number of maximal fragments (cf. eq 7.1) among all the k fragments, $1 \leq m \leq k$, and let $\sigma_1, ..., \sigma_m$ be the index for the maximal fragments.

Applying any of the equations 7.20-7.23 for all the m maximal fragments we obtain the following m values (for example, by eq 7.20):

$$PD(Fr_{i,j}^{\sigma_1}), PD(Fr_{i,j}^{\sigma_2}), ..., PD(Fr_{i,j}^{\sigma_m})$$

The matrix entry associated to the pair (i,j) is the mean value:

$$PD_{i,j} = \frac{\sum_{t=1}^{m} PD(Fr_{i,j}^{\sigma_t})}{m} \qquad (7.24)$$

The resulting matrices are in general *unsymmetric* but they can be symmetrized as shown in Chap. 2. The symbols for the fragmental property matrices will be detailed below.

7.3.4 Fragmental Property Indices

Fragmental property indices are calculated at any fragmental property matrices given by eqs 7.20-7.24. Four types of index operators are defined: $P\_$, $P2$, $E\_$, $E2$ according to the relations:

$$P\_(M) = \tfrac{1}{2} \Sigma\Sigma [\mathbf{M}]_{i,j} \quad ; \qquad P2(M) = \tfrac{1}{2} \Sigma\Sigma [\mathbf{M}]_{i,j} [\mathbf{M}]_{j,i};$$
$$E\_(M) = \tfrac{1}{2} \Sigma\Sigma [\mathbf{M}]_{i,j} [\mathbf{A}]_{i,j} \quad ; \qquad E2(M) = \tfrac{1}{2} \Sigma\Sigma [\mathbf{M}]_{i,j} [\mathbf{M}]_{j,i} [\mathbf{A}]_{i,j} \qquad (7.25)$$

where $\mathbf{M}$ is any property matrix, symmetric or unsymmetric.

7.3.5 Symbolism of the Fragmental Property Matrices and Indices

The name of *fragmental property matrices* is of the general form:
ABcDdEfffffG   (7.26)

where:

$\mathbf{A} \in \{\mathbf{D}, \mathbf{R}\}$; D = Dense; R = Rare;

$\mathbf{B} \in \{\mathbf{T}, \mathbf{G}\}$; T = Topological; G = Geometric;

$\mathbf{c} \in \{\mathbf{f}, \mathbf{j}, \mathbf{s}\}$; f = CF-type; j = CJ-type; s = Sz-type;

$\mathbf{Dd} \in \{\mathbf{Di}, \mathbf{De}\}$; Di = Distance; De = Detour;

$E \in \Phi$ (i.e. $E \in \{M, E, C, P\}$ where $M = mass$; $E = electronegativity$; $C = cardinality$; $P =$ other atomic property - implicitly, *partial charge*; explicitly, a property given by manual input);

$fffff \in \Omega$ (i.e. $fffff \in \{\_p\_, \_1/p\_, \_d\_, \_1/d\_, \_p.d\_, \_p/d\_, \_p/d2, p2/d2\}$ with the known meaning given in eq 7.19);

$G \in \Psi$ (i.e. $G \in \{S, P, A, G, H\}$ with the known meaning from eq 7.18).

The name of *fragmental property indices* is of the general form:

$$ABcDdEfffffGii \tag{7.27}$$

where:

$ii \in \{P\_, P2, E\_, E2\}$ with the known meaning from eq 7.25.

If an operator, such as $f(x)=1/x$ (inverse operator) or $f(x)=ln(x)$, is applied the indices are labeled as follows:

$$lnABcDdEfffffGii := \quad ln(ABcDdEfffffGii);$$

$$1/ABcDdEfffffGii := \quad \frac{1}{ABcDdEfffffGii} \tag{7.28}$$

For example, index *lnDGfDeM_p_SP_* is the logarithm of index *DGfDeM_p_SP_* computed on the property matrix **DGfDeM_p_S**. The model used is dense, geometric, on fragment of type *CF*, with the cutting path being detour. The chosen property is mass, the descriptor for property is even the property (mass) and the sum operator counts the vertex descriptors.

7.3.6 Some Particular Fragmental Property Models

Let i, j be two vertices in $V(G)$ and $Fr_{i,j}$ any fragment referred to i with respect to j.

Fragmental Mass

In evaluating the fragmental mass, the chosen property is $\Phi = M$, descriptor $\Omega = p$, superposition $\Psi = S$, *and* the model is *rare topological, RT*. The fragmental mass descriptor takes the form:

$$PD(Fr_{i,j}) = \sum_{v \in Fr_{i,j}} M_v \tag{7.29}$$

It models the molecular mass of the fragment. The *name* of the associated property matrix is **RTcDdM_p_S**, with the known meaning for c and Dd.

If $c = s$ and $Dd = Di$ then **RTsDiM_p_S**, it models the molecular mass of the Szeged Distance Fragments (equation 7.5). If $c = f$ and $Dd = Di$ then the matrix

RTfDiM_p_S collects mean values (see eq 7.24) of mass of all the fragments belonging to i (with respect to j) according to the CF criterion (eqs 7.1, 7.4).

Fragmental Electronegativity

The well known equalizing principle of electronegativity E, is here considered: the fragment electronegativity is the geometric mean of electronegativities of the s atoms joined to form that fragment (see also Section 7.3.1).

Let the property $\Phi = E$ (electronegativity); descriptor $\Omega = p$; superposition $\Psi = G$; the model is *rare topological, RT*. The fragmental electronegativity descriptor of $Fr_{i,j}$ is:

$$PD(Fr_{i,j}) = |Fr_{i,j}| \sqrt{\prod_{v \in Fr_{i,j}} E_v} \tag{7.30}$$

It models the electronegativity of the fragment. The *name* of the property matrix associated with it is **RTcDdE_p_G** . Note that E_v is the group electronegativity for vertex v calculated with formula:

$$E_v = \sqrt[\sum_{j \in \Gamma_v} b(v,j)]{\prod_{j \in \Gamma_v} E_{aj}^{b(v,j)}} \tag{7.31}$$

where $b(v, j)$ is the *conventional bond order* between v and j (e.g. 1, 1.5, 2, 3 for single, aromatic, double and triple bonding, respectively), E_a is the atomic electronegativity (Sanderson) and $j \in \Gamma_v$ is any atom (H atoms included) consisting the group Γ_v.

Fragmental Numbers

The property $\Phi = C$ (cardinality) was introduced for recovering some graph-theoretical quantities and/or graph theoretical analogue indices (see below).

For descriptor $\Omega = p$, superposition $\Psi = \{P, A, G\}$, and the model *rare topological, RT,* the cardinal numbering descriptor of $Fr_{i,j}$ is:

$$PD(Fr_{i,j}) = \prod_{v \in Fr_{i,j}} 1 = \frac{\sum_{v \in Fr_{i,j}} 1}{|Fr_{i,j}|} = |Fr_{i,j}| \sqrt{\prod_{v \in Fr_{i,j}} 1} = 1 \tag{7.32}$$

The arithmetic mean A, geometric mean G and product P applied to 1 (value for vertex property) leave it unchanged. The mean value for all fragments belonging to i vs. j (CJ and CF only) is also 1. All matrices **RTcDdC_p_P, RTcDdC_p_A** and **RTcDdC_p_G** have all their entries unity, except the main diagonal elements that are zero.

The indices $RTcDdC\_p\_PP\_$, $RTcDdC\_p\_AP\_$, $RTcDdC\_p\_GP\_$ give the number of edges in the complete graph having the same number of vertices N, as the considered molecular graph:

$$RTcDdC\_p\_PP\_ = RTcDdC\_p\_AP\_ = RTcDdC\_p\_GP\_ = N(N-1)/2$$

Similarly, the indices calculated on edge, $RTcDdC\_p\_PE\_$, $RTcDdC\_p\_AE\_$, $RTcDdC\_p\_GE\_$ give the number of edges in the molecular structure.

Uniform Field Gravity

Let the property $\Phi = M$, descriptor $\Omega = p/d2$, superposition $\Psi = S$ and *rare geometrical* model.

The uniform gravity descriptor of $Fr_{i,j}$ is calculated by:

$$PD(Fr_{i,j}) = \sum_{v \in Fr_{i,j}} \frac{M_v}{d_{v,j}^2} \tag{7.33}$$

It models the value of the gravitational field induced by the fragment $Fr_{i,j}$ in the point j. Values given by (7.33) are collected in the matrix **RGsDdM_p/d2S** while averaged values are considered in **RGfDdM_p/d2S** and **RGjDdM_p/d2S** matrices.

Non-Uniform Field Gravity

Let the property $\Phi = M$, descriptor $\Omega = p/d2$, superposition $\Psi = S$ and *dense geometrical* model. The *distance* (vs. j) of the *center of equivalent fragmental gravity* of $Fr_{i,j}$ is:

$$d_{ECP(Fr_{i,j}),j} = \sqrt{\left(\sum_{v \in Fr_{i,j}} M_v\right) \Bigg/ \left(\left(\sum_{v \in Fr_{i,j}} \frac{M_v}{d_{v,j}^2} \cdot \frac{\vec{d}_{v,j}}{d_{v,j}}\right) \cdot \left(\sum_{v \in Fr_{i,j}} \frac{M_v}{d_{v,j}^2} \cdot \frac{\vec{d}_{v,j}}{d_{v,j}}\right)\right)^{\frac{1}{2}}} \tag{7.34}$$

It models the distance at which a point mass equal to the fragment mass $\sum_{v \in Fr_{i,j}} M_v$

should be located vs. j such that the gravitational field induced by $Fr_{i,j}$ in j be equal to the field induced by all atoms of the fragment. The associated matrix is of the form **DGcDdM_p/d2S**.

Uniform Electrostatic field

Let the property $\Phi = P$ (Q_P implicitly, in the Cluj Program), descriptor $\Omega = p/d2$, superposition $\Psi = S$ and *rare geometrical* model. The uniform electrostatic field descriptor of $Fr_{i,j}$ is:

$$PD(Fr_{i,j}) = \sum_{v \in Fr_{i,j}} \frac{Q_{Pv}}{d_{v,j}^2} \qquad (7.35)$$

It models the value of electrostatic field induced by the fragment in j. The property matrix is of the form: **RGcDdP_p/d2S**.

Non-Uniform Electrostatic Field

For the property $\Phi = P$ (Q_P implicitly), descriptor $\Omega = p/d2$, superposition $\Psi = S$ and *dense geometrical* model, the *distance* (vs. j) of the *center of equivalent electrostatic field* of $Fr_{i,j}$ is:

$$d_{ECP(Fr_{i,j}),j} = \sqrt{ \sum_{v \in Fr_{i,j}} Q_{Pv} \Bigg/ \left(\left(\sum_{v \in Fr_{i,j}} \frac{Q_{Pv}}{d_{v,j}^2} \cdot \frac{\vec{d}_{v,j}}{d_{v,j}} \right) \cdot \left(\sum_{v \in Fr_{i,j}} \frac{Q_{Pv}}{d_{v,j}^2} \cdot \frac{\vec{d}_{v,j}}{d_{v,j}} \right) \right)^{\frac{1}{2}} } \qquad (7.36)$$

It models the distance at which a point charge equal to the fragment charge $\displaystyle\sum_{v \in Fr_{i,j}} Q_{Pv}$

be located vs. j such that the electrostatic field induced by it in j be equal to the field induced by the all atoms of the fragment. The associated matrix is of the form: **DGcDdP_p/d2S**.

Uniform Field Gravitational Potential

It is obtained for the property $\Phi = M$, descriptor $\Omega = p/d$, superposition $\Psi = S$ and *rare geometrical* model. The property descriptor of $Fr_{i,j}$ is:

$$PD(Fr_{i,j}) = \sum_{v \in Fr_{i,j}} \frac{M_v}{d_{v,j}} \qquad (7.37)$$

It models the value of the gravitational potential induced by the fragment in j. The property matrix is of the form: **RGcDdM_p/d_S**.

***Non*-Uniform Field-Type Gravitational Potential**

For the property $\Phi = M$; descriptor $\Omega = p/d$; superposition $\Psi = S$; dense *geometrical* model, the *distance* (vs. j) of the *center of equivalent fragmental gravity* of $Fr_{i,j}$ is:

$$d_{ECP(Fr_{i,j}),j} = \sqrt{ \left(\sum_{v \in Fr_{i,j}} M_v \right) \Bigg/ \left(\left(\sum_{v \in Fr_{i,j}} \frac{M_v}{d_{v,j}} \cdot \frac{\vec{d}_{v,j}}{d_{v,j}} \right) \cdot \left(\sum_{v \in Fr_{i,j}} \frac{M_v}{d_{v,j}} \cdot \frac{\vec{d}_{v,j}}{d_{v,j}} \right) \right)^{\frac{1}{2}} } \qquad (7.38)$$

It models the distance at which a point mass equal to the fragment mass ($\sum_{v \in Fr_{i,j}} M_v$) should be located vs. j such that the gravitational potential induced by it in j be equal to the potential induced by the all atoms of the fragment. The associated matrix is of the form **DGcDdM_p/d_S**.

Uniform Field Coulombian Potential

It is obtained for the property $\Phi = P$ (Q_p implicitly), descriptor $\Omega = p/d$, superposition $\Psi = S$ and *rare geometrical* model. The electrostatic potential descriptor of $Fr_{i,j}$ is:

$$PD(Fr_{i,j}) = \sum_{v \in Fr_{i,j}} \frac{Q_{Pv}}{d_{v,j}} \tag{7.39}$$

It models the value of the electrostatic potential induced by the fragment in j. The property matrix is of the form: **RGcDdP_p/d_S**.

Non-Uniform Field Electrostatic Potential

For the property $\Phi = P$ (Q_P implicitly); descriptor $\Omega = p/d$; superposition $\Psi = S$ and *dense geometrical* model, the *distance* (vs. j) of the *center of equivalent electrostatic potential* of $Fr_{i,j}$ is:

$$d_{ECP(Fr_{i,j}),j} = \sqrt{\left(\sum_{v \in Fr_{i,j}} Q_{Pv}\right) \Big/ \left(\left(\sum_{v \in Fr_{i,j}} \frac{Q_{Pv}}{d_{v,j}} \cdot \frac{\vec{d}_{v,j}}{d_{v,j}}\right) \cdot \left(\sum_{v \in Fr_{i,j}} \frac{Q_{Pv}}{d_{v,j}} \cdot \frac{\vec{d}_{v,j}}{d_{v,j}}\right)\right)^{\frac{1}{2}}} \tag{7.40}$$

It models the distance at which a point charge equal to the fragment charge ($\sum_{v \in Fr_{i,j}} Q_{Pv}$) should be located vs. j such that the electrostatic potential induced by it in j be equal to the potential induced by all the atoms of the fragment. The associated matrix is of the form **DGcDdP_p/d_S**.

Fragmental Numbers and Graph-Theoretical Matrices of CJ-, CF- and Sz – Type

Let the property $\Phi = C$, descriptor $\Omega = p$, superposition $\Psi = S$ and *rare topological* model. Value of cardinal numbering descriptor for $Fr_{i,j}$ is:

$$PD(Fr_{i,j}) = \sum_{v \in Fr^{\sigma_i}_{i,j}} 1 = \left| Fr^{\sigma_i}_{i,j} \right| \tag{7.41}$$

It models the number of atoms in the fragment. The associated matrices are of the form: **RTcDdC__p__S**. Note that these matrices are exactly the graph-theoretical matrices corresponding to the Cluj and Szeged criteria (see eqs 7.1-7.6):

$$\textbf{RTfDiC\_\_p\_\_S} = \textbf{CFD}; \qquad \textbf{RTfDeC\_\_p\_\_S} = \textbf{CF}\Delta$$
$$\textbf{RTjDiC\_\_p\_\_S} = \textbf{CJD}; \qquad \textbf{RTjDeC\_\_p\_\_S} = \textbf{CJ}\Delta$$
$$\textbf{RTsDiC\_\_p\_\_S} = \textbf{SZD}; \qquad \textbf{RTsDeC\_\_p\_\_S} = \textbf{SZ}\Delta$$

In all the above presented models, j appears as a *virtual probe atom* . In the opposite to the CoMFA approach, whose descriptors are calculated as interactions of the molecule with external grid probe atoms, our approach makes use of internal probe atoms: the property of fragment $Fr_{i,j}$ is viewed as the interaction of atoms forming the fragment $Fr_{i,j}$ with the atom j (with no chemical identity, however).

Model Degeneration and Computational Features

The degeneration in the above models may occur in cases when the values of property are not diverse enough, like is case of cardinality (see *Fragmental Numbers,* this Section). Another degeneration is in the case: **RTfDiC__p__H = TfDiC_1/p_S**.

The fragmental analysis was made by the aid of four original 16-bit windows computer programs. First program, *ClujTeor* calculates *topological descriptors* of *Cluj* and *Szeged* type and generates the fragments for the molecules. Second, *ClujProp* calculates the *fragmental properties*. The third one, *StatMon* makes *monovariate regressions* and sorts indices according to the correlation score. The forth program, *StatQ* performs *multi-linear regression* (2-variate, 4-variate, etc.) and saves on disk the best couples of indices. The total number of indices is given by: 2560(see Sect. 7.1) × 3(i.e. x, $ln(x)$, $1/x$) × 3(i.e. the cutting methods: CJ, CF, Sz) × 2(i.e. the path criteria: Di, De) = 46080. Note that in most cases, the degeneration induced by property values and operators lead to a total number of distinct indices around 19,000. In bivariate regression, the first 2^{14}-1=16383 indices recording the best scores in monovariate regression are considered.

7.4. STUDY OF CORRELATION

To illustrate the quality of the family of fragmental property indices in correlation a set of 17 chemical structures from the class of substituted 3-(Phthalimidoalkyl)-pyrazolin-5-ones was selected.

7.4.1 Structure of substituted 3-(Phthalimidoalkyl)-pyrazolin-5-ones

The structure of the selected chemical compounds is given in Figures 7.1.(a, b).

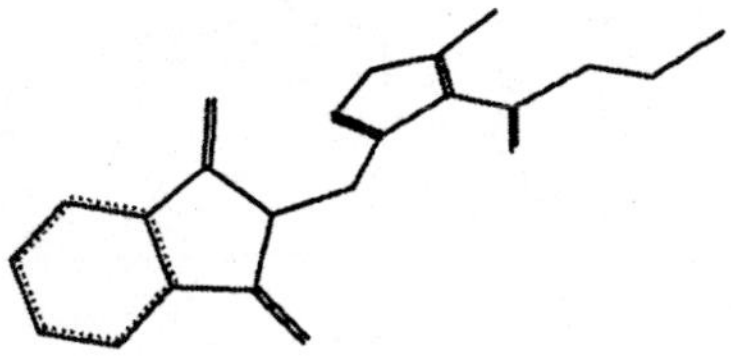

1

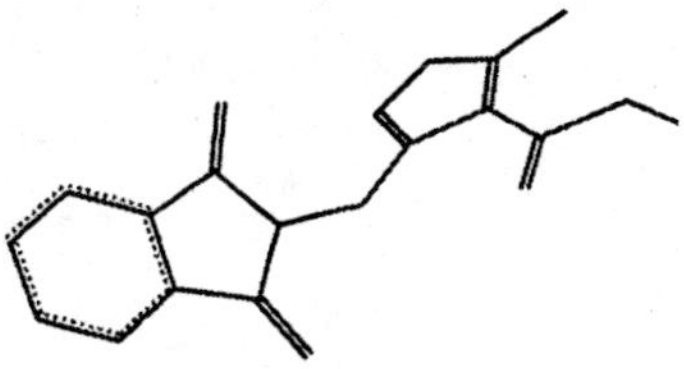

2

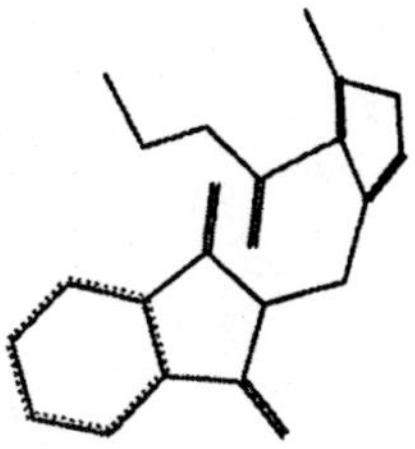

3

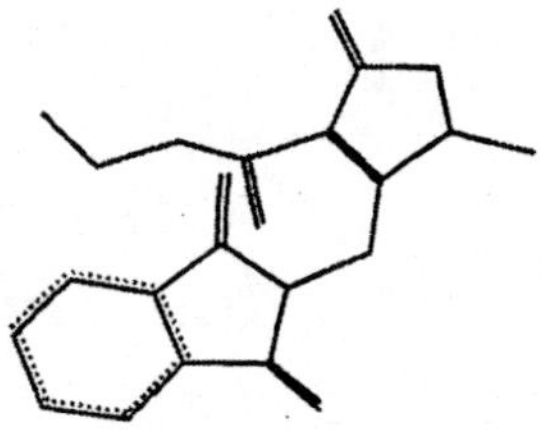

4

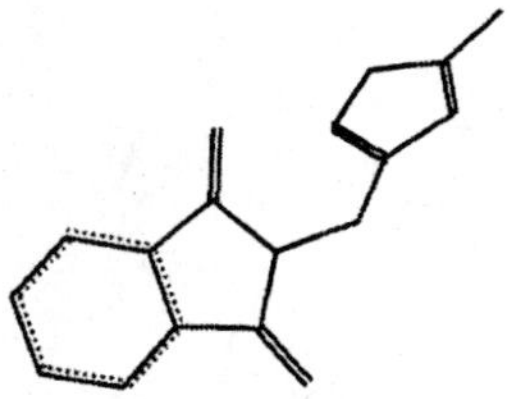

5

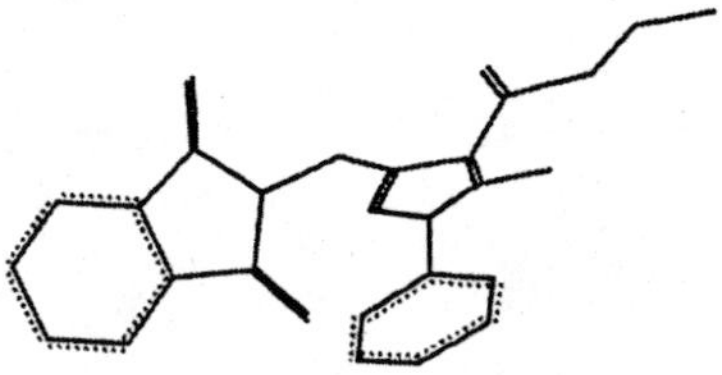

6

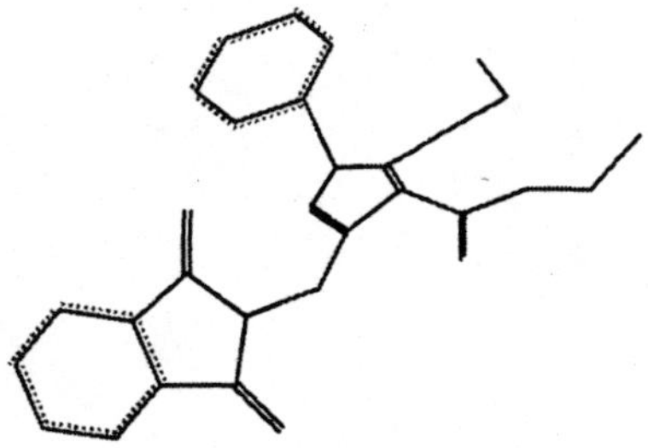

7

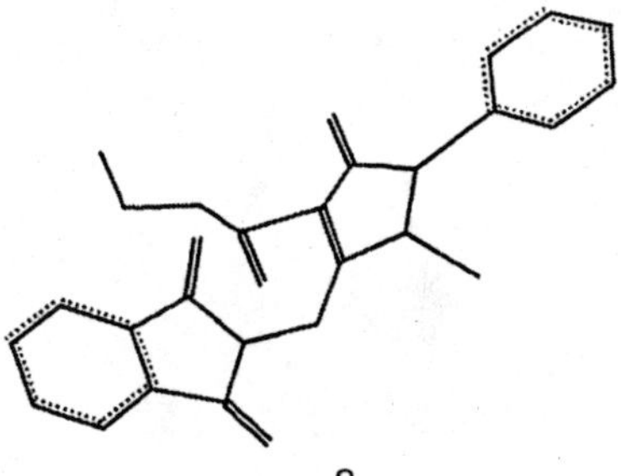

8

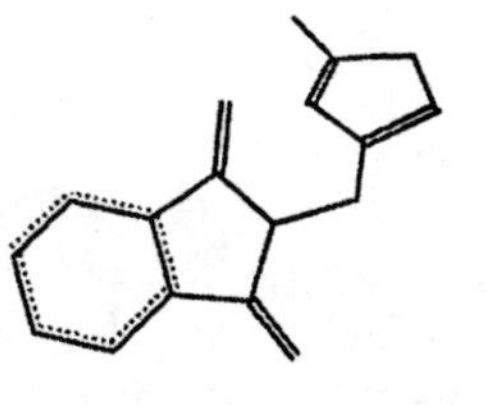

9

Figure 7.1.a. Structure of 17 substituted 3-(Phthalimidoalkyl)-pyrazolin-5-ones; molecules 1 to 9

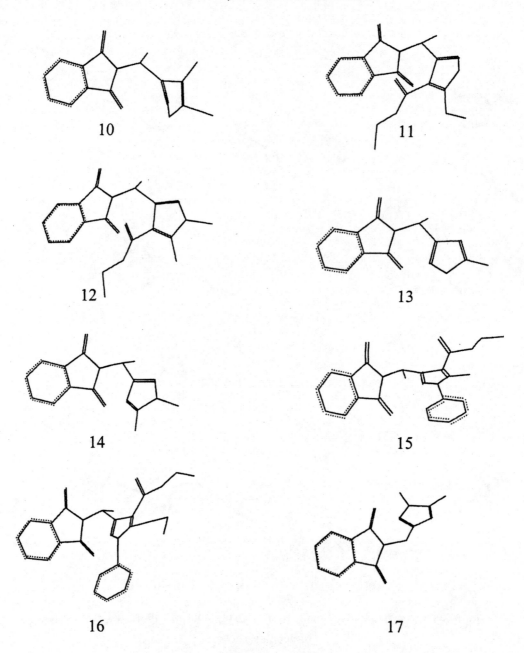

Figure 7.1.b. Structure of 17 substituted 3-(Phthalimidoalkyl)-pyrazolin-5-ones; molecules 10 to 17.

7.4.2 Properties of substituted 3-(Phthalimidoalkyl)-pyrazolin-5-ones

The sum of one-electron energy calculated at the Extended-Huckel level was the first molecular property taken in correlation. A second molecular property was the biological activity of the above listed pyrazolin-5-ones, namely the inhibitory activity (in

%) of a solution of 0.05 mg/ml pyrazolin-5-one on *Lepidium sativum L.* (Cresson). The The data are listed in Table 7.1.

Table 7.1. The Sum of One-Electron Energy Calculated at Single Point Semi-Empirical Extended-Huckel and the Inhibitory Activity on *Lepidium sativum L.* (Cresson) for 17 Substituted 3-(Pthalimidoalkyl)-Pyrazolin-5-Ones*

No.	Compound	Energy (kcal/mol)	Inhibition (%) (0.05 mg/ml)
1	Gly-Pyr	-50978.12	28.4
2	Gly-Pyr-O-Me	-48531.44	28.0
3	Gly-Pyr-1-N-Me	-50863.04	30.4
4	Gly-Pyr-2-N-Me	-53416.95	27.7
5	Gly-Pyr-Decarb	-38604.68	14.3
6	Gly-Ph-Pyr	-62330.33	68.3
7	Gly-Ph-Pyr-O-Me	-64752.65	49.4
8	Gly-Ph-Pyr-2-N-Me	-64751.09	65.2
9	Gly-Ph-Pyr-Decarb	-38588.46	46.9
10	Ala-Pyr	-43209.47	29.3
11	Ala-Pyr-O-Me	-55729.99	28.9
12	Ala-Pyr-1-N-Me	-55832.12	32.6
13	Ala-Pyr-Decarb	-41020.54	12.2
14	Ala-Pyr-1-N-Me-Decarb	-43743.36	18.2
15	Ala-Ph-Pyr	-64701.39	71.7
16	Ala-Ph-Pyr-O-Me	-67104.63	50.6
17	Gly-Pyr-1-N-Me-Decarb	-41057.45	15.1

* Values of inhibition are taken from [48i] and values of energy are calculated by HyperChem program (HyperCube Inc.)

7.4.3 QSPR Analysis for Energy

Monovariate Regression for Energy
For the first five best indices in monovariate correlation, the equation of the model is:
Predicted energy = $b_0 + b_1 \cdot \ln Index$ (7.42)
The indices and their values are shown in Table 7.2.

Table 7.2 Values of the Bests Five Indices in Monovariate Regression

Index	1	2	3	4	5
Name	$DGjDeC\_1/p\_SP\_$	$DGfDeC\_1/p\_SP\_$	$DGjDiC\_1/p\_SP\_$	$RTsDiC\_1/p\_GP2$	$RTsDiC\_\_p\_\_SP\_$
r	**-0.999466**	-0.999463	-0.999439	-0.999414	-0.999405
b_0	89016.51763	89128.37361	89375.15450	89382.16518	81477.18650
b_1	-25198.26985	-25214.11504	-25150.19421	-25379.59389	-16830.75255
1	257.98245	258.17130	263.03785	253	2644.0
2	235.67776	235.86768	240.61058	231	2296.5
3	262.04099	262.37413	268.90715	253	2644.0
4	287.68696	287.93773	296.79757	276	3001.0
5	155.73903	155.92757	159.59795	153	1233.5
6	414.86839	415.11870	425.81775	406	5376.5
7	445.15796	445.41579	456.60416	435	5953.5
8	448.66651	448.91642	460.66835	435	5941.5
9	156.12878	156.32527	160.50819	153	1233.5
10	193.88140	194.13758	197.96803	190	1691.0
11	312.32217	312.59997	321.23457	300	3406.0
12	312.17034	312.44917	321.26878	300	3390.0
13	174.41853	174.62444	178.39353	171	1449.0
14	194.01648	194.23557	198.60151	190	1702.0
15	444.70485	444.96963	455.79605	435	5946.0
16	476.02635	476.29883	487.62160	465	6560.0
17	174.26687	174.46487	178.66246	171	1461.0

The best single variable QSPR (boldface in Table 7.2) was

Predicted energy = 89016.5 − 25198.3·ln$DGjDeC\_1/p\_SP\_$ (7.43)

Statistics for the best regression are given in Table 7.3.

Table 7.3. Statistics for eq 7.43

	r	St Err.of r	B	St Err.of B	s	$F(1,15)$	$t(15)$	p-level
Intercept			89016.5	1193.28			74.59	.00000
ln$DGjDeC\_1/p\_SP\_$	-0.999466	0.00843	-25198.3	212.61	338.5	14046	-118.5	.00000

For the meaning of statistic parameters, the reader is invited to consult Chapter 9 of this book.

Figure 7.2. shows the plot of the calculated energy (cf eq 7.43) vs. the natural logarithm of the $DGjDeC\_1/p\_SP\_$ index.

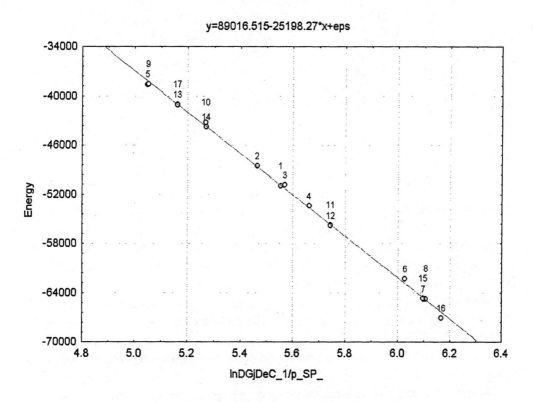

Figure 7.2. The best monovariate regression (cf eq 7.43)

Values of ln$DGjDeC\_1/p\_SP\_$, energy and predicted energy (cf eq 7.43) are included in Table 7.4.

Table 7.4 Regression Results: ln$Index$, Energy and Predicted Energy, cf eq 7.43.

No	ln$DGjDeC\_1/p\_SP\_$	Energy	Predicted Energy
1	5.55289	-50978.12	-50906.74219
2	5.46246	-48531.44	-48628.16016
3	5.56850	-50863.04	-51300.07031
4	5.66187	-53416.95	-53652.88281
5	5.04818	-38604.68	-38188.92969
6	6.02796	-62330.33	-62877.67969
7	6.09843	-64752.65	-64653.34766
8	6.10628	-64751.09	-64851.17187
9	5.05068	-38588.46	-38251.91016
10	5.26725	-43209.47	-43708.98437
11	5.74403	-55729.99	-55723.23437
12	5.74355	-55832.12	-55710.98047
13	5.16146	-41020.54	-41043.28906
14	5.26794	-43743.36	-43726.53516
15	6.09741	-64701.39	-64627.68359
16	6.16547	-67104.63	-66342.74219
17	5.16059	-41057.45	-41021.36719

Bivariate Regression for Energy

The first 16383 indices in monovariate regression are input for bivariate correlation. The algorithm searches for the best correlation for every pair (i, j) of indices ($1 \leq i < j \leq 16383$), and writes on disk the new best found correlation.

Here, the first best found three pairs of indices for bivariate correlation. Indices are labeled with their monovariate rank score. The pairs are $biv1$(1, 11717), $biv2$(95, 1414) and $biv3$(108, 1383).

Case $biv3$(108,1383) was the best found correlation within the *fragmental property family*.

Note that the best scored index in monovariate correlation is not present in the pair of best bivariate correlation ($1 \notin \{108,1383\}$). This fact suggests that the best scored index in monovariate correlation does not explain at the best the property, when coupled with another index belonging to the family. Selection of the pairs of indices for bivariate correlation must be done among all the set (1...16383). Experimental it is proved that there is no method less time consuming (such as an orthogonalization procedure) in obtaining the best scored pair of indices.

The bivariate scores are shown below:

$biv1$ (1 = ln$DGjDeC\_1/p\_SP\_$, 11717 = ln$RGsDeMp2/d2SE2$)

$biv1$ = -25958.9·ln$DGjDeC\_1/p\_SP\_$ + 1220.67·ln$RGsDeMp2/d2SE2$ + 80244.9

$$(7.44)$$

Correlation coefficient: Energy vs $biv1$, $r = 0.999570$

Values of indices, energy and $biv1$ (predicted energy cf eq 7.44) are included in Table 7.5.

Table 7.5 Indices, Energy and $biv1$ (Predicted Energy cf eq 7.44)

No	lnDGjDeC_1/p_SP_	lnRGsDeMp2/d2SE2	Energy	$biv1$
1	5.55289	10.78950	-50978.12	-50731.59240
2	5.46246	10.41587	-48531.44	-48840.30885
3	5.56850	10.56440	-50863.04	-51411.57213
4	5.66187	10.60332	-53416.95	-53787.89129
5	5.04818	10.40413	-38604.68	-38100.29205
6	6.02796	11.07533	-62330.33	-62714.98246
7	6.09843	10.82098	-64752.65	-64854.73277
8	6.10628	11.27766	-64751.09	-64501.06933
9	5.05068	10.41900	-38588.46	-38588.46280
10	5.26725	10.81842	-43209.47	-43281.26493
11	5.74404	10.90985	-55729.99	-55546.56394
12	5.74355	10.60678	-55832.12	-55903.89805
13	5.16146	10.44110	-41020.54	-40995.68193
14	5.26794	10.44431	-43743.36	-43756.00117
15	6.09741	11.00936	-64701.39	-64598.33663
16	6.16547	10.82029	-67104.63	-66595.96010
17	5.16059	10.41325	-41057.45	-41007.09916

$biv2$ (95 = $DTjDeP\_1/d\_SE\_$, 1414 = $1/DTsDiP\_\_d\_\_AE\_$)
$biv2 = -385.13 \cdot DTjDeP\_1/d\_SE\_ + 796417 \cdot 1/DTsDiP\_\_d\_\_AE\_ - 37166$

$$(7.45)$$

Correlation coefficient: energy vs $biv2$, $r = 0.999884$

Values of indices, energy and $biv2$ (predicted energy cf eq 7.45) are included in Table 7.6.

Table 7.6 Indices, Energy and $biv2$ (Predicted Energy cf eq 7.45)

No	DTjDeP_1/d_SE_	1/DTsDiP__d__AE_	Energy	biv2
1	59.70321	0.01174	-50978.12	-50813.46614
2	55.55278	0.01270	-48531.44	-48448.36827
3	59.70321	0.01174	-50863.04	-50813.46614
4	64.99726	0.01129	-53416.95	-53207.83788
5	38.10040	0.01653	-38604.68	-38673.81954
6	81.74904	0.00801	-62330.33	-62268.72347
7	87.25182	0.00776	-64752.65	-64592.17922
8	88.15976	0.00786	-64751.09	-64855.46663
9	38.10040	0.01653	-38588.46	-38588.43460
10	47.66825	0.01510	-43209.47	-43496.59129
11	70.34408	0.01082	-55729.99	-55640.48566
12	70.83694	0.01093	-55832.12	-55742.02042
13	43.00040	0.01598	-41020.54	-41002.88913
14	47.31508	0.01481	-43743.36	-43593.46139
15	88.34726	0.00788	-64701.39	-64917.41882
16	94.01670	0.00763	-67104.63	-67299.73910
17	42.21508	0.01527	-41057.45	-41261.34229

$biv3$ (108 = $RTfDeM\_p/d2SP\_$, 1383 = $1/DTsDeE\_1/p\_SE\_$)

$biv3 = -54.019 \cdot RTfDeM\_p/d2SP\_ + 697864.87 \cdot 1/DTsDeE\_1/p\_SE\_ - 43266.3$

(7.46)

Correlation coefficient: energy vs $biv3$, $r = 0.999934$

Values of indices, energy and $biv3$ (predicted energy cf eq 7.46) are included in Table 7.7.

7.4.4. Conclusions for Energy Analysis

1. The *best index* in monovariate regression does not provide the best explanation for the *measured* property when coupled (in a bivariate correlation) with any other index belonging to the discussed family.

2. The best bivariate correlation is not obtained as the *best orthogonal indices*, (see the Randić's *DCA*, Section 9.6.) but only as the *best couple of indices*, resulted by the trial of the whole family.

3. The constant high correlation (r > 0.999) between the best indices and the quantum mechanically calculated energy provided by the semi-empirical Extended-Huckel approach demonstrates the *quantum nature* of *FPI*.

Table 7.7. Indices, Energy and $biv3$ (Predicted Energy cf eq 7.46)

No	$RTfDeM\_p/d2SP\_$	$1/DTsDeE\_1/p\_SE\_$	Energy	$biv3$
1	291.65807	0.01169	-50978.12	-50861.14698
2	253.84179	0.01204	-48531.44	-48578.47928
3	286.32871	0.01101	-50863.04	-51050.62598
4	318.26541	0.01013	-53416.95	-53390.70245
5	141.99293	0.01752	-38604.68	-38707.64398
6	435.14085	0.00641	-62330.33	-62302.08286
7	475.71099	0.00600	-64752.65	-64773.71407
8	475.03007	0.00600	-64751.09	-64737.52944
9	141.99293	0.01752	-38588.46	-38588.45484
10	195.34755	0.01526	-43209.47	-43166.88331
11	358.32197	0.00986	-55729.99	-55742.37228
12	349.41745	0.00932	-55832.12	-55639.52197
13	164.77147	0.01577	-41020.54	-41162.38985
14	188.35158	0.01428	-43743.36	-43476.32911
15	474.16792	0.00599	-64701.39	-64698.52071
16	516.35251	0.00563	-67104.63	-67231.41652
17	164.07111	0.01579	-41057.45	-41107.89635

Figure 7.3. illustrates the plot of energy (quantum mechanically calculated) vs $biv3$ (predicted energy cf eq 7.46).

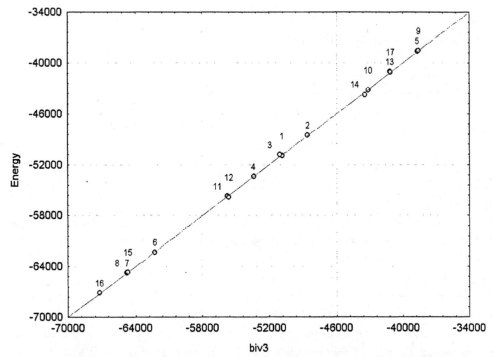

Figure 7.3. The plot: energy vs $biv3$ (predicted energy cf eq 7.46)

Statistics for the regression: energy vs $biv3$ are given in Table 7.8.

Table 7.8 Statistics for the Regression: Energy vs $biv3$.

	r	St Err.of r	B	St Err.of B	s	$F(1,15)$	$t(15)$	p-level
Intercept			-0.0000	157.23			-0.0000	1.0000
biv3	0.99993	0.00296	1.0000	0.003	119.14	113500	336.904	0.0000

4. Sum of one-electron-energies for the set of 17 molecules is best modeled by $biv3$ (i.e. the calculated energy by eq 7.46).

5. An insight of eq 7.46 (i.e. $biv3$), reveals the dependency of this energy by the molecular topology (topological models) and the nature of atoms (mass and electronegativity).

6. Let $\Sigma abs(b_1 \cdot x_1)/(\Sigma(abs(b_1 \cdot x_1) + \Sigma abs(b_2 \cdot x_2))$ be a measure of individual contribution of indices in variation of $biv3$. The value given by $\Sigma|-54 \cdot RTfDeM\_p/d2SP\_|/(\Sigma|-54 \cdot RTfDeM\_p/d2SP\_| + \Sigma|697864 \cdot 1/DTsDeE\_1/p\_SE\_|)$ = 0.68035 says that about 68% sum of one-electron-energy is a measure of field (p/d2 in the expression of $RTfDeM\_p/d2SP\_$).

7. The preferred operator in monovariate regression is ln (all the best 5 indices, see eq 7.42 and Table 7.2).

7.4.5 QSAR Analysis for Inhibition

Monovariate Regression for Inhibition
For the first seven best indices in monovariate regression, the equation of the model is:

$$\text{Predicted inhibition} = b_0 + b_1 \cdot Index \qquad (7.47)$$

the index values of which are shown in Table 7.9 and statistics in Table 7.10.

Bivariate Regression for Inhibition
The first best found three pairs of indices in bivariate correlation are presented. Indices are labeled with their monovariate scores. The pairs are: $biv1(1, 11961)$, $biv2(235, 4052)$ and $biv3(235, 7783)$. The second index for the bivariate correlation was chosen from the 1..16383 best scored monovariate indices.

The case $biv3(235, 7783)$ was the best found correlation in the algorithm selection. For algorithm details see section $Bivariate\ Regression\ for\ Energy$.

Note that, as in the case of energy, the best scored index in monovariate correlation is not present in the pair of best bivariate correlation ($1 \notin \{235, 7783\}$). Selection of pairs of indices for bivariate correlation must be done among all the family (1...16383).

Table 7.9 Inhibition and Values of the Best Seven Indices in Monovariate Regression

No	Inhib	1	2	3	4	5	6	7
1	28.4	64089	11.068	1.5603E-05	-345.37	82.636	4775.5	637.31
2	28	64448	11.074	1.5516E-05	-221.58	76.173	3596.9	569.26
3	30.4	68490	11.134	1.4601E-05	-286.06	84.060	4846.4	772.27
4	27.7	65346	11.087	1.5303E-05	-213.29	105.128	6000.1	966.24
5	14.3	64947	11.081	1.5397E-05	-161.27	77.538	3085.2	491.68
6	68.3	77978	11.264	1.2824E-05	-538.20	154.502	13086.9	1807.18
7	49.4	72755	11.195	1.3745E-05	-455.55	116.674	8640.4	1314.55
8	65.2	77294	11.255	1.2938E-05	-588.78	143.425	11578.5	1742.13
9	46.9	65165	11.085	1.5346E-05	-255.93	78.655	3166.4	504.32
10	29.3	65341	11.087	1.5304E-05	-278.17	85.016	4816.9	643.25
11	28.9	65547	11.091	1.5256E-05	-186.94	96.383	7273.2	1037.66
12	32.6	66652	11.107	1.5003E-05	-272.94	91.621	6019.0	970.02
13	12.2	65588	11.091	1.5247E-05	-229.75	79.968	3410.3	546.83
14	18.2	65333	11.087	1.5306E-05	-147.46	78.974	3586.2	574.37
15	71.7	77537	11.259	1.2897E-05	-643.15	140.442	11386.1	1669.85
16	50.6	73461	11.205	1.3613E-05	-648.32	112.870	8445.2	1320.70
17	15.1	65119	11.084	1.5356E-05	-102.56	77.259	3264.8	520.13

Table 7.10 Name of the Best Seven Indices and their Monovariate Correlation

No	Index	r	b_0	b_1
1	$RGsDeCp2/d2SE2$	0.899523	-194.68	0.003370
2	$lnRGsDeCp2/d2SE2$	0.897333	-2612.4	237.92
3	$1/RGsDeCp2/d2SE2$	0.894772	281.7	-16741138.9
4	$DGjDiP\_p/d\_AP\_$	0.894351	4.9767	-0.095527
5	$RGsDeM\_p/d2SE\_$	0.888106	-28.384	0.654080
6	$RGsDeEp2/d2SE\_$	0.887772	3.6751	0.005184
7	$RGsDeMp2/d2SE\_$	0.885762	1.8078	0.036454

The best monovariate QSAR was

$$\text{Predicted inhibition} = -194.68 + 0.003370 \cdot RGsDeCp2/d2SE2 \tag{7.48}$$

Statistics for the best scored index $RGsDeCp2/d2SE2$ (cf eq 7.48) are given in Table 7.11.

Table 7.11 Statistics for the Best Scored Index $RGsDeCp2/d2SE2$ (cf eq 7.48).

	r	St Err. r	B	St Err. B	s	$F(1,15)$	$t(15)$	p-level
Intercept			-194.68	29.04			-6.704	.000007
$RGsDeCp2/d2SE2$	0.89952	0.1128	0.00337	0.00042	8.591	63.59	7.975	.000001

The plot of the inhibition vs. the index $RGsDeCp2/d2SE2$ is shown in Figure 7.4.

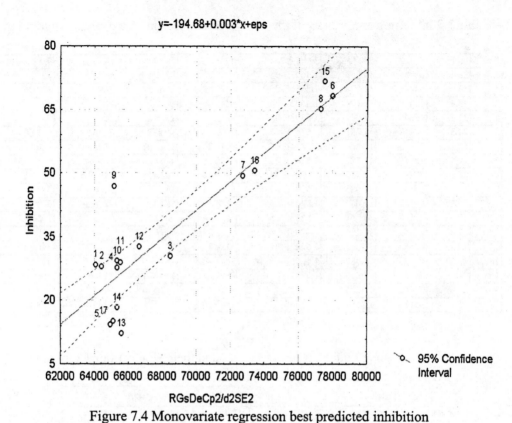

Figure 7.4 Monovariate regression best predicted inhibition

The bivariate correlations are as follows:

$biv1$ (1 = $RGsDeCp2/d2SE2$, 11961 = $1/DGjDeP\_p/d2GE2$)
$biv1$ = 0.003054·$RGsDeCp2/d2SE2$ − 1719.3·$1/DGjDeP\_p/d2GE2$ − 138.6

$$(7.49)$$

Correlation coefficient: inhibition vs $biv1$. r = 0.988830

Table 7.12 Values of Indices, Inhibition and $biv1$ (Predicted Inhibition, cf eq 7.49).

No	$RGsDeCp2/d2SE2$	$DGjDeP\_p/d2GE2$	INHIBITION	$biv1$
1	64088.68725	55.64699	28.4	26.265
2	64447.53674	64.68759	28.0	31.679
3	68490.38729	43.68903	30.4	31.253
4	65345.82007	52.33072	27.7	28.147
5	64947.11517	37.00776	14.3	13.326
6	77978.39585	52.22715	68.3	66.667
7	72755.40201	52.00063	49.4	50.570
8	77293.73666	52.54815	65.2	64.776
9	65165.30346	40.14345	46.9	46.900
10	65340.85566	48.57654	29.3	25.593
11	65546.50225	48.61544	28.9	26.249
12	66652.22261	51.43069	32.6	31.562
13	65587.90626	38.58451	12.2	17.182
14	65332.92136	38.58311	18.2	16.401
15	77536.67347	57.10956	71.7	68.132
16	73460.71348	59.88603	50.6	57.078
17	65119.29156	38.30207	15.1	15.422

$biv2$ (235 = $DGsDiPp2/d2SE2$, 4052 = $\ln DTsDiE\_p/d\_HE2$)

$$biv2 = 0.775 \cdot DGsDiPp2/d2SE2 + -30.994 \cdot \ln DTsDiE\_p/d\_HE2 + 30.782$$

(7.50)

Correlation coefficient: inhibition vs $biv2$, $r = 0.993240$

$biv3$ (235 = $DGsDiPp2/d2SE2$, 7783 = $\ln RGsDiEp2/d2HE2$)

$biv3 = 0.9372 \cdot DGsDiPp2/d2SE2 - 10.058 \cdot \ln RGsDiEp2/d2HE2 + 14.488$

(7.51)

Correlation coefficient: inhibition vs $biv3$, $r = 0.993770$

7.4.6 Conclusions for Inhibition Analysis

1. The *best index* in monovariate regression does not offer the best explanation for the *measured* property when coupled with any other index belonging to this family in a bivariate correlation.

2. The best bivariate correlation is the *best couple of indices*, resulted by the trial in the whole family of fragmental property indices.

Table 7.13. Values of Indices, Inhibition, $biv2$ (Predicted Inhibition, cf eq 7.50) and $biv3$ (Predicted Inhibition, cf eq 7.51).

No	DGsDiPp2/d2SE2	DTsDiE_p/d_HE2	RGsDiEp2/d2HE2	Inhibition	$biv2$	$biv3$
1	-14.80225	0.78665	0.05815	28.4	26.748	29.229
2	-16.55124	0.79529	0.05973	28.0	25.054	27.321
3	-15.07688	0.71684	0.07599	30.4	29.416	26.281
4	-12.22525	0.69343	0.06758	27.7	32.655	30.133
5	-17.15632	1.20376	0.29558	14.3	11.738	10.668
6	14.24503	0.45129	0.02181	68.3	66.482	66.317
7	-3.66908	0.44415	0.01870	49.4	53.093	51.078
8	15.21789	0.49053	0.03332	65.2	64.652	62.966
9	-16.87848	1.20376	0.27948	46.9	46.903	46.902
10	2.91247	1.17026	0.35940	29.3	28.167	27.511
11	-10.87153	0.79379	0.07637	28.9	29.514	30.172
12	-11.10275	0.73055	0.06304	32.6	31.909	31.885
13	-13.62422	1.27247	0.30716	12.2	12.755	13.592
14	-15.50583	1.03057	0.13654	18.2	17.832	19.984
15	22.24726	0.49099	0.02304	71.7	70.071	73.268
16	-1.97165	0.48236	0.02024	50.6	51.851	51.871
17	-17.44972	0.96513	0.13847	15.1	18.359	18.021

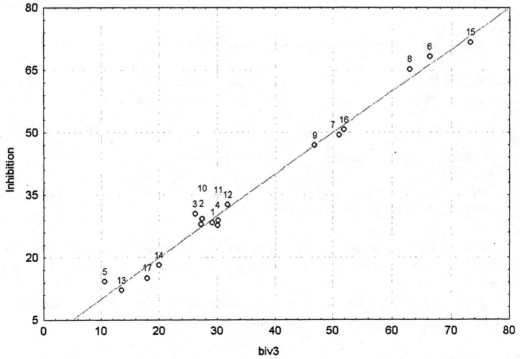

Figure 7.5. The plot: inhibition vs $biv3$ (predicted inhibition cf eq 7.51)

The plot of inhibition vs $biv3$ (predicted inhibition cf eq 7.51) is shown in Figure 7.5.

Statistics for the regression: inhibition vs $biv3$ (predicted inhibition cf eq 7.51) are given in Table 7.14.

Table 7.14. Statistics for the Regression: Inhibition vs $biv3$

	r	St Err. r	B	St Err. B	s	$F(1,15)$	$t(15)$	p-level
Intercept			0.00037	1.1778			0.00031	.9997
biv3	0.99377	0.028771	0.99999	0.0289	2.19	1193	34.54	.0000

3. The constant high correlation ($r > 0.88$) between the best indices and the mitodepressive activity on *Lepidium Savitium L. (Cresson)* demonstrate ability of this family of indices to estimate the biological activity of the considered set of chemical structures.

4. An inspection onto eq 7.51 suggests that the mitodepressive activity on *Lepidium Savitium L. (Cresson)* is dependent on the geometric feature of molecules, the nature of atoms (electronegativity) and the electrostatic field of atoms induced by their partial charges.

5. The geometric models are dominant both in monovariate and bivariate regression (7 of the best 7 among the monovariate regressions and 5 of the best 6 in the bivariate regressions).

7.4.7 Correlation between Energy and Inhibition

The plot: inhibition vs energy (Figure 7.6) reveals that between the two properties no good correlation exists: $r = 0.77898363$. It implies that these properties cannot be modeled by the same indices. Our results clearly showed that the inhibition is best modeled by geometric models whereas topological models better describe the energy.

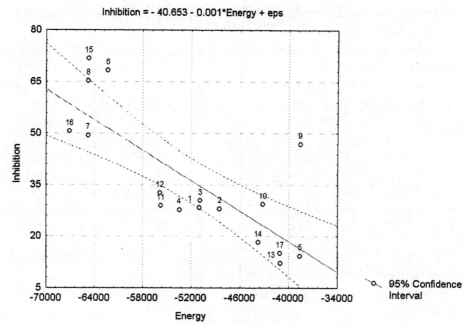

Figure 7.6. The plot: inhibition *vs.* energy (quantum mechanically calculated – see text).

The low correlation between inhibition and the sum of one-electron energy demonstrates that the inhibition is not dependent on the energy.

7.4.8 General Conclusions to Correlating Studies

1.   Fragmental property indices take into account the chemical nature of atoms (mass and electronegativity), various kinds of interactions between the fragments of molecules and the 3D geometry of molecular structures.

2.   There exist an analogy between *CoMFA* and *FPI*: both of them calculate the interaction of a chemical structure (or substructure) with a *probe atom* in the 3D space. The property of fragment $Fr_{i,j}$ is viewed as the interaction of atoms forming the fragment $Fr_{i,j}$ with the atom j . The major difference is that *CoMFA* uses external probe atoms (with defined chemical identity) whereas *FPI* considers internal probe atoms with no chemical identity. Only the fragments (i.e. substructures) are chemically well defined.

3.   Bivariate correlations with indices belonging to the fragmental property index family can offer good quality models for quite diverse molecular properties such as the inhibition of mitodepressive activity on *Lepidium Savitium L.* ($r > 0.99$) as well as the sum of one-electron energy calculated at the Extended-Huckel level ($r > 0.9999$). These results demonstrate the correlating ability of this family of indices.

4.   The *best couple of indices* are found by performing all combinations of two indices bivariate regressions within the family. At such a large pool of indices, the two-dimensional description (i.e. bivariate correlation), providing a direct structural

interpretation of a molecular property, appears to be one of the most powerful methods in the characterization of molecular structures.

REFERENCES

1. Filizola M.; Rosell G.; Guerrero A.; Pérez J. J.; Conformational Requirements for Inhibition of the Pheromone Catabolism in Spodoptera Littoralis. *Quant. Struct.-Act. Relat.* **1998**, *17*, 205-210.

2. Lozoya E.; Berges M.; Rodríguez J.; Sanz F.; Loza M. I.; Moldes V. M.; Masauer C. F.; Comparison of Electrostatic Similarity Approaches Applied to a Series of Kentaserin Analogues with $5-HT_{2A}$ Antagonistic Activity. *Quant. Struct.-Act. Relat.* **1998**, *17*, 199-204.

3. Winkler D. A.; Burden F. R.; Holographic QSAR of Benzodiazepines. *Quant. Struct.-Act. Relat.* **1998**, *17*, 224-231.

4. Wikler D. A.; Burden F. R.; Watkins A. J. R, Atomistic Topological Indices Applied to Benzodiazepines using Various Regression Methods. *Quant. Struct.-Act. Relat.* **1998**, *17*, 14-19.

5. Zbinden P.; Dobler M.; Folkers G.; Vedani A.; PrGen: Pseudoreceptor Modeling Using Receptor-mediated Ligand Alignment and Pharmacophore Equilibration. *Quant. Struct.-Act. Relat.* **1998**, *17*, 122-130.

6. Cramer R. D. III, Patterson D. E., Bunce J. D.; Comparative Molecular Field Analysis (CoMFA). 1. Effect of Shape on Binding of Steroids to Carrier Proteins. *J. Am. Chem. Soc.* **1988**, *110*, 5959-5967.

7. Unity Program for SIMCA (Soft Independent Modeling Class Analogy). *Tripos Associates*, St. Louis, MO.

8. Merz, A.; Rognan, D.; Folkers, G. 3D QSAR Study of N2-Phenylguanines as Inhibitors of Herpes Simplex Virus Thymide Kinase, Antiviral and Antitumor Research, *http:\\www.pharma.ethz.ch /text/research/tk/qsar.html*.

9. Gurba P. E.; Parham M. E.; Voltano J. R. Comparison of QSAR Models Developed for Acute Oral Toxicity (LD_{50}) by Regression and Neural Network Techniques. Conference on Computational Methods in Toxicology – April, **1998**, Holiday Inn/I-675, Dayton, Ohio, USA, abstract available at *http://www.ccl.net /ccl/toxicology/abstracts/abs9.html*.

10. HyperChem, Molecular Modelling System, *Hypercube Inc.*

11. Molconn-Z, http://www.eslc.vabiotech.com/molconn.

12. Waller C. L.; Wyrick, S. D.; Park, H. M.; Kemp, W. E.; Smith, F. T. Conformational Analysis, Molecular Modeling, and Quantitative Structure-Activity Relationship Studies of Agents for the Inhibition of Astrocytic Chloride Transport. *Pharm. Res.* **1994**, *11*, 47-53.

13. Horwitz J. P.; Massova, I.; Wiese, T.; Wozniak, J.; Corbett, T. H.; Sebolt-Leopold, J. S.; Capps, D. B.; Leopold, W. R. Comparative Molecular Field Analysis of in

Vitro Growth Inhibition of L1210 and HCT-8 Cells by Some Pyrazoloacridines. *J. Med. Chem.* **1993**, *36*, 3511-3516.

14. McGaughey G. B.; MewShaw R. E.; Molecular Modeling and the Design of Dopamine D_2 Partial Agonists, (presented at the Charleston Conference, March, **1998**), *Network Science*, http://www.netsci.org/Science/Compchem/feature20.html.

15. Myers A. M.; Charifson, P. S.; Owens, C. E.; Kula, N. S.; McPhall, A. T.; Baldessarini, R. J.; Booth, R. G.; Wyrick, S. D. Conformational Analysis, Pharmacophore Identification, and Comparative Molecular Field Analysis of Ligands for the Neuromodulatory .sigma.3 Receptor. *J. Med. Chem.* **1994**, *37*, 4109-4117.

16. Chuman H.; Karasawa M.; Fujita T. A Novel Three-Dimensional QSAR Procedure: Voronoi Field Analysis. *Quant. Struct.-Act. Relat.* **1998**, *17*, 313-326.

17. Kellogg G. E.; Semus S. F.; Abraham D. J. HINT: a New Method of Empirical Hydrophobic Field Calculation for CoMFA. *J. Comput.-Aided Mol. Des.* **1991**, *5*, 545-552.

18. Kim K. H. Use of the Hydrogen-Bond Potential Function in Comparative Molecular Field Analysis (CoMFA): An extension of CoMFA. *Quant. Struct.-Act. Relat.* **1993**, *12*, 232-238.

19. Durst G. L. Comparative Molecular Field Analysis (CoMFA) of Herbicidal Protoporphyrinogen Oxidase Inhibitors using Standard Steric and Electrostatic Fields and an Alternative LUMO Field. *Quant. Struct.-Act. Relat.* **1998**, *17*, 419-426.

20. Waller C.L.; Marshall G. R. Three-Dimensional Quantitative Structure-Activity Relationship of Angiotensin-Converting Enzyme and Thermolysin Inhibitors. II. A Comparison of CoMFA Models Incorporating Molecular Orbital Fields and Desolvation Free Energy Based on Active-Analog and Complementary-Receptor-Field Alignment. *J. Med. Chem.* **1993**, *36*, 2390-2403.

21. Waller C. L.; Kellogg G. E. Adding Chemical Information of CoMFA Models with Alternative 3D QSAR Fields. *Network Science*, **1996**, *Jan*, http://www.netsci.org/Science/Compchem/feature10.html.

22. Pajeva I. L., Wiese M. A Comparative Molecular Field Analysis of Propafenone-type Modulators of Cancer Multidrug Resistance. *Quant. Struct.-Act. Relat.* **1998**, *17*, 301-312.

23. Klebe G.; Abraham U. On the Prediction of Binding Properties of Drug Molecules by Comparative Molecular Field Analysis. *J. Med. Chem.* **1993**, *36*, 70-80.

24. Czaplinski K.-H. A.; Grunewald G. L. A Comparative Molecular Field Analysis Derived Model of Binding of Taxol Analogs to Microtubes. *Bioorg. Med. Chem. Lett.* **1994**, *4*, 2211-2216.

25. Waller C. L.; Oprea, T. I.; Giolitti, A.; Marshall, G. R. Three-Dimensional QSAR of Human Immunodeficiency Virus. (1). Protease Inhibitors. 1. A Determined Alignment Rules. *J. Med. Chem.* **1993**, *36*, 4152-4160.

26. Akagi T. Exhaustive Conformational Searches for Superimposition and Three-Dimensional Drug Design of Pyrethroids. *Quant. Struct.-Act. Relat.* **1998**, *17*, 565-570.

27. Thompson E. The use of Substructure Search and Relational Databases for Examining the Carcinogenic Potential of Chemicals. Conference on Computational Methods in Toxicology – April, **1998**, Holiday Inn/I-675, Dayton, Ohio, USA, abstract available at *http://www.ccl.net /ccl/toxicology/abstracts/tabs6.html.*

28. Todeschini R.; Lasagni M.; Marengo E. New Molecular Descriptors for 2D and 3D Structures. Theory, *J. Chemometrics*, **1994**, *8*, 263-272.

29. Todeschini R.; Gramatica P.; Provenzani R.; Marengo E. Weighted Holistic Invariant Molecular (WHIM) descriptors. Part2. There Development and Application on Modeling Physico-chemical Properties of Polyaromatic Hydrocarbons. *Chemometrics and Intelligent Laboratory Systems*, **1995**, *27*, 221-229.

30. Todeschini R.; Vighi M.; Provenzani R.; Finizio A.; Gramatica P. Modeling and Prediction by Using WHIM Descriptors in QSAR Studies: Toxicity of Heterogeneous Chemicals on Daphnia Magna. *Chemosphere*, **1996**, *8*, 1527.

31. Bravi G.; Gancia E.; Mascagni P.; Pegna M.; Todeschini R.; Zaliani A. MS-WHIM. New 3D Theoretical Descriptors Derived from Molecular Surface Properties: A Comparative 3D QSAR Study in a Series of Steroids. *J. Comput.-Aided Mol. Des.* **1997**, *11*, 79-92.

32. Zaliani A.; Gancia E. MS-WHIM Scores for Amino Acids: A New 3D-Description for Peptide QSAR and QSPR Studies. *J. Chem. Inf. Comput. Sci.* **1999**, *39*, 525-533.

33. Sears, F. W.; Zemansky, M. W.; Young, H. D. University Physics, fifth edition, **1976**, *Addison – Wesley Publishing Company*, USA-Canada, Catalog Card no. 75-20989.

34. Famini, G. R.; Wilson, L. Y. Using Theoretical Descriptors in Quantitative Structure Activity Relationship and Linear Free Energy Relationships. *Network Science*, **1996**, *Jan*, http:\\www.netsci.org /Science /Compchem/future08.html.

35. Sun, L.; Weber, S. G. Prediction of Molecular Recognition – Enhanced Phenobarbital Extraction Based on Solvatochromic Analysis. *Journal of Molecular Recognition*, **1998**, *11*, 28-31.

36. Collection of Journal of Chemical Information and Computer Sciences, ISSN 00952338, Am. Chem. Soc. Section Chemical Computation.

37. Golender, V.; Vesterman, B.; Vorpagel, E. APEX-3D Expert System for Drug Design. *Network Science*, **1996**, *Jan*, http:\\www.netsci.org/Science/Compchem/feature09.html.

38. Landau, L. D.; Lifshitz, E. M. Statistical Physics, 3-rd edition, revised by Lifshitz, E. M and Pitaevkim, L. P., Nauka, Moscow, **1978**, Chap. Phase Transition of Rank Two and Critical Phenomena.

39. Rose, V. S.; Wood, J. Generalized Cluster Significance Analysis and Stepwise Cluster Significance Analysis with Conditional Probabilities. *Quant. Struct.-Act. Relat.* **1998**, *17*, 348-356.

40. Labute, P. QuaSAR-Binary: A New Method for the Analysis of High Thoughput Screening Data. *Network Science*, **1998**, *May*, http:\\www.netsci.org/Science/Compchem/feature21.html.

41. Young, H. D.*Statistical Treatment of Experimental Data*. McGraw-Hill, New York, **1962**.

42. Reif, F. *Fundamentals of Statisticals and Thermal Physics*, McGraw-Hill, Chap.1 New York, **1965**.

43. Diudea M.V.; Silaghi-Dumitrescu I.; Valence Group Electronegativity as a Vertex Discriminator. *Rev. Roumaine Chim.* **1989**, *34*, 1175-1182.

44. Diudea M.V.; Kacso I. E.; Topan M. I. Molecular Topology. 18. A Qspr/Qsar Study by Using New Valence Group Carbon-Related Electronegativities. *Rev. Roum. Chim.* **1996**, *41*, 141-157.

45. Crawford, F. S. Jr.; *Waves*, Berkeley Physics Course, Newton, Massachusetts, vol. 3, **1968**.

46. Purcell, E. M. *Waves*, Berkeley Physics Course, Newton, Massachusetts, vol. 3, **1965**.

47. Atkins, P. W. *Physical Chemistry*, fifth edition, Oxford University Press (Oxford, Melbourne, Tokyo), **1994**.

48. Nikolić, S.; Medić-Sarić, M.; Matijević-Sosa; J. A QSAR Study of 3-(Phtalimidoalkyl)-pyrazolin-5-ones, *Croat. Chem. Acta*, **1993**, *66*, 151-160.

Chapter 8.

SYMMETRY AND SIMILARITY

The investigation of molecular structure involves research on its constitution (i.e. the number and chemical identity of atoms and bonds joining them) and configuration in 3D-space.

Molecules show various types of geometrical symmetry. The symmetry is reflected in several molecular properties, such as dipole moments, IR vibrations, ^{13}C - NMR signals etc., properties which are dependent on the spatial structure of molecules. The molecular topology reveals a different type of symmetry: the *topological symmetry* (i.e. constitutional symmetry). It is defined in terms of *connectivity*, as a constitutive principle of molecules and expresses the equivalence relationships between elements of graph: vertices, bonds or larger subgraphs. It makes use of groups theory formalism in modeling an N - dimensional space. The geometrical aspects are disregarded.

Similarity (or relatedness) of molecular structures expresses the common features occurring within a set of molecules. It is established on the ground of various criteria and procedures. Both symmetry and similarity provide equivalence classes: the first one at the level of molecular graph and its subgraphs while the last one among the members of a whole set of molecules. The two notions are interrelated, as will be detailed in the following.

8.1. ISOMORPHISM AND AUTOMORPHISM

Let $G = (V, E)$ and $G' = (V', E')$ be two graphs, with $|V| = |V'|$, and a function f, *mapping* the vertices of V onto the vertices belonging to the set V', $f : V \rightarrow V'$. That is, the function f makes a one-to-one correspondence between the vertices of the two sets. The two graphs are called *isomorphic*, $G \quad G'$, if there exists a mapping f that preserves adjacency (i.e. if $(i, j) \in E$, then $(f(i), f(j)) \in E'$). In searching isomorphicity, labeled graphs are compared. In the chemical field, such a study will answer if two molecular graphs represent one and the same chemical compound.

Let the mapping be a permutation P, represented in a two-row notation[1] as:

$$P = \begin{pmatrix} 1 & 2 \dots i \dots N \\ p_1 & p_2 \dots p_i \dots p_N \end{pmatrix}$$

which shows that vertex 1 gets permuted to vertex p_1, vertex 2 to vertex p_2, vertex i to vertex p_i and so on. The permutation that leaves the graph unchanged is called the *permutation identity* and denoted P_{11}. Some permutations preserve the adjacency and some others not. The former type provides an isomorphism of a graph with itself, which is called an *automorphism*.

Let $Aut(G) = (P_{11}, P_{1i}, P_{1j}...)$ be the set of automorphisms of a graph G and $\otimes$ a *binary operation* (i.e. a *composition* rule) defined on that set. $Aut(G)$ is called an *automorphism group* if the following conditions are satisfied: [2,3]

1.   For any two permutations P_{1i}, $P_{1j} \in Aut(G)$ there exists a *unique element*, $P_{1k} \in Aut(G)$, such that $P_{1k} = P_{1i} \otimes P_{1j}$.

2.   The operation is *associative*: $P_{1i} \otimes P_{1j} \otimes P_{1k} = P_{1i} \otimes (P_{1j} \otimes P_{1k}) = (P_{1i} \otimes P_{1j})$ $\otimes P_{1k}$, for all P_{1i}, P_{1j} and $P_{1k} \in Aut(G)$.

3.   The set $Aut(G)$ contains a *unique permutation P_{11}* , called *permutation identity*, such that $P_{1i} \otimes P_{11} = P_{11} \otimes P_{1i} = P_{1i}$, for all $P_{1i} \in Aut(G)$.

4.   4. For every permutation $P_{1i} \in Aut(G)$ there exists an *inverse*, $P^{-1}{}_{1i} \in Aut(G)$ that obey the relation: $P_{1i} \otimes P^{-1}{}_{1i} = P^{-1}{}_{1i} \otimes P_{1i} = P_{11}$

A permutation can be described by a *permutation matrix* $\mathbf{P}$, whose elements $[\mathbf{P}]_{ij}$ = 1 if vertex i is permuted to vertex j and $[\mathbf{P}]_{ij} = 0$ otherwise. The permutation identity, $\mathbf{P}_{11}$, is a diagonal matrix whose elements equal unity.

In matrix form, an isomorphism can be expressed as: [4,5]

$$\mathbf{A}(G_2) = \mathbf{P}_{12}^{-1}\mathbf{A}(G_1)\mathbf{P}_{12} \tag{8.1}$$

where $\mathbf{A}(G_1)$ and $\mathbf{A}(G_2)$ are the adjacency matrices of the two isomeric graphs and $\mathbf{P}$ is the permutation matrix. Since the $\mathbf{P}$ matrix is orthogonal, eq 8.1 can be written as:

$$\mathbf{A}(G_2) = \mathbf{P}_{12}^{T}\mathbf{A}(G_1)\mathbf{P}_{12} \tag{8.2}$$

relation in which $\mathbf{P}^T$ is the transpose of matrix $\mathbf{P}$.

In case of an automorphism the relation 8.2 becomes: [4]

$$\mathbf{A}(G) = \mathbf{P}^T\mathbf{A}(G)\mathbf{P} \tag{8.3}$$

Figure 8.1 illustrates the above notions. It can be seen that a permutation P in a two-row notation is easily written in its matrix form. In this Figure, P_{113} leads to an isomorphism (cf. eq 8.2) while P_{12} provides an automorphism (cf. eq 8.3). Furthermore, condition 1 is satisfied, as shown in the multiplicative table and any $\mathbf{P}$ matrix admits an inverse (see above and condition 4); the permutation P_{11} leaves the graph unchanged (condition 3) and finally, the composition rule $\otimes$, which is just the matrix multiplication)

$G_{8.1}$
$P_{11} = \begin{pmatrix} 1 & 2 & 3 & 4 & 5 \\ 1 & 2 & 3 & 4 & 5 \end{pmatrix}$

$G_{8.2}$
$P_{12} = \begin{pmatrix} 1 & 2 & 3 & 4 & 5 \\ 3 & 1 & 2 & 4 & 5 \end{pmatrix}$

$G_{8.3}$
$P_{13} = \begin{pmatrix} 1 & 2 & 3 & 4 & 5 \\ 1 & 3 & 2 & 4 & 5 \end{pmatrix}$

$G_{8.4}$
$P_{14} = \begin{pmatrix} 1 & 2 & 3 & 4 & 5 \\ 2 & 1 & 3 & 4 & 5 \end{pmatrix}$

$G_{8.5}$
$P_{15} = \begin{pmatrix} 1 & 2 & 3 & 4 & 5 \\ 3 & 2 & 1 & 4 & 5 \end{pmatrix}$

$G_{8.6}$
$P_{16} = \begin{pmatrix} 1 & 2 & 3 & 4 & 5 \\ 2 & 3 & 1 & 4 & 5 \end{pmatrix}$

$G_{8.13}$
$P_{113} = \begin{pmatrix} 1 & 2 & 3 & 4 & 5 \\ 1 & 4 & 2 & 3 & 5 \end{pmatrix}$

$G_{8.7}$
$P_{17} = \begin{pmatrix} 1 & 2 & 3 & 4 & 5 \\ 1 & 2 & 3 & 5 & 4 \end{pmatrix}$

$G_{8.8}$
$P_{18} = \begin{pmatrix} 1 & 2 & 3 & 4 & 5 \\ 3 & 1 & 2 & 5 & 4 \end{pmatrix}$

$G_{8.9}$
$P_{19} = \begin{pmatrix} 1 & 2 & 3 & 4 & 5 \\ 1 & 3 & 2 & 5 & 4 \end{pmatrix}$

$G_{8.10}$
$P_{110} = \begin{pmatrix} 1 & 2 & 3 & 4 & 5 \\ 2 & 1 & 3 & 5 & 4 \end{pmatrix}$

$G_{8.11}$
$P_{111} = \begin{pmatrix} 1 & 2 & 3 & 4 & 5 \\ 3 & 2 & 1 & 5 & 4 \end{pmatrix}$

$G_{8.12}$
$P_{112} = \begin{pmatrix} 1 & 2 & 3 & 4 & 5 \\ 2 & 3 & 1 & 5 & 4 \end{pmatrix}$

$A(G_{8.1} - G_{8.12})$:
$$\begin{pmatrix} 0 & 0 & 0 & | & 1 & 1 \\ 0 & 0 & 0 & | & 1 & 1 \\ 0 & 0 & 0 & | & 1 & 1 \\ \hline 1 & 1 & 1 & | & 0 & 0 \\ 1 & 1 & 1 & | & 0 & 0 \end{pmatrix}$$

$A(G_{8.13})$:
$$\begin{pmatrix} 0 & 0 & 1 & 0 & 1 \\ 0 & 0 & 1 & 0 & 1 \\ 1 & 1 & 0 & 1 & 0 \\ 0 & 0 & 1 & 0 & 1 \\ 1 & 1 & 0 & 1 & 0 \end{pmatrix}$$

Isomorphism : $\quad \mathbf{P}_{113}^{T} \, \mathbf{A}\,(G_{8.1}) \, \mathbf{P}_{113} = \mathbf{A}\,(G_{8.13})$

$\mathbf{P}_{113}^{T}$ $\qquad$ $\mathbf{A}\,(G_{8.1})$ $\qquad$ $\mathbf{P}_{113}$ $\qquad$ $\mathbf{A}\,(G_{8.13})$

$$\begin{pmatrix} 1 & 0 & 0 & 0 & 0 \\ 0 & 0 & 1 & 0 & 0 \\ 0 & 0 & 0 & 1 & 0 \\ 0 & 1 & 0 & 0 & 0 \\ 0 & 0 & 0 & 0 & 1 \end{pmatrix} \times \begin{pmatrix} 0 & 0 & 0 & 1 & 1 \\ 0 & 0 & 0 & 1 & 1 \\ 0 & 0 & 0 & 1 & 1 \\ 1 & 1 & 1 & 0 & 0 \\ 1 & 1 & 1 & 0 & 0 \end{pmatrix} \times \begin{pmatrix} 1 & 0 & 0 & 0 & 0 \\ 0 & 0 & 0 & 1 & 0 \\ 0 & 1 & 0 & 0 & 0 \\ 0 & 0 & 1 & 0 & 0 \\ 0 & 0 & 0 & 0 & 1 \end{pmatrix} = \begin{pmatrix} 0 & 0 & 1 & 0 & 1 \\ 0 & 0 & 1 & 0 & 1 \\ 1 & 1 & 0 & 1 & 0 \\ 0 & 0 & 1 & 0 & 1 \\ 1 & 1 & 0 & 1 & 0 \end{pmatrix}$$

Automorphism : $\mathbf{P}_{12}^{T} \, \mathbf{A}\,(G_{8.1}) \, \mathbf{P}_{12} = \mathbf{A}(G_{8.2}) = \mathbf{A}(G_{8.1})$

$\mathbf{P}_{12}^{T}$ $\qquad$ $\mathbf{A}(G_{8.1})$ $\qquad$ $\mathbf{P}_{12}$ $\qquad$ $\mathbf{A}(G_{8.2}) = \mathbf{A}(G_{8.1})$

$$\begin{pmatrix} 0 & 1 & 0 & 0 & 0 \\ 0 & 0 & 1 & 0 & 0 \\ 1 & 0 & 0 & 0 & 0 \\ 0 & 0 & 0 & 1 & 0 \\ 0 & 0 & 0 & 0 & 1 \end{pmatrix} \times \begin{pmatrix} 0 & 0 & 0 & 1 & 1 \\ 0 & 0 & 0 & 1 & 1 \\ 0 & 0 & 0 & 1 & 1 \\ 1 & 1 & 1 & 0 & 0 \\ 1 & 1 & 1 & 0 & 0 \end{pmatrix} \times \begin{pmatrix} 0 & 0 & 1 & 0 & 0 \\ 1 & 0 & 0 & 0 & 0 \\ 0 & 1 & 0 & 0 & 0 \\ 0 & 0 & 0 & 1 & 0 \\ 0 & 0 & 0 & 0 & 1 \end{pmatrix} = \begin{pmatrix} 0 & 0 & 0 & 1 & 1 \\ 0 & 0 & 0 & 1 & 1 \\ 0 & 0 & 0 & 1 & 1 \\ 1 & 1 & 1 & 0 & 0 \\ 1 & 1 & 1 & 0 & 0 \end{pmatrix}$$

Matrix Multiplication Table of $Aut(G_{8.1})$

	p_{11}	p_{12}	p_{13}	p_{14}	p_{15}	p_{16}	p_{17}	p_{18}	p_{19}	p_{110}	p_{111}	p_{112}
p_{11}	p_{11}	p_{12}	p_{13}	p_{14}	p_{15}	p_{16}	p_{17}	p_{18}	p_{19}	p_{110}	p_{111}	p_{112}
p_{12}	p_{12}	p_{16}	p_{14}	p_{15}	p_{13}	p_{11}	p_{18}	p_{112}	p_{110}	p_{111}	p_{19}	p_{17}
p_{13}	p_{13}	p_{15}	p_{11}	p_{16}	p_{12}	p_{14}	p_{19}	p_{111}	p_7	p_{112}	p_{18}	p_{110}
p_{14}	p_{14}	p_{13}	p_{12}	p_{11}	p_{16}	p_{15}	p_{110}	p_{19}	p_{18}	p_{17}	p_{112}	p_{111}
p_{15}	p_{15}	p_{14}	p_{16}	p_{12}	p_{11}	p_{13}	p_{111}	p_{110}	p_{112}	p_{18}	p_{17}	p_{19}
p_{16}	p_{16}	p_{11}	p_{15}	p_{13}	p_{14}	p_{12}	p_{112}	p_{17}	p_{111}	p_{19}	p_{110}	p_{18}
p_{17}	p_{17}	p_{18}	p_{19}	p_{110}	p_{111}	p_{112}	p_{11}	p_{12}	p_{13}	p_{14}	p_{15}	p_{16}
p_{18}	p_{18}	p_{112}	p_{110}	p_{111}	p_{19}	p_{17}	p_{12}	p_{16}	p_{14}	p_{15}	p_{13}	p_{11}
p_{19}	p_{19}	p_{111}	p_{17}	p_{112}	p_{18}	p_{110}	p_{13}	p_{15}	p_{11}	p_{16}	p_{12}	p_{14}
p_{110}	p_{110}	p_{19}	p_{18}	p_{17}	p_{112}	p_{111}	p_{14}	p_{13}	p_{12}	p_{11}	p_{16}	p_{15}
p_{111}	p_{111}	p_{110}	p_{112}	p_{18}	p_{17}	p_{19}	p_{15}	p_{14}	p_{16}	p_{12}	p_{11}	p_{13}
p_{112}	p_{112}	p_{17}	p_{111}	p_{19}	p_{110}	p_{18}	p_{16}	p_{11}	p_{15}	p_{13}	p_{14}	p_{12}

Figure 8.1. Isomorphic ($G_{8.1}$ and $G_{8.13}$) and automorphic ($G_{8.1}$ - $G_{8.12}$) graphs and matrix multiplication table, cf. eqs 8.1 - 8.3.

is associative (condition 2). Thus, $Aut(G_{8.1})$, with its 3!2!=12 automorphic permutations, is a group.

Thus, these permutations lead either to isomorphic or automorphic labeled graphs, $G(Lb)$. A graph having N vertices can be labeled in $N!$ ways, thus resulting $N!$ different $G(Lb_i)$; $i = 1,2,...N!$ but representing one and the same *abstract graph* (as proposed by Klin and Zefirov).[6] Among these $G(Lb_i)$, only the automorphic ones preserve the connectivity (and the adjacency matrix) in the original graph. Any graph possesses at least one automorphism, e.g. that induced by the permutation identity, P_{11}.

Given a graph $G=(V, E)$ and a group $Aut(G)$, two vertices, $i, j \in V$ are called *equivalent* if there is a group element, $aut(n_i) \in Aut(G)$, such that $j \; aut(n_i) \; i$ (i.e. an automorphic permutation that transforms one to the other - a permutation that is edge invariant). The set of all vertices j obeying the above *equivalence relation* (see also Sect. 8.4) is called the *orbit of vertex i*, V_{ni}. Synonyms are: *automorphic partition, class of equivalence*. Vertices belonging to the same equivalence class can not be differentiated by graph-theoretical parameters.[4]

Suppose $V_{n1}, V_{n2},...V_{nm}$ are the m disjoint *automorphic partitions* of the set of vertices, $V(G)$ (with $|V(G)| = N = n_1 + n_2 + ... + n_m$ vertices):

$$V = V_{n_1} \bigcup V_{n_2} \bigcup ...\bigcup V_{n_m} \tag{8.4}$$

$$V_{n_i} \bigcap V_{n_j} = \phi \tag{8.5}$$

The group of automorphisms, $Aut(G)$,

$$Aut(G) = aut(n_1) \times aut(n_2) \times...\times aut(n_m) \tag{8.6}$$

(with $aut(n_i)$ being a group element containing $n_i!$ permutations) is a subgroup (of $n_1!n_2!...n_m!$ permutations), of the complete permutation group, of $N!$ elements, $Per(G)$.[1, 4, 7]

The quotient set $V/Aut(G)$ is often called the *orbit space*.[8] It describes all symmetry properties of a graph.[4]

A search for $Aut(G)$ may provide a *canonical code*. A code, $Cd(G,Lb)$ of a labeled graph, $G(Lb)$, is a string derived from the graph by a set of rules. It is a description of $G(Lb)$ which allows the (labeled) graph reconstruction. Codes are useful in computer structure storage and retrieval procedures as well as in enumeration and generation of isomers.

Two codes may be compared and ordered (by either a lexicographical or a numerical relation): they may differ or may be identical, $Cd(G,Lb_1) = Cd(G,Lb_2)$, situation in which the corresponding labeling are equal: $Lb_1 = Lb_2$. It comes out that, if two vertex labelings are identical, $V(Lb_1) = V(Lb_2)$, the corresponding vertices are automorphic.

A rigorous search for $Cd_{can}(G,Lb)$, has to construct all $N!$ permutations, to generate and compare all corresponding codes, $Cd(G,Lb_i)$; $i = 1,2,...N!$. Finally, a *maximal*, $Cd_{Mcan}(G,Lb)$, (*or a minimal*, $Cd_{mcan}(G,Lb)$) *canonical code* is selected along with the automorphism partitions. The process of generating $Cd_{can}(G,Lb)$ by investigating automorphism permutations is called *canonical code generation by automorphism permutation, CCAP*.[9] The identification of *topological symmetry* allows reduction of the number of tests ($N!$) by avoiding the generation of non-automorphic permutations.[4, 10, 11]

Consider a vertex invariant, $In = In_1, In_2,...,In_N$, which assigns a value In_i to vertex i. Two vertices, i and j, showing $In_i = In_j$ belong to the same *atomic invariant class, AIC*. The process of vertex partitioning in *AIC* induced by a given *In* is called *graph invariant atom partitioning, GIAP*. The partitioning of vertices into m classes, with $n_1, n_2,...n_m$ vertices in each class, is taken as a basis in generating the canonical code. Note that *GIAP* may by different from the orbits of automorphism since no vertex invariant is known so far to always discriminate two non-equivalent vertices in any graph. The classes of vertices are *ordered* with some rules, vertices in the first class being labeled by 1, 2, ...n_1, vertices in the second class by $n_1 + 1$, $n_1 + 2$, ..., $n_1 + n_2$, and so on.

A reliable algorithm for canonical coding would obligatory include two steps:[9]

       (i) *GIAP:* computes a discriminant atom invariant and provides an *initial atom partitioning* along with a *GIAP labeling* ;

       (ii) *CCAP:* generates codes and identifies the *canonical code* (by exploring all permutations over the *GIAP* classes); from the *canonical labeling*, (i.e., those providing the canonical code) true *orbits of automorphism* are identified.

Thus, the *GIAP* results can be used as ground for both *canonical coding* and search for $Aut(G)$, as shown above. For some applications, such as the numbering of ^{13}C-NMR signals,[12] the knowledge of $Aut(G)$ is not necessary; only the automorphic partitions are quite sufficient. Other major chemical applications of topological symmetry are: (i) chemical documentation system; storage and retrieval of chemical compounds in structure databases and (ii) computer generation of chemical structures, involved in molecular and synthesis design as well as in structure elucidation search.

Several procedures for canonical coding (or only *GIAP* procedures) were developed. [4, 5, 13-38] Among these, the Morgan algorithm[13] was the first and the best known, in the original form (*EC* -Extended *Connectivity* algorithm – used by the *CAS* in the chemical registry system) or as its extensions (*SEMA* - *S*tereochemically *E*xtended *M*organ *A*lgorithm[17, 39]). Balaban et al.[24] have proposed a variant, which provides automorphic partitions by hierarchic ordering (and numbering) of vertices (*HOC* - *H*ierarchically *O*rdered extended *C*onnectivities). The *HOC* algorithm also considers the stereochemical information[24] and is followed by a *CCAP* procedure. The ordering provided by *HOC* for the carbon atoms in some polycyclic aromatic hydrocarbons was shown to parallel the experimental 1H-NMR chemical shifts of their attached hydrogen atoms.[40] Balasubramanian developed algorithms for generating the equivalence classes in edge-weighted graphs[1, 7] as well as in 3D-molecular structures[1] and proposed applications in NMR and ESR spectroscopy.[41, 42] Among the more recent *GIAP* procedures, the *MOLORD* (*MOL*ecular *ORD*ering) performed by Diudea et al. [43] is presented.

8.2. TOPOLOGICAL SYMMETRY BY *MOLORD* ALGORITHM

The *MOLORD* algorithm is built on the ground of iterative line graphs, L_n,[43-45] that will be discussed before the algorithm.

8.2.1. Line Graphs

The points of the *line graph*, $L(G)$, represent lines of G and two points of $L(G)$ are adjacent if the corresponding lines of G are incident to a common point.[46] By repeating this procedure n times, the *iterative line graph*, L_n ; $n = 0, 1, 2, \ldots$ (with $n = 0$ for the original graph, G) can be obtained. Figure 8.2 illustrates the line graphs L_n for $G_{8.14}$ (2-Methylbutane); $n = 0 - 3$.

The number of vertices, N_{n+1} and edges Q_{n+1} in L_{n+1} is given by relations:[43-45]

$$N_{n+1} = Q_n \tag{8.7}$$

$$Q_{n+1} = -Q_n + (1/2)\sum_{i \in L_n}(k_i)^2 \tag{8.8}$$

$$Q_{n+1} = \sum_{i \in L_n}\binom{k_i}{2} = \sum_{i \in L_n}k_i(k_i-1)/2 = B_n \tag{8.9}$$

where k_i is the vertex degree and B_n - Bertz' s branching index,[45] which is the exact number of edges in the L_{n+1} line graph.

In regular graph (i.e. graphs in which all vertices have the same degree), the number of edges Q_{n+1} can be calculated by a recursive relation, derived from eq 8.8 or eq 8.9 by substituting the value for the vertex degree (see also[45]) :

$$k_n = 2Q_n / N_n = 2Q_n/Q_{n-1} \tag{8.10}$$
$$Q_{n+1} = -Q_n + 2Q_n^2 / Q_{n-1} \tag{8.11}$$

The number of edges in L_{n+1} can also be calculated by:

$$Qn+1 = (1/2)\ kn+1\ Nn+1 \tag{8.12}$$

Since in regular graphs:

$$kn+1 = 2\ (kn\ -1) \tag{8.13}$$

and taking into account eq 8.7, eq 8.12 becomes

$$Q_{n+1} = Q_n\ (k_n - 1) \tag{8.14}$$

From relations (8.13) and (8.14) k_n and Q_n can be expressed in terms of the starting parameters, k_0 and Q_0 (i.e. the degree and number of edges in the initial graph, L_0).

$$k_n = 1 + 2^n k_0 - \sum_{e=0}^{n} 2^e = 2^n k_0 - 2^{(n+1)} + 2 \tag{8.15}$$

$$Q_n = Q_0 \prod_{i=0}^{n-1} (k_i - 1) = Q_0 \prod_{i=0}^{n-1} (2^i k_0 - 2^{(i+1)} + 1) \tag{8.16}$$

In case of multigraphs, the (multiple) line graphs[47] will account for the bond orders.

8.2.2. MOLORD Algorithm

The MOLORD algorithm[43] characterizes vertices or subgraphs (of various size) of the initial graph by means of invariants derived from the topology of line graphs, L_0 (=G), L_1, ... L_m.

Some notations need to be introduced.

Vertices $i_n \in L_n$ (i.e. the current line graph) denote pairs of vertices i.e. lines in the lower-order line graph, L_{n-1}:

$$i_n = (j_{n-1}, k_{n-1}) \tag{8.17}$$

where the two points j and k are necessarily connected by an edge in L_{n-1}. One can write that $j_{n-1} \in i_n$ and $k_{n-1} \in i_n$. The relatedness of vertices (subgraphs) in process of iteration can be expressed by:

$$\delta(i_n, i_{n+1}) = \begin{vmatrix} 1 & if & (i_n \in i_{n+1}) \\ 0 & otherwise \end{vmatrix} \tag{8.18}$$

The definition can be easily extended for any two arbitrary ranks n and $m \geq n$, stating that $\delta(i_n, i_m) = 1$ only if the vertex i_n appears in at least one of the subgraphs defining vertex i_m. On going back to L_0, it can be seen that i_n denotes a subgraph consisting of n edges, in L_0.

The algorithm consists of the following four steps:

Step 1 : computes local, $I(i_n)$, and global, $GI(L_n)$ classical invariants on each L_n within the set of line graphs L_0 to L_m :

$$GI(L_n) = \sum_{i_n} I(i_n) \tag{8.19}$$

Step 2 : evaluates a partial local invariant $PI_m(i_n)$ of a vertex i_n, with respect to the mth order line graph, L_m :

$$PI_m(i_n) = \frac{GI(L_n)}{GI(L_m)} \sum_{i_m} I(i_m) \delta(i_n, i_m) \tag{8.20}$$

Here, $I(i_m)$ denotes a certain local invariant of vertex i_m, with respect to the topology of graph L_m. Furthermore, the partial invariant of i_n with respect to L_m is calculated by summing up all the local invariants $I(i_m)$ of those vertices in L_m which are *related* to i_n, according to the m-n successive line graphs, $L_n, \ldots L_m$. The ratio $GI(L_n) / GI(L_m)$ is used as a normalizing factor meant to ensure that the resulting PI values can be compared with each other, irrespective of the current L_m for which they are evaluated.

Step 3 : computes a synthetic local invariant of vertex i_n, in a series of successive line graph, $L_n, \ldots, L_m$:

$$SI_m(i_n) = \sum_{k=n}^{m} PI_k(i_n) f^{(n-k)} \tag{8.21}$$

Subscript m in $SI_m(i_n)$ indicates the last line graph (L_m) taken into account. The factor f can be used to give different weight to the contributions arising from line graphs of various ranks (usually 10 unless otherwise specified) . Note that in case $n = m$, the synthetic invariant $SI_m(i_n)$ is reduced to the classical invariant $I(i_n)$.

Step 4 : evaluates the final expression for the global synthetic index of a graph, L_n:

$$GSI_m(L_n) = \sum_{i_n} SI_m(i_n) \tag{8.22}$$

The *MOLORD* algorithm offers a *GIFP* (Graph Invariant Fragment Partitioning) and a (decreasing) ordered *GIFP* labeling according to a certain invariant (i.e. topological index). The spectrum of local values, $SI_m(i_n)$ (per fragments of various size) and global values, $GSI_m(L_n)$, can be used both for partitioning purposes and correlating studies. The algorithm is exemplified on 2-Methylbutane $G_{8.14}$, (Figure 8.2). The line graphs, L_n; $n = 0$-3 are given along with the corresponding **LDS** matrices within an output list with including some detailed calculations. The focused data are marked by gray and/or boldface letters/numbers (see below).

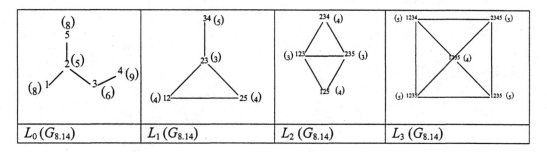

| $L_0(G_{8.14})$ | $L_1(G_{8.14})$ | $L_2(G_{8.14})$ | $L_3(G_{8.14})$ |

Figure 8.2. Line graphs L_n; $n = 1$-3, of 2-Methylbutane ($G_{8.14}$).

The corresponding *DS* are given in brackets.

An Example of *MOLORD* algorithm:

Graph $G_{8.14}$; matrices **LDS**; values derived for $I = X(LDS)$; $t_i = 1$.

Current rank of line graph: 0
LDS(L_0); (k_{i0}); (line graph evolution)
  5 vertices & 4 edges in line graph

  1 (1) :   8   5   14   9   (1)
  2 (3) :   5   22   9   0   (2)
  3 (2) :   6   14   16   0   (3)
  4 (1) :   9   6   5   16   (4)
   5 (1) :   8   5   14   9   (5)

Global operator value, $GI(L_0)$:   1.25903433

Fragments of 1 atoms after 0 line graph:

1- Fragment: 2   Atoms: 2   Bonds: I: 0.5746136
2- Fragment: 3   Atoms: 3   Bonds: I: 0.3256480
3- Fragment: 5   Atoms: 5   Bonds: I: 0.1242019
4- Fragment: 1   Atoms: 1   Bonds: I: 0.1242019
5- Fragment: 4   Atoms: 4   Bonds: I: 0.1103690

** Sum of fragmental indices (I):   1.259034329

Current rank of line graph: 1
LDS(L_1); (k_{i1});   (line graph evolution)
  4 vertices & 4 edges in line graph

1 (2) :   4   7   5 (1   2)
2 (3) :   3 13   0 (2   3)
3 (2) :   4   7   5 (2   5)
4 (1) :   5   3   8 (3   4)

 Global operator value, $GI(L_1)$:   2.13992226

Fragments of 1 atoms after 1st line graph:

1- Fragment: 2   Atoms: 2   Bonds: I: 0.6888219
2- Fragment: 3   Atoms: 3   Bonds: I: 0.3937350
3- Fragment: 1   Atoms: 1   Bonds: I: 0.1531101
4- Fragment: 5   Atoms: 5   Bonds: I: 0.1531101
5- Fragment: 4   Atoms: 4   Bonds: I: 0.1220641

** Sum of fragmental indices (I):   1.510841194
Note:

$$0.6888219 = 0.5746136 + \left\{ \left[2\left(\frac{4.0705}{2}\right)^{-1} + \left(\frac{3.1300}{3}\right)^{-1} \right] \frac{1.2590329}{2.1399223} \right\} 10^{-1}$$

$$SI_1(2) = PI_0(2)\,10^{0-0} \quad + \quad \{PI_1(2)\}\,10^{0-1}$$

Fragments of 2 atoms after 1st line graph:

1- Fragment: 2   Atoms: 2 3   Bonds: 2   I: 0.9584665
2- Fragment: 3   Atoms: 2 5   Bonds: 3   I: 0.4913401
3- Fragment: 1   Atoms: 1 2   Bonds: 1   I: 0.4913401
4- Fragment: 4   Atoms: 3 4   Bonds: 4   I: 0.1987755

** Sum of fragmental indices (I):   2.139922257

Current rank of line graph: 2
LDS(L_2); (k_{i2});   (line graph evolution)
  4 vertices & 5 edges in line graph

  1 (3) :   3 11 0 (1 2 3)
  2 (2) :   4 6 4 (1 2 5)
  3 (3) :   3 11 0 (2 3 5)
  4 (2) :   4 6 4 (2 3 4)

Global operator value, $GI(L_2)$:   2.91438507

Fragments of 1 atoms after 2nd line graphs:

  1- Fragment: 2   Atoms: 2   Bonds: I: 0.7014123
  2- Fragment: 3   Atoms: 3   Bonds: I: 0.4041974
  3- Fragment: 5   Atoms: 5   Bonds: I: 0.1594053
  4- Fragment: 1   Atoms: 1   Bonds: I: 0.1594053
  5- Fragment: 4   Atoms: 4   Bonds: I: 0.1241920

** Sum of fragmental indices (I):   1.548612224

Fragments of 2 atoms after 2nd line graphs:

  1- Fragment: 2   Atoms: 2 3   Bonds: 2   I: 1.1362917
  2- Fragment: 1   Atoms: 1 2   Bonds: 1   I: 0.5983362
  3- Fragment: 3   Atoms: 2 5   Bonds: 3   I: 0.5983362
  4- Fragment: 4   Atoms: 3 4   Bonds: 4   I: 0.2349425

** Sum of fragmental indices (I):   2.567906708

   Note:

$$0.5983362 = 0.4913401 + \left\{ \left[\left(\frac{3.1100}{3} \right)^{-1} + \left(\frac{4.0604}{2} \right)^{-1} \right] \frac{2.1399223}{2.91438507} \right\} 10^{-1}$$

$$SI_2(1,2) = PI_1(1,2)\, 10^{1-1} \quad + \quad \{PI_2(1,2)\} 10^{1-2}$$

Fragments of 3 atoms after 2nd line graph:

  1- Fragment: 3   Atoms: 2 3 5   Bonds: 2 3   I: 0.9646302

2- Fragment: 1   Atoms: 1 2 3   Bonds: 1   2   I: 0.9646302
3- Fragment: 4   Atoms: 2 3 4   Bonds: 2   4   I: 0.4925623
4- Fragment: 2   Atoms: 1 2 5   Bonds: 1   3   I: 0.4925623

** Sum of fragmental indices (I):   2.914385068

Current rank line graph: 3
**LDS(L_3); (k$_{i3}$);   (line graph evolution)
  5 vertices & 8 edges in line graph

  1 (3) :   5 14 5 (1 2 3 5)
  2 (4) :   4 20 0 (1 2 3 5)
  3 (3) :   5 14 5 (1 2 3 4)
  4 (3) :   5 14 5 (1 2 3 5)
  5 (3) :   5 14 5 (2 3 4 5)

Global operator value, $GI(L_3)$:   3.28678422

Fragments of 1 atoms after 3rd line graph:

  1- Fragment: 2   Atoms: 2   Bonds: I: 0.7026713
  2- Fragment: 3   Atoms: 3   Bonds: I: 0.4054564
  3- Fragment: 1   Atoms: 1   Bonds: I: 0.1604408
  4- Fragment: 5   Atoms: 5   Bonds: I: 0.1604408
  5- Fragment: 4   Atoms: 4   Bonds: I: 0.1246391

** Sum of fragmental indices (I):   1.553648361

Fragments of 2 atoms after 3rd line graph:

  1- Fragment: 2   Atoms: 2 3   Bonds: 2   I: 1.1576909
  2- Fragment: 3   Atoms: 2 5   Bonds: 3   I: 0.6159358
  3- Fragment: 1   Atoms: 1 2   Bonds: 1   I: 0.6159358
  4- Fragment: 4   Atoms: 3 4   Bonds: 4   I: 0.2425418

** Sum of fragmental indices (I):   2.632104376

Fragments of 3 atoms after 3rd line graph:

  1- Fragment: 1   Atoms: 1 2 3   Bonds: 1 2   I: 1.2043210
  2- Fragment: 3   Atoms: 2 3 5   Bonds: 2 3   I: 1.2043210
  3- Fragment: 2   Atoms: 1 2 5   Bonds: 1 3   I: 0.6805053
  4- Fragment: 4   Atoms: 2 3 4   Bonds: 2 4   I: 0.5960578

** Sum of fragmental indices (I):   3.685205051

Fragments of 4 atoms after 3rd line graph:

1- Fragment: 1   Atoms: 1 2 3 5   Bonds: 1 2 3   I: 2.1195826
2- Fragment: 3   Atoms: 2 3 4 5   Bonds: 2 3 4   I: 0.5836008
3- Fragment: 2   Atoms: 1 2 3 4   Bonds: 1 2 4   I: 0.5836008

** Sum of fragmental indices (I):   3.286784221

Note:

$$2.1195826 = \left\{ \left[2\left(\frac{5.1405}{3}\right)^{-1} + \left(\frac{4.2000}{4}\right)^{-1} \right] \frac{3.28678422}{3.28678422} \right\} 10^{3-3}$$

$$SI_3(1,2,3,5) = \left\{ \left[\sum_{(1,2,3,5)} \delta((1,2,3,5),(1,2,3,5))\, I(1,2,3,5) \right] \frac{GI(L_3)}{GI(L_3)} \right\} f^{(m-m)}$$

The index used in the above example is defined by:

$$I(i_n) = X(LDS)_{i_n} = \left(\frac{t_i}{k_i} \sum_{j=0}^{ecc_i} [\mathbf{LDS}]_{ij} 10^{-zj} \right)^{-1} \tag{8.23}$$

where t_i/k_i is a weighting factor (i.e. electronegativities per degree - in Figure 8.2, $t =$ 1), ecc_i is the eccentricity of vertex i; z is the maximal number of bites of an entry in the matrix **LDS**.

If in the first step of *MOLORD*, the values $I(i_n)$ are normalized by $max\ I(i_n) \in L_n$ and in step 2, the scaling factor $GI(L_n)\ /GI(L_m)$ is omitted, a variant called *MOLCEN* is obtained.[48] This algorithm provides centric ordering of vertices (see below), with values in the range [0 - 1]; value 1 is assigned for the central vertices.

8.3. INTRAMOLECULAR ORDERING

Under this topic, we include both the identification of *GIFP classes* and *fragment ordering*, by the following criteria: (i) of centrality; (ii) of centrocomplexity and (iii) lexicographic (see also Chap. Topological Indices).

8.3.1. Criteria of Centrality

The *center* of a graph is the set of vertices, $\{i\} \in V(G)$, which obey the relation:

$$ecc_i = r(G) \tag{8.24}$$

where

$$ecc_i = max\ D_{i,j \in V(G)} \tag{8.25}$$
$$r(G) = min\ ecc_{i \in V(G)} = min\ max\ D_{i,j \in V(G)} \tag{8.26}$$

ecc_i being the eccentricity of the vertex i while $r(G)$ the *radius* of the graph. In other words, the central vertices have their eccentricity equal to the radius of the graph, which, in turn, is the *minimal maximal* distance in the graph. The *diameter*, $d(G)$, is, in the opposite, the *maximal eccentricity* in the graph:

$$d(G) = max\ ecc_{i \in V(G)} = max\ max\ D_{i,j \in V(G)} \tag{8.27}$$

Any tree has either a center or a dicenter.[46, 49, 50] Note that the requirement (8.24) is only necessary but not sufficient. The finding of the graph center, in cycle-containing structures, is not always a simple task. In this respect, Bonchev *et al.*[49] have proposed the distance-based criteria, *1D-3D*, as follows:

1D: minimum vertex eccentricity: min ecc_i

2D: minimum vertex distance sum: min $\sum_j D_{ij}$

3D: minimum number of occurrence of the largest distance: min $[LC]_{ij,max}$
(see Chap. Topological Matrices, Sect. Layer Matrices). *If the largest distance occurs for several vertices, the next largest distance* (i.e. $[LC]_{ij,max-1}$) *is considered, and so on.*

Criteria *1D-3D* are applied hierarchically. The algorithm which implements these criteria is called *IVEC*.[50] It finds the center of a graph and its orbits of *GIFP*, which are ordered from the center to the *periphery* (i.e. the vertices having *max ecc_i*). The centrality ordering given by *IVEC* is illustrated on a set of polycyclic graphs,[50] included in Table 8.1. On the same set, the *MOLCEN* algorithm[48] (working by indices *C(LK)* and *X(LK)* - see Chap. Topological Indices) finds the same ordering, with only slight differences.

Table 8.1. *IVEC* and *MOLCEN* Ordering (According to the Values of Indices $C(LK)$ and $X(LK)$ Calculated on L_0 - L_2).

Graph		Vertices
	IVEC	(1), (2), (3), (4), (5), (6)
	C(LK)	(1), (2), (3), (4), (5), (6)
	X(LK)	(1), (4), (2), (3), (5), (6)
		Edges
	IVEC	(12), (14), (23), (15), (34), (45), (26)
	C(LK)	(14), (12), (15), (34), (23), (45), (26)
	X(LK)	(14), (12), (15), (34), (45), (23), (26)
		Vertices
	IVEC	(1:2), (3), (4;5), (6)
	C(LK)	(1:2), (3), (4;5), (6)
	X(LK)	(1:2), (3), (4;5), (6)
		Edges
	IVEC	(12), (13;23), (14;45), (36), (45)
	C(LK)	(12), (13;23), (14;45), (36), (45)
	X(LK)	(12), (13;23), (14;45), (36), (45)
		Vertices
	IVEC	(1), (2), (3;4), (5;6)
	C(LK)	(1), (2), (3;4), (5;6)
	X(LK)	(1), (2), (3;4), (5;6)
		Edges
	IVEC	(12), (13;14), (25;26), (34)
	C(LK)	(12), (13;14), (25;26), (34)
	X(LK)	(12), (13;14), (25;26), (34)
		Vertices
	IVEC	(1), (2), (3;4), (5), (6)
	C(LK)	(1), (2), (3;4), (5), (6)
	X(LK)	(1), (2), (3;4), (5), (6)
		Edges
	IVEC	(12), (13;14), (15), (23;24), (35;45), (26)
	C(LK)	(12), (13;14), (15), (23;24), (35;45), (26)
	X(LK)	(12), (13;14), (15), (23;24), (35;45), (26)
		Vertices
	IVEC	(1;2), (3), (4), (5), (6)
	C(LK)	(1;2), (3), (4), (5), (6)
	X(LK)	(1;2), (3), (4), (5), (6)
		Edges
	IVEC	(12), (13;23), (14;24), (34), (15;25), (36)
	C(LK)	(12), (13;23), (14;24), (34), (15;25), (36)
	X(LK)	(12), (13;23), (14;24), (34), (15;25), (36)
		Vertices

IVEC	(1;2;3), (4), (5), (6)	
C(LK)	(1;2;3), (4), (5), (6)	
X(LK)	(1;2;3), (4), (5), (6)	

Edges

IVEC	(12;13;23), (14;24;34), (15;25;35), (46)
C(LK)	(12;13;23), (14;24;34), (15;25;35), (46)
X(LK)	(12;13;23), (14;24;34), (15;25;35), (46)

Vertices

IVEC	(1), (2), (3), (4), (5), (6)
C(LK)	(1), (2), (3), (4), (5), (6)
X(LK)	(1), (4), (2), (3), (5), (6)

Edges

IVEC	(12), (13), (23), (15), (14), (24), (35), (26)
C(LK)	(12), (13), (23), (14), (24), (15), (26), (35)
X(LK)	(12), (13), (23), (14), (15), (24), (26), (35)

Vertices

IVEC	(1), (2), (3), (4), (5), (6)
C(LK)	(1), (2), (3), (4), (5), (6)
X(LK)	(1), (4), (2), (3), (5), (6)

Edges

IVEC	(12), (13), (14), (23), (15), (25), (34), (26), (46)
C(LK)	(12), (13), (23), (14), (15), (25), (34), (26), (46)
X(LK)	(12), (13), (23), (14), (15), (25), (26), (34), (46)

Vertices

IVEC	(1), (2;3), (4;5), (6)
C(LK)	(1), (2;3), (4;5), (6)
X(LK)	(1), (2;3), (4;5), (6)

Edges

IVEC	(12;13), (23), (14;15), (24;35), (26;36), (45)
C(LK)	(12;13), (23), (14;15), (24;35), (26;36), (45)
X(LK)	(12;13), (23), (14;15), (24;35), (26;36), (45)

Vertices

IVEC	(1;2), (3), (4), (5), (6)
C(LK)	(1;2), (3), (4), (5), (6)
X(LK)	(1;2), (3), (4), (5), (6)

Edges

IVEC	(12), (13;23), (14;24), (15;25), (34), (36), (56)
C(LK)	(12), (13;23), (14;24), (34), (15;25), (36), (56)
X(LK)	(12), (13;23), (14;24), (34), (15;25), (36), (56)

In layer matrices, particularly in **LDS**, the $1D$ criterion[49] is scanned by the column counter, j; the $2D$ criterion is included in the column $j=0$ (the distance sum being just the property collected by this matrix). The $3D$ criterion is somewhat nondecisive. It is known that there are graphs having pairs of vertices with the same *distance degree* *sequence, DDS*:[51-53] 17, 24, 29, 25, 26, 23, 9. Figure 8.3 illustrates such graphs, which are labeled in a canonical ordering given just by **LDS** matrix.[53]

These graphs show identical global sequence, *DDS*. Moreover, vertices labeled 15 and 16 show the same sequence DDS_i : 4, 4, 2, 4, 3, in both graphs. It is obvious that the two vertices can not be discriminated by the 1*D*-3*D* criteria. More powerful is the matrix **LDS** and index $C(LDS)_i$ which separate these vertices, both intra- and intermolecularly. Figure 8.3 shows matrices **LDS** along with the canonical-**LDS** and central ordering induced by $C(LDS)_i$. It can be seen that the central ordering reverses the canonical-**LDS** one, with a single inversion (vertex 15 before vertex 16 in $G_{8.15}$).

The matrix **LDS** degenerates very rarely in trees but there are cyclic regular graphs which show degenerate **LDS**.[53] When included in the frame of *MOLORD* algorithm, **LDS** succeeded in separating the *GIFP* classes for subgraphs even larger than one edge. Figure 8.4 exemplifies such a performance in case of cuneane ($G_{8.17}$). It illustrates the fact that the geometrical symmetry implies the topological symmetry; the reciprocal is, however, not always true.

The finding of the center of a graph is of interest in the chemical nomenclature, or in coding of chemical structures (see also[9, 50]) or also in correlating some physico-chemical properties (e.g. centric indices and octanic number).[54, 55]

8.3.2. Criteria of Centrocomplexity

If in a molecular graph , a "center of importance" is defined, the reminder substructures can be ordered with respect to this center. Such a criterion was called "centrocomplexity"[53] and it takes into account the chemical nature of vertices and edges in molecules.

8.3.2.1. Accounting for the Nature of Heteroatoms

Kier and Hall[56] have extended the validity of Randić index[57] χ (see chap. Topological Indices) to heteroatom-containing molecules. They introduced the δ_i valences in the construction of the analogue index χ^v:

$$\delta_i^v = Z_i^v - h_i \tag{8.28}$$

where Z_i^v is the number of valence electrons of atom i and h_i is the number of hydrogen atoms attached to atom i . For atoms belonging to the third period of Periodic Table, δ_i^v is calculated by:

$$\delta_i^v = \frac{Z_i^v - h_i}{Z_i - Z_i^v} \tag{8.29}$$

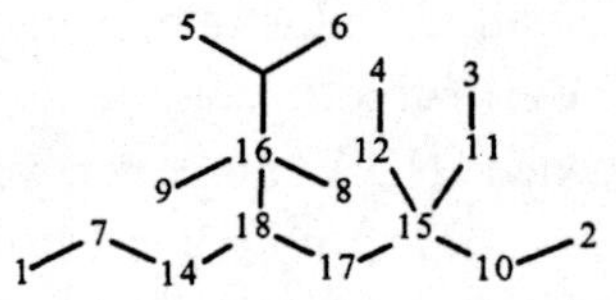

$G_{8.15}$

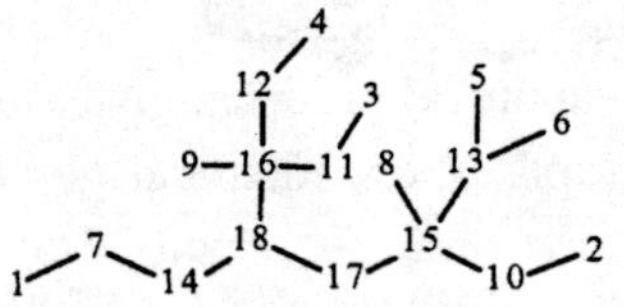

$G_{8.16}$

DDS: 17.24.29.25.26.23.9

Canonical **LDS** ($G_{8.15}$)

									Central	$C(LDS)_i \, 10^{-2}$
1	85	69	55	43	94	240	343	237	18	0.92893
2	79	63	49	171	201	104	260	239	17	0.98765
3	79	63	49	171	201	104	260	239	16	0.98765
4	79	63	49	171	201	104	260	239	15	0.98765
5	77	61	126	173	100	118	274	237	14	0.99034
6	77	61	126	173	100	118	274	237	13	0.99034
7	77	140	43	94	240	343	237	0	12	1.61963
8	69	49	169	254	118	274	237	0	11	1.68283
9	65	49	169	254	118	274	237	0	10	1.68283
10	65	128	171	201	104	260	239	0	9	1.70661
11	63	128	171	201	104	260	239	0	8	1.70661
12	63	128	171	201	104	260	239	0	7	1.70661
13	61	203	173	100	118	274	237	0	6	1.71299
14	55	112	179	240	343	237	0	0	5	2.83069
15	49	234	280	104	260	239	0	0	4	2.98095
16	49	234	254	118	274	237	0	0	3	2.96090
17	45	92	293	497	239	0	0	0	2	4.98333
18	43	149	309	428	237	0	0	0	1	5.03388
			$C(LDS)10^2$							36.48041

Canonical LDS ($G_{8.16}$)

									Central	$C(LDS)_i \, 10^{-2}$
1	85	69	55	43	94	240	347	233	18	0.93163
2	79	63	49	171	197	104	260	243	17	0.98305
3	79	63	49	171	179	118	274	233	16	0.98405
4	79	63	49	171	179	118	274	233	15	0.98405
5	77	61	126	173	122	104	260	243	14	0.99260
6	77	61	126	173	122	104	260	243	13	0.99260
7	77	140	43	94	240	347	233	0	12	1.62391
8	69	49	169	276	104	260	243	0	11	1.68925
9	65	49	169	258	118	274	233	0	10	1.68981
10	65	128	171	197	104	260	243	0	9	1.69970
11	63	128	171	179	118	274	233	0	8	1.70115
12	63	128	171	179	118	274	233	0	7	1.70115
13	61	203	173	122	104	260	243	0	6	1.71633
14	55	112	179	240	347	233	0	0	5	2.83731
15	49	234	276	104	260	243	0	0	4	2.97033
16	49	234	258	118	274	233	0	0	3	2.97154
17	45	92	293	493	243	0	0	0	2	4.97244
18	43	149	309	432	233	0	0	0	1	5.04486
			$C(LDS)10^2$							36.48577

Figure 8.3 .Canonical and central ordering (cf. **LDS** and $C(LDS)_i$, respectively) of $G_{8.15}$ and $G_{8.16}$

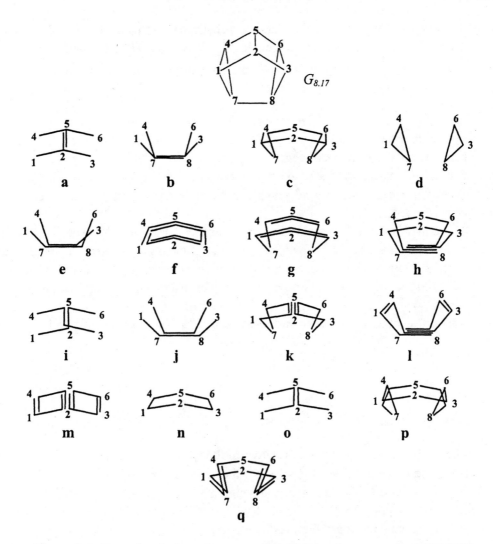

Figure 8.4. Central ordering of cuneane (cf. $SI_3(i_3)$ values - see the *MOLCEN* algorithm)

where Z_i is the atomic number of atom i. Analogue heteroatom accounting was made by Balaban.[58]

Diudea and Silaghi[59] have proposed group electronegativity valences, denoted EVG and defined by:

$$ESG_i = (ESA_i\, ESH^{hi})^{1/(1+hi)} \tag{8.30}$$

$$h_i = (8 - GA_i) - v_i \tag{8.31}$$

$$EVG_i = (ESG_i)^{1/(1+vi)} \tag{8.32}$$

where GA_i is the number of column in the Periodic Table for the atom A belonging to the vertex (i.e. group) i. ESA and ESH denote the Sanderson electronegativities for the atom

A and hydrogen, respectively. The number of hydrogen atoms attached to the group i is denoted by h_i while v_i stands for the degree of i. When $v_i > (8\text{-}GA_i)$, then $h_i = 0$. In case of multiple bonds, $v_i = \sum_j b_{ij}$, where b_{ij} is the conventional bond orders around the vertex i.

Note that these group electronegativities obey the electronegativity equalizing principle within the group i (see eq 8.30) and per molecule, each group is considered bonded to neighbors with electronegativity 1.[59]

The EVG_i values were used in the construction of the DS index (see Chap. Topological Indices) that showed good correlation with several physico-chemical and biological properties.[59]

A variant of EVG parameters was further developed.[60] The EC valence electronegativities are based on the idea of the modification of covalent radius of an atom by its hybridization state.[61] Such a modification is reflected in the electronegativity values corresponding to the considered state. The following scenario defined the EC parameters:

(i) - covalent radii relative to carbon atom (0.772 ANG) are calculated by eq. 8.33-8.35 :

$$rc_{ni} = rc_{1i} + \Delta rc_{ni} \tag{8.33}$$
$$rc_{1i} = r_{1i} / 0.772 \tag{8.34}$$
$$\Delta rc_{ni} = (r_{ni} - r_{1i}) / 7.72 \tag{8.35}$$

where : rc is the atomic radius relative to the carbon atom; n is the row and i is the column in the Periodic Table; Δrc stand for the "excess of relative radius".

(ii) - values EC, for the atoms belonging to the nth row of Periodic Table are calculated by dividing the group electronegativities ESG_i to the mean relative length, mlc, of the bonds around the considered vertex/group i :

$$EC_{ni} = (ESG_{ni} / mlc_{ni}) / EC_C \tag{8.36}$$
$$EC_C = 2.746 / 1.4996 \tag{8.37}$$
$$mlc_{ni} = ml_C \, rc_{ni} \tag{8.38}$$

EC values are listed in Table 8.2. Two Randić-type indices were constructed by using the EC values (see Chap. Topological Indices). They showed good correlation with some physico-chemical properties.[59, 60]

Table 8.2. *EC* Electronegativities.

-Br	1.2447	-NH$_2$	1.0644
-CBr$_3$	1.1266	-NO	1.4063
-CCl$_3$	1.1932	-NO$_2$	1.4861
-CF$_3$	1.3260	-O-	1.4634
-CH$_2$-	0.9622	-OCH$_3$	1.1248
-CH$_2$Br	1.0110	-OH	1.2325
-CH$_2$Cl	1.0305	-P(CH$_3$)$_2$	0.9351
-CH$_2$F	1.0674	-P<	0.8988
-CH$_2$I	0.9744	-PCH$_3$-	0.9314
-CH$_2$OH	1.0228	-PH-	0.9124
-CH$_2$SH	0.9804	-PH$_2$	0.9170
-CH$_3$	0.9575	-PHCH$_3$	0.93053
-CH<	0.9716	-S-	1.1064
-CH=CH$_2$	1.0381	-SCH$_3$	1.0073
-CH=O	1.1596	2PO	0.1222
-CHBr$_2$	1.0672	3P=O	1.3333
-CHCl$_2$	1.1089	=C=	1.1581
-CHF$_2$	1.1897	=CH-	1.0441
-CHI$_2$	0.9914	=CH$_2$	1.0891
-CI$_3$	1.0088	=N-	1.3147
-COOH	1.2220	=NH	1.2474
-Cl	13717	=O	1.6564
-C≡	1.1476	=P-	0.9658
-C≡N	1.2377	=S	1.2523
-F	1.6514	>C<	1.0000
-H	0.9175	>C=	1.0747
-I	1.0262	>C=O	1.2397
-N(CH$_3$)$_2$	1.0292	-NHCH$_3$	1.0379
-N<	1.2234	≡CH	1.2142
-NH-	1.1021	≡N	1.5288

8.3.2.2. X(L^{e}M) Descriptors

The descriptors $X(L^eM)$ are built on layer matrices: **LDS**, **L^{e}W**, etc. The chemical nature of atoms is considered by means of the parameter t_i (see Chap. Topological Indices).

Figure 8.5 offers an example of centrocomplexity ordering (and separating of automorphism groups) in which the *important* property is the valence/degree of vertices. The graph G$_{8.18}$ shows vertices 3 and 6, those are endospectral (i.e. have the same sequence of ^eW_i parameters- see Figure): these vertices can be distinguished by means of **L^{1}W** and index $X(L^1W)_i$, respectively.

(a)   (b) **L^{1}W** ($G_{8.18}$):

1	1 3 3 2 3 3 2 2 1
2	3 4 2 3 3 2 2 1 0
3*	2 5 5 3 2 2 1 0 0
4	2 5 6 4 2 1 0 0 0
5	3 5 4 5 3 0 0 0 0
6*	2 5 5 3 3 2 0 0 0
7	2 4 4 3 2 3 2 0 0
8	2 3 2 3 3 2 3 2 0
9	1 2 2 2 3 3 2 3 2
10	1 3 3 2 3 3 2 2 1
11	1 3 4 4 5 3 0 0 0

$G_{8.18}$

$^{e}W_{(3;6)}$: 2, 5, 9 ,21, 39, 88, 168, 370, 721, 1560

$X(L^1W)_3 = 2.553221$

$X(L^1W)_6 = 2.553320$

Figure 8.5. (a) Endospectral vertices (3 and 6 -marked with *) in the graph $G_{8.18}$. (b) Matrix L^1W and the index $X(L^1W)_i$, which separates these vertices.

Perception of heteroatom, by means of $X(LDS)$ index and *MOLORD* algorithm, is illustrated in Figure 8.6. and Tables 8.3 and 8.4, for a set of cuneanes. Values are listed in decreasing ordering of centrocomplexity.

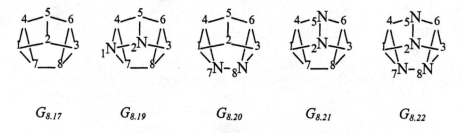

$G_{8.17}$ $G_{8.19}$ $G_{8.20}$ $G_{8.21}$ $G_{8.22}$

Figure 8.6 Cuneane and heterocuneanes.

8.3.2.3. $^{e}W_M$ *and* $^{e}E_M$ *Descriptors*

The descriptors $^{e}W_M$ represent walk degrees weighted by the property collected by the square matrix M.[62, 63] They can be calculated by the $^{e}W_M$ algorithm (see Chap. Topological Matrices). If the algorithm runs on the matrix C (of connectivities) then the resulting $^{e}W_C$ *naturally* take into account the multiple bond. If in the first step the EC values are setting as diagonal elements, the resulting descriptors are the *weighted electronegativities*, $^{e}E_M$, of rank e:[55]

Table 8.3. *MOLORD* Ordering of Cuneanes (Figure 8.6);
Values $SI_m(i_0)$ and $GSI_m(L_0)$; $f=10$; $I = X(LDS)$.

GRAPH		$G_{8.17}$		$G_{8.19}$		$G_{8.20}$		$G_{8.21}$
$G_{8.22}$								
Values $SI_0(i_0)$: vertices								
2 0.337702	2	0.353210	7	0.353210	2	0.353210	2	0.353210
5 0.337702	5	0.337702	8	0.353210	5	0.353210	5	0.353210
7 0.337702	7	0.337702	2	0.337702	7	0.337702	7	0.353210
8 0.337702	8	0.337702	5	0.337702	8	0.227702	8	0.353210
1 0.310649	1	0.324915	1	0.310649	1	0.310649	1	0.310649
3 0.310649	3	0.310649	3	0.310649	3	0.310649	3	0.310649
4 0.310649	4	0.310649	4	0.310649	4	0.310649	4	0.310649
6 0.310649	6	0.310649	6	0.310649	6	0.310649	6	0.310649
Values $GSI_0(L_0)$:								
2.593402	2.623177		2.624419		2.624419		2.655436	
Values $SI_1(i_0)$: vertices								
2 0.404711	2	0.422259	7	0.420017	2	0.422272	2	0.422311
5 0.404711	5	0.405217	8	0.420017	5	0.422272	5	0.422311
7 0.402482	7	0.402963	2	0.404749	7	0.402493	7	0.420029
8 0.402482	8	0.402481	5	0.404749	8	0.402493	8	0.420029
1 0.374424	1	0.390656	1	0.374943	1	0.374943	1	0.375461
3 0.374424	3	0.374931	3	0.374943	3	0.374943	3	0.375461
4 0.374424	4	0.374883	4	0.374943	4	0.374943	4	0.375461
6 0.374424	6	0.374426	6	0.374943	6	0.374943	6	0.375461
Values $GSI_1(L_0)$:								
3.112083	3.147812		3.149303		3.149303		3.186523	

Table 8.4. *MOLORD* Ordering of Cuneanes (Figure 8.6);
Values $Si_m(i_1)$ and $GSI_m(L_1)$; $f=10$; $I = X(LDS)$.

GRAPH	$G_{8.17}$		$G_{8.19}$		$G_{8.20}$		$G_{8.21}$		$G_{8.22}$
Values $SI_1 (i_1)$: edges									
(2, 5)	0.268012	(1, 2)	0.280276	(7, 8)	0.280261	(2, 5)	0.280320	(2, 5)	0.280320
(1, 2)	0.267970	(2, 5)	0.274097	(2, 5)	0.268012	(1, 2)	0.274054	(7, 8)	0.280261
(2, 3)	0.267970	(2, 3)	0.274054	(1, 2)	0.267970	(2, 3)	0.274054	(1, 2)	0.274054
(4, 5)	0.267970	(4, 5)	0.267970	(2, 3)	0.267970	(4, 5)	0.274054	(2, 3)	0.274054
(5, 6)	0.267970	(5, 6)	0.267970	(4, 5)	0.267970	(5, 6)	0.274054	(4, 5)	0.274054
(7, 8)	0.267956	(7, 8)	0.267956	(5, 6)	0.267970	(7, 8)	0.267956	(5, 6)	0.274054
(1, 7)	0.254624	(1, 7)	0.260405	(1, 7)	0.260405	(1, 7)	0.254624	(1, 7)	0.260405
(3, 8)	0.254624	(3, 8)	0.254624	(3, 8)	0.260405	(3, 8)	0.254624	(3, 8)	0.260405
(4, 7)	0.254624	(4, 7)	0.254624	(4, 7)	0.260405	(4, 7)	0.254624	(4, 7)	0.260405
(6, 8)	0.254624	(6, 8)	0.254624	(6, 8)	0.260405	(6, 8)	0.254624	(6, 8)	0.260405
(1, 4)	0.242557	(1, 4)	0.248064	(1, 4)	0.242557	(1, 4)	0.242557	(1, 4)	0.242557
(3, 6)	0.242557	(3, 6)	0.242557	(3, 6)	0.242557	(3, 6)	0.242557	(3, 6)	0.242557
Values $GSI_1 (L_1)$:									
	3.111455		3.147218		3.146885	3.146099			3.183529
Values $SI_2 (i_2)$: edges									
(2, 5)	0.320767	(1, 2)	0.334647	(7, 8)	0.334616	(2, 5)	0.334692	(2, 5)	0.334693
(1, 2)	0.320733	(2, 5)	0.327854	(1, 2)	0.320936	(1, 2)	0.327828	(7, 8)	0.334628
(2, 3)	0.320733	(2, 3)	0.327824	(2, 3)	0.320936	(2, 3)	0.327828	(1, 2)	0.328032
(4, 5)	0.320733	(4, 5)	0.321131	(4, 5)	0.320936	(4, 5)	0.327828	(2, 3)	0.328032
(5, 6)	0.320733	(5, 6)	0.320937	(5, 6)	0.320936	(5, 6)	0.327828	(4, 5)	0.328032
(7, 8)	0.320704	(7, 8)	0.320907	(2, 5)	0.320771	(7, 8)	0.320717	(5, 6)	0.328032
(1, 7)	0.305766	(1, 7)	0.312525	(1, 7)	0.312524	(1, 7)	0.305978	(1, 7)	0.312737
(3, 8)	0.305766	(4, 7)	0.306145	(3, 8)	0.312524	(3, 8)	0.305978	(3, 8)	0.312737
(4, 7)	0.305766	(3, 8)	0.305970	(4, 7)	0.312524	(4, 7)	0.305978	(4, 7)	0.312737
(6, 8)	0.305766	(6, 8)	0.305771	(6, 8)	0.312524	(6, 8)	0.305978	(6, 8)	0.312737
(1, 4)	0.293140	(1, 4)	0.299612	(1, 4)	0.293518	(1, 4)	0.293543	(1, 4)	0.293918
(3, 6)	0.293140	(3, 6)	0.293340	(3, 6)	0.293518	(3, 6)	0.293543	(3, 6)	0.293918
Values $GSI_2 (L_1)$:									
	3.733746		3.776662		3.776261	3.777719			3.820234

$$\mathbf{M} + {}^e\mathbf{E} = {}^e\mathbf{E}_\mathbf{M} \tag{8.39}$$

$$[{}^{e+1}\mathbf{E}_\mathbf{M}]_{ii} = \left[\prod_j ([{}^e\mathbf{E}_\mathbf{M}]_{jj})^{[\mathbf{M}]_{ij}} \right]^{1/\sum_j [\mathbf{M}]_{ij}} \; ; \; [{}^1\mathbf{E}_\mathbf{M}]_{jj} = EC \quad values^{60} \tag{8.40}$$

$$[{}^e\mathbf{E}_\mathbf{M}]_{ij} = [{}^e\mathbf{E}_\mathbf{M}]_{ij} = [\mathbf{M}]_{ij} \tag{8.41}$$

where $\mathbf{M}$ is the matrix used for *weighting* ${}^e\mathbf{E}_\mathbf{M}$, and ${}^e\mathbf{E}$ is the diagonal matrix of atomic electronegativities. Summing the two matrices results in the matrix ${}^e\mathbf{E}_\mathbf{M}$ whose elements are defined by eqs 8.40 and 8.41. Finally, $[{}^e\mathbf{E}_\mathbf{M}]_{ii}$ is assigned to ${}^e E_{M,i}$ (see above). Note that relation (8.40) is in agreement with the equalizing principle of atomic electronegativities .

Descriptors $^eE_{M,i}$ can be used as independent parameters or in association with ^eW_M parameters, to give the parameters $^eW_{ME,i}$:

$$^eW_{ME,i} = {}^eW_{M,i}\,{}^eE_{M,i} \tag{8.42}$$

The $^eW_{ME,i}$ descriptors allow the perception of both heteroatom and multiple bond in graphs. Figure 8.7 illustrates such descriptors for the graph $G_{8.14}$.

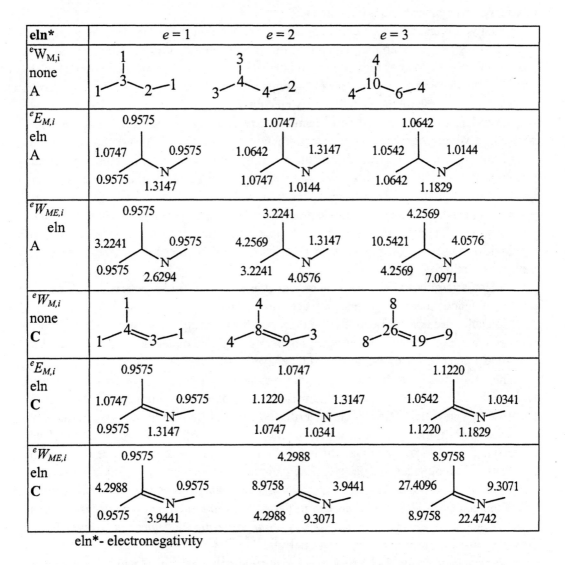

eln*- electronegativity

Figure 8.7. Heteroatom and multiple bond perception in $G_{8.14}$

8.4. MOLECULAR SIMILARITY

"Similarity is one of the most instantly recognizable and universally experienced abstractions known to humankind".[64]

Because of its fundamental role in a large variety of situations and fields, the similarity concept has strongly attracted the interest of scientific world. It is reflected in the occurrence of several English synonyms: relatedness, equivalence, proximity, closeness, resemblance, isomorphism, etc.

Usually, things, facts or concepts are classified (i.e. partitioned) into groups or categories according to simple perceptions or more elaborated criteria. Members of such groups will possess one or more common attributes. Similarity is always with respect to some particular characterization of groups. If the similarity is well behaved mathematically these members will satisfy an equivalence relationship (see below).

Several *levels* of similarity in chemistry are actually recognized:[65] (1) *Chemical similarity*, which compare and group chemical systems with respect to various macroscopic properties such as melting point, refraction index, chromatographic retention index, etc. (2) *Molecular similarity*, which involves the comparison and grouping of individual molecules according to their 2D and 3D structural information and property information, such as dipole moments and charge density. (3) *Intramolecular similarity*, which compare and group intramolecular entities, such as molecular orbitals or topological fragments (see Section 8.1). This Section is focused on the concept of molecular similarity.

Molecular similarity, like molecular branching, is an intuitive notion.[64, 66] A unique and unambiguous measure of similarity does not exist.[65]

Molecules are nonrigid entities that preserve their identity under small deformations, such as vibrations or rotations at some temperature. Thus, molecules can be viewed as topological objects,[67] mathematically well behaved.

The descriptions of molecules used in *molecular similarity analysis* are named *molecular descriptions*. A simple enumeration of atoms, or a fragment location, or an electrostatic potential surface characterization of the molecule can be termed as molecular description.[65] In many cases, the molecular description is a *vector of numbers*, quantifying some local, or global attributes such as the presence or absence of a certain fragment, a topological index, etc. Each element of vector is called a *molecular descriptor*. All molecular descriptions induce a *partitioning into equivalence classes* on a set of molecules. It is now appropriate to define concepts such as *equivalence relation, equivalence class, mapping, matching, partial ordering* and *proximity*, which are frequently used in molecular similarity analysis.

Equivalence relation. Let S be a set of molecular structures and R a binary relation on S relating pairs of its elements. If $x, y \in S$ are thus related, xRy will be written. The relation R is an *equivalence relation* if some properties are satisfied:

1. xRx, for all $x \in S$ (reflexivity)
2.   If xRy, then yRx (symmetry)

3. If xRy and yRz, then xRz (transitivity)

The set of all elements $y \in S$, such that xRy, represent the *equivalence class of x*. By imposing an equivalence relation R on a set S results in partitioning S into disjoint subsets, each subset being an equivalence class under R. This set of subsets is denoted by S/R (i.e. the *quotient set, S modulo R*).[8]

Let f be a function *mapping* the elements of S onto the elements of any other set Y. That is, for any $x \in S$, f assigns a corresponding value $y = f(x)$ in Y. This correspondence can be written as $f: S \rightarrow Y$. If Y is the set of descriptions, the mapping function associates a molecular description with each molecule in S. Those molecules in S are equivalent which are mapped to the same molecular description. Such a function f may be a labeling (or a coding) or simply a measuring process. It can be shown that various molecular descriptions associated with their algebraic representation form a group.[8, 67]

A *matching* can be achieved by overlapping two molecules. An overlapping can indicate the common features shared by two molecules or by two molecular descriptions.

A *partial ordering* refers to some *local ordering* induced by *local covering* (i.e. substructure matching) within the molecules belonging to the set S. Such an ordering can be illustrated by a Hasse diagram.[68]

Mathematically, the *ordering relation* requires the antisymmetry property (2'): If xRy and yRx then x = y, instead of the symmetry property (2) (see above). Randić[69] reported a partial ordering of alkane isomers with respect to the path numbers p_2 and p_3. Other graph-theoretical descriptors, such as topological indices, sequences of descriptors, etc., may be used in the characterizing and subsequently partial ordering and clustering of molecular structures. (see Sections 8.3 and 8.5). Molecules may also be ranked with respect to some experimental property. Compounds closely positioned in a derived ordering are expected to have close (i.e. similar) properties.

Proximity is basically expressed by two categories: *similarity* and *dissimilarity*. *Similarity* expresses the relatedness of two molecules, with a large number if their molecular descriptions are closely related and with a number going to zero in case they are unrelated.[65] The ratio of the count of matched atoms and bonds to the corresponding count for the whole molecule, multiplied by the analogous ratio for a comparing molecule has been proposed[70] as a similarity measure between two molecules. Such measures have the correlation property (*zero* for no correlation and *one* for full correlation).

Dissimilarity expresses the relatedness of two molecules, with a number close to zero when their molecular descriptions are closely related and with a large number if they are unrelated. For example, the number of atoms and bonds that cannot be matched up in overlapping two molecules may be a measure of dissimilarity between two molecules. This particular dissimilarity measure[71] constitutes a metric (see below) and is also referred to as *chemical distance*.[38, 53, 72-75]

Similarity and dissimilarit are both included in the more general term *proximity*. Four main types of proximity coefficients have been reported.

Distance coefficients usually assume a Minkowski metric within an *m*-dimensional space:[53, 69, 71, 76]

$$D(x,y) = \left[\sum_{i=1}^{m} |x_i - y_i|^z \right]^{1/z} \tag{8.43}$$

where $x = (x_1, x_2, ... x_m)$ and $y = (y_1, y_2, ... y_m)$ are the two structures of m points. Such coefficients are extensively used owing to their geometrical interpretation: when $z = 1$, the city-block distance (or the Manhattan distance D_M) is obtained and when $z = 2$, the Euclidean distance D_E results.[77, 78] Randić[69] evaluated the Euclidean distance on a set of monoterpenes by using path sequences as descriptors. Basak et. al.[76] performed a *PCA* (*Principal Component Analysis*) study on a set of 3692 molecules by using a pool of 90 topological indices and D_E as a measure of dissimilarity. For other examples, see Sect. 8.5.

Any proximity measure is a metric if it satisfies the triangle inequality: (1) $D(x,y) = 0$ for $x = y$; (2) $D(x,y) = D(y,x)$ and (3) $D(x,z) \leq D(x,y) + D(y,z)$.

Association coefficients are used when binary variables are involved.

Correlation coefficients measure the degree of statistical correlation between two molecules or their descriptions.

Probabilistic coefficients count the distribution of frequencies of occurrence of some common features in a dataset.[79]

A molecular description is in essence a mapping from a set S of structures onto a set Y of molecular descriptions. This mapping, together with some concepts of matching, partial ordering and proximity, defines a *molecular similarity space.*[65]

Similarity procedures thus produce a partitioning of sets of molecules into disjoint subsets or clusters based on their similarity. The procedures are classified as hierarchical or nonhierarchical depending on whether relationships can be established between the clusters.[80]

The clustering process is achieved in three stages: (1) the selection of appropriate variables for the molecular description, (2) the weighting of these variable and (3) the definition of the similarity measure. The choice among a variety of possibilities depends very much on the nature of the molecules under study but is, ultimately, a personal preference of each researcher.

Complementarity is another form of similarity, which needs the use of some *shape descriptors.*[67, 71, 81-86]

In case of flexible molecules, the similarity analysis requires a conversion from 2D to 3D molecular structures to which analogue considerations may be addressed. For such a purpose, a computer program, which takes into account the torsion angles and Euclidean distances, is needed. Some programs are actually available: CONCORD (University of Texas at Austin and TRIPOS Associates[87]), ChemModel (Chemical Design Ltd.), ALLADIN (Martin et al.[88]), etc.

8.5. INTERMOLECULAR ORDERING

Ordering of a set of molecules with respect to certain graph theoretical descriptors follows approximately the same criteria as the *intramolecular ordering*, with the difference that here global descriptors are used.

8.5.1. Criteria of Centrality

Let us consider a set of isomers. Their global sequence, *DDS*, can be ordered according to the *1D-3D* criteria,[49, 50] this time applied "intermolecularly".[89] Tables 8.5 and 8.6 lists the distance sequences and central ordering (C_{ord}), in increasing order, of heptanes and octanes, respectively. For comparison, the global value $C(L3DS)$ (calculated by using 3D distances in optimized geometry) was considered. It can be seen that a single inversion: 3EC5; 22M2C5 (in heptanes) and 3E3M5; 223M3C5 (in octanes), appeared between the two central orderings.

Table 8.5. Distance Degree Sequence (*DDS*) of Heptanes, Lexicographic (X_{ord}) and Central Ordering (C_{ord}, cf. 1D-3D Criteria), Compared with the Indices $X(L^1W)$, DM^1 and $C(L3DS)$, Respectively.

DDS						X_{ord}	$X(L^1W)$	DM^1	C_{ord}	$C(L3DS)$
6	5	4	3	2	1	C7	14.39506	13.42462	C7	0.58938
6	6	4	3	2	0	2MC6	14.61504	14.76562	2MC6	0.74061
6	6	5	3	1	0	3MC6	14.63682	15.08212	3MC6	0.78580
6	6	6	3	0	0	3EC5	14.65860	15.36658	24M2C5	0.99623
6	7	4	4	0	0	24M2C5	14.83680	16.36313	3EC5	1.03608
6	7	6	2	0	0	23M2C5	14.87640	16.94921	22M2C5	1.02187
6	8	4	3	0	0	22M2C5	15.05460	17.94975	23M2C5	1.06819
6	8	6	1	0	0	33M2C5	15.09420	18.48528	33M2C5	1.14982
6	9	6	0	0	0	223M3C4	15.31200	20.54701	223M3C4	1.34805

Table 8.6. Distance Degree Sequence (DDS) of Octanes, Lexicographic (X_{ord}) and Central Ordering (C_{ord}, cf. 1D-3D Criteria), Compared with the Indices $X(L^1W)$, DM^1 and $C(L3DS)$, Respectively.

DDS							X_{ord}	$X(L^1W)$	DM^1	C_{ord}	$C(L3DS)$
7	6	5	4	3	2	1	C8	16.83951	15.61028	C8	0.44479
7	7	5	4	3	2	0	2MC7	17.05950	17.02015	2MC7	0.56848
7	7	6	4	3	1	0	3MC7	17.08148	17.56044	3MC7	0.60339
7	7	6	5	2	1	0	4MC7	17.08346	17.56044	4MC7	0.63236
7	7	7	5	2	0	0	3EC6	17.10544	17.91494	25M2C6	0.71165
7	8	5	4	4	0	0	25M2C6	17.27968	18.60840	22M2C6	0.73761
7	8	6	5	2	0	0	24M2C6	17.30344	19.20822	3EC6	0.75937
7	8	7	4	2	0	0	23M2C6	17.32324	19.60890	24M2C6	0.76325
7	8	8	4	1	0	0	34M2C6	17.34502	20.00839	23M2C6	0.76779
7	8	8	5	0	0	0	3E2MC5	17.34700	20.10744	34M2C6	0,81452
7	9	5	4	3	0	0	22M2C6	17.49946	20.51561	33M2C6	0.82412
7	9	7	4	1	0	0	33M2C6	17.54302	21.42983	224M3C5	1.01891
7	9	8	4	0	0	0	234M3C5	17.56480	22.07279	3E2MC5	1.04532
7	9	9	3	0	0	0	3E3MC5	17.58460	22.13693	234M3C5	1.05845
7	1	5	6	0	0	0	224M3C5	17.72320	22.80578	3E3MC5	1.11623
7	1	8	3	0	0	0	223M3C5	17.78260	24.14856	223M3C5	1.08177
7	1	9	2	0	0	0	233M3C5	17.80240	24.49869	233M3C5	1.13486
7	1	9	0	0	0	0	2233M4C4	18.23800	29.75000	2233M4C	1.39893

8.5.2. Criteria of Centrocomplexity

8.5.2.1. X(L^{e}M) Descriptors

Descriptors of the type $X(L^eM)$ succeeded in separating pairs of *recalcitrant* isomers (i.e. which can not be discriminated by classical spectral parameters).

In simple cases, the ordering induced by the matrix $\mathbf{L^1W}$ and the corresponding index $X(L^1W)$ is sufficient. The ordering supplied by the above descriptors is identical to the lexicographic ordering of DDS (see also[90]) in the sets of heptanes and octanes (Tables 8.5 and 8.6.), or to that induced by the super-index[71] DM^1 (the same Tables).

There are graphs with pairs of vertices showing oscillating values of ^eW_i,(when e increases) . In such cases, higher elongation, e, is needed for discrimination . Figure 8.8 shows a pair of such graphs ($G_{8.23}$ and $G_{8.24}$) in which vertices: 3 and 6'; 6 and 3'; 12 and 12' are isospectral . Using layer matrices $\mathbf{L^2W}$ allows the discrimination of both the mentioned vertices and the two graphs.[53]

(a)

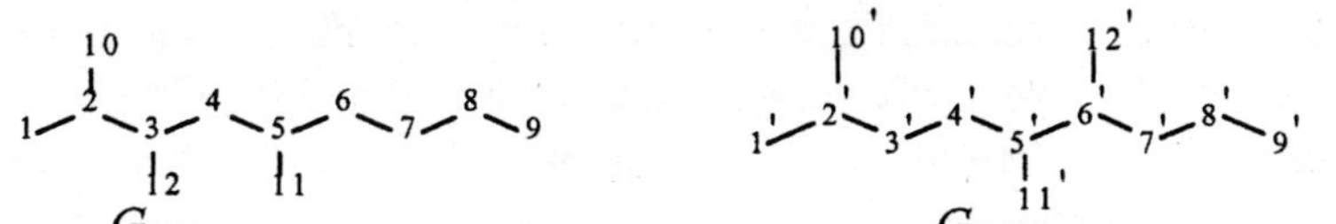

$G_{8.23}$ $G_{8.24}$

$^eW_i : 3; 6'$:   3, 6, 14, 29, 66, 136, 310, 633, 1449, 2937, 6747
   6; 3' :   2, 5, 9, 22, 40, 92, 180, 432, 816, 1941, 3717
   12; 12':   1, 3, 6, 14, 29, 66, 136, 310, 633, 1449, 2937
^eW :   11, 24, 49, 106, 222, 479, 1014, 2186, 4651, 10023, 21380

L^2W matrices:

	$L^2W(G_{8.23})$									**$L^2W(G_{8.24})$**							
1	3	5	9	9	5	8	4	3	2	3	4	8	5	6	9	8	3	2
2	5	12	9	5	8	4	3	2	0	4	11	5	6	9	8	3	2	0
3	6	14	11	8	4	3	2	0	0	5	9	12	9	8	3	2	0	0
4	6	11	16	10	3	2	0	0	0	5	11	13	14	3	2	0	0	0
5	5	14	10	11	8	0	0	0	0	6	14	13	7	8	0	0	0	0
6	5	9	12	8	8	6	0	0	0	6	14	11	7	4	6	0	0	0
7	4	8	7	9	6	8	6	0	0	5	9	11	8	5	4	6	0	0
8	3	6	5	5	9	6	8	6	0	3	7	6	9	8	5	4	6	0
9	2	3	4	5	5	9	6	8	6	2	3	5	6	9	8	5	8	6
10	3	5	9	9	5	8	4	3	2	3	4	8	5	6	9	8	3	2
11	3	5	11	10	11	8	0	0	0	3	6	11	13	7	8	0	0	0
12	3	6	11	11	8	4	3	2	0	3	6	11	11	7	4	6	0	0
$X(L^2W)$:	48.9915008066362410									48.9915008066421810								

(b) *Vertex ordering:*

$G_{8.23}$:

$X(L^1W)_i$	3	5	2	4	6	7	8	12	11	(1	10)	9
$X(L^2W)_i$	3	4	5	2	6	7	12	8	11	(1	10)	9
$X(L^3W)_i$	3	5	2	4	6	7	12	8	11	(1	10)	9
$X(L^{10}W)_i$	3	4	5	2	6	12	11	7	(1	10)	8	9
$X(L^{12}W)_i$	3	4	2	5	6	12	11	7	(1	10)	8	9
$X(L^{13}W)_i$	3	5	4	2	6	12	7	(1	10)	11	8	9
$X(L^{14}W)_i$	3	4	2	5	6	12	11	(1	10)	7	8	9
eigenvector	3	4	5	2	6	12	11	(1	10)	7	8	9

$G_{8.24}$:

$X(L^1W)_i$	5	6	2	4	3	7	8	11	12	(1	10)	9
$X(L^2W)_i$	5	6	4	3	7	2	8	11	12	(1	10)	9
$X(L^3W)_i$	5	6	4	2	3	7	8	11	12	(1	10)	9
$X(L^{10}W)_i$	5	6	4	7	3	2	12	11	8	(1	10)	9
$X(L^{12}W)_i$	5	6	4	7	3	12	2	11	8	(1	10)	9
$X(L^{13}W)_i$	6	5	4	7	3	2	11	12	8	(1	10)	9
$X(L^{14}W)_i$	5	6	4	7	3	12	11	2	8	(1	10)	9
eigenvector	5	6	4	7	3	11	12	2	8	(1	10)	9

Figure 8.8. (a) Isospectral graphs, ^eW_i and ^eW sequences, matrices L^2W; (b) Vertex ordering of $G_{8.23}$ and $G_{8.24}$ cf. normalized $NX(L^eW)_i$ and normalized first eigenvector values.

It is useful that local values $X(L^eW)_i$ be normalized by dividing to the corresponding global values (actually $NX(L^eW)_i$ values - see Tables 8.7 and 8.8). At large values of e, the NX values are superposed over the *vertex weights* (i.e. $^eW_i /2^eW$ - see[91]) and also over the coefficients of normalized first eigenvector . Tables 8.7 and 8.8 offer NX data only for even values of e (for which the ordering is closer to that induced by the normalized first eigenvector - see Figure 8.8(b)). This result is in agreement with the suggestion of Bonchev et.al.[91] to rather consider the closed walks (i.e. $^eSRW_i$ values). However, our results indicate a better correlation (0.995) with ^eW_i values (of even e values) than with $^eSRW_i$ values (0.977).[53]

Table 8. 7. Normalized $NX(L^eW)_i$ Values and Their Correlation with the Coefficients of Normalized First Eigenvector, of $G_{8.23}$.

	eigenvector	normalized eigenvector $\times 10^2$	$NX(L^eW)_i \times 10^2$			
			$(L^{10}W)$	$(L^{12}W)$	$(L^{14}W)$	$(L^{12}SRW)$
1	0.1847	5.9903	5.5734	5.6128	5.6477	3.5974
2	0.3966	12.8628	12.5986	12.8050	12.9548	15.7588
3	0.4823	15.6423	14.6547	14.7643	14.8401	21.3600
4	0.4145	13.4434	13.8359	13.8774	13.9006	15.0778
5	0.4078	13.2261	12.7587	12.7736	12.7670	16.1178
6	0.2712	8.7958	9.6846	9.5313	9.4184	8.4983
7	0.1747	5.6660	5.7881	5.6671	5.5935	4.6879
8	0.1039	3.3698	4.0015	3.8479	3.7435	2.3098
9	0.0484	1.5697	1.7165	1.6325	1.5862	0.6503
10	0.1847	5.9903	5.5734	5.6128	5.6477	3.5974
11	0.1899	6.1590	6.6061	6.5539	6.5110	3.6474
12	0.2246	7.2844	7.2298	7.3235	7.3896	4.7207
			$r = 0.99351$	0.99492	0.99566	0.97727
			$s = 0.08678$	0.07389	0.19199	0.18375

8.5.2.2. ^eW_M *Descriptors*

Numbers ^eW_M (i.e., descriptors Wiener of higher rank)[63] have proved a highly discriminating capability. In this respect four graphs were selected: $G_{8.15}$: $G_{8.16}$;[52] and $G_{8.25}$: $G_{8.26}$,[92] (Figure 8.9). These graphs show degenerate DDS. Moreover these graphs show identical sequences for several ^1W_M numbers. The immediate consequence is the degeneracy of the corresponding Wiener-type numbers. Results are listed in Table 8.9.

Table 8. 8. Normalized $NX(L^eW)_i$ Values and Their Correlation with the Coefficients of Normalized First Eigenvector, of $G_{8.24}$.

	eigenvector	normalized eigenvector $\times 10^2$	$NX(L^eW)_i \times 10^2$			
			$(L^{10}W)$	$(L^{12}W)$	$(L^{14}W)$	$(L^{12}SRW)$
1	0.1039	3.3699	4.0015	3.8479	3.7435	2.1647
2	0.2230	7.2327	7.3449	7.2302	7.1495	8.3657
3	0.2712	8.7961	9.6846	9.5313	9.4184	8.4982
4	0.3594	11.6567	11.2019	11.2105	11.2110	11.7223
5	0.5005	16.2331	16.4405	16.5834	16.6681	22.0785
6	0.4823	15.6423	14.6547	14.7643	14.8401	21.3601
7	0.3106	10.0739	9.9940	10.0990	10.1873	9.4761
8	0.1847	5.9903	5.5735	5.6128	5.6477	3.7423
9	0.0860	2.7893	2.7642	2.7755	2.7977	0.9095
10	0.1039	3.3699	4.0015	3.8479	3.7435	2.1645
11	0.2331	7.5603	7.1302	7.1759	7.2038	4.8208
12	0.2246	7.2844	7.2298	7.3235	7.3896	4.7207
			$r = 0.99351$	0.99492	0.99566	0.97211
			$s = 0.07084$	0.03612	0.11162	1.16896

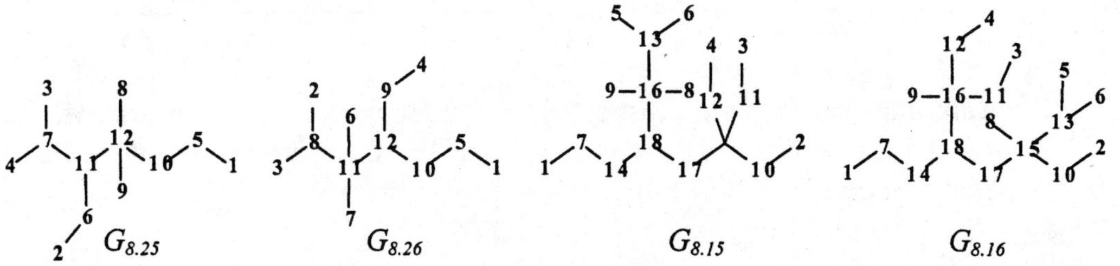

$G_{8.25}$ $G_{8.26}$ $G_{8.15}$ $G_{8.16}$

Figure 8.9. Pairs of graphs with degenerate DDS: $G_{8.25}$ and $G_{8.26}$: 11, 15, 16, 16, 5, 3 $G_{8.15}$ and $G_{8.16}$: 17, 24, 29, 25, 26, 23, 9

Pair $G_{8.25}$: $G_{8.26}$ and $G_{8.15}$: $G_{8.16}$ show degeneracy among the topological indices based on distances in graph (see Table 8.10.). Numbers ^1W_M are also degenerate (even for some walk numbers of rank 2 : $^2W_{De}$ and $^2W_{We}$ but not for $^2W_{Dp}$ and $^2W_{Wp}$). The walk numbers of rank 3, ^3W_M, succeeded in separating both of these pairs of isomers.

Table 8.9. Numbers ^eW_M (of Rank 1 - 3) for the Graphs $G_{8.25}$, $G_{8.26}$, $G_{8.15}$ and $G_{8.16}$

	e	$G_{8.25}$	$G_{8.26}$	$G_{8.15}$	$G_{8.16}$
$^eW_{De}$	1	196	196	583	583
	2	6692	6692	39173	39173
	3	227288	227252	2625203	2625299
$^eW_{We}$	1	196	196	583	583
	2	10686	10686	70137	70097
	3	592184	592292	9051023	9066815
$^eW_{He}$	1	29.33333	29.35001	55.23572	55.23572
	2	149.82250	150.01529	353.43560	353.43560
	3	762.56399	764.13899	2258.69928	2258.66741
$^eW_{Dp}$	1	450	450	1638	1638
	2	38171	38119	329089	329089
	3	3186855	3176484	65720352	65729760
$^eW_{Wp}$	1	450	450	1638	1638
	2	45940	45946	464101	463865
$^eW_{Hp}$	1	20.74287	20.76191	35.48334	35.48334
	2	79.35440	79.53998	154.70112	154.70112
	3	300.62484	301.76187	671.90912	671.89416
$^eW_{W(A,De,De)}$	1	3780	3491	33851	33896
	2	2979036	2373482	155875988	155261932

The walk numbers ^eW_M are constructed on any topological square matrix M. The Schultz-type indices, particularly those path-calculated on the matrix combination: D_e, A, M, (e.g. entries 13 and 15 - Table 8.10) show good discriminating ability.

8.5.3. Distance Measure by C- and X-Type Descriptors

Diudea[53] evaluated the Manhattan distance, D_M, by using local descriptors of centrality and centrocomplexity (C- and X-type, respectively) derived on layer matrices $\mathbf{L^eM}$. The set of testing graphs ($G_{8.27}$-$G_{8.30}$) is that in Figure 8.10 (see also[51]). It can be seen that these graphs are built from *semi-hexes* (denoted A and B) ranged in the following sequence : (a) *ABAB*, (b) *ABBA* and (c) *BAAB*. It was proved (by circular permutations[53]) that there are only four distinct combinations: *a-a, b-a, b-c* and *b-b*.

Table 8.10. Distance-Based Indices of the Graphs $G_{8.25}$, $G_{8.26}$, $G_{8.15}$ and $G_{8.16}$

	Index	$G_{8.25}$	$G_{8.26}$	$G_{8.15}$	$G_{8.16}$
1	W	196	196	583	583
2	H_{De}	29.33333	29.35	55.23571	55.23571
3	$IP(CJD)$	450	450	1638	1638
4	$IE(CJD)$	196	196	583	583
5	$IP(RCJD)$	26.47508	26.47508	54.26032	54.26245
6	$IP(SZD)$	1253	1310	7286	7264
7	$IE(SZD)$	196	196	583	583
8	$IP(RSZD)$	6.68064	6.48620	8.05156	7.74600
9	$IP(SCH_{(A,A,CJD)})$	3833	3833	14438	14430
10	$IE(SCH_{(A,A,CJD)})$	75	75	294	294
11	$IP(SCH_{(A,A,SZD)})$	6369	6346	33499	33429
12	$IE(SCH_{(A,A,SZD)})$	75	75	294	294
13	$IP(SCH_{(De,A,CJD)})$	514937	514001	7292966	7293518
14	$IE(SCH_{(De,A,CJD)})$	42928	42841	416098	416098
15	$IP(SCH_{(De,A,SZD)})$	1714455	1824190	38021230	38048434
16	$IE(SCH_{(De,A,SZD)})$	216629	212995	3206482	3208168

(a)

$G_{8.26}$

$G_{8.27}$

$ABAB:ABAB$
$a \quad\quad a$

$G_{8.28}$

$ABBA:ABAB$
$b \quad\quad a$

$G_{8.29}$

$ABBA:BAAB$
$b \quad\quad c$

$G_{8.30}$

$ABBA:ABBA$
$b \quad\quad b$

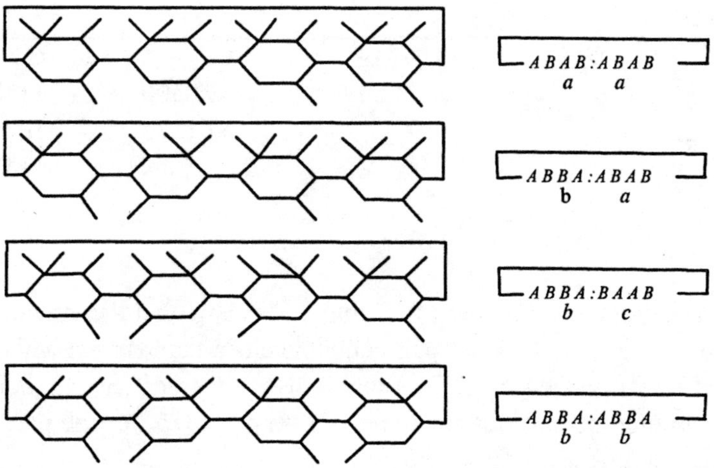

(b)

e	1	2	3	4	5	6	7	8	9
$2DDS(G_{8.27}-G_{8.30})$	88	152	200	216	248	224	168	136	96
$2\,{}^eWS(G_{8.27})$	88	240	616	1648	4312	11440	30088	79592	209704
$2\,{}^eWS(G_{8.28})$	88	240	616	1648	4312	11440	30088	79592	209712
$2\,{}^eWS(G_{8.29})$	88	240	616	1648	4312	11440	30088	79592	209712
$2\,{}^eWS(G_{8.30})$	88	240	616	1648	4312	11440	30088	79592	209720

Figure 8.10. (a) Graphs with degenerate **LC** and **L^{1}W** matrices. (b) *DDS* and *${}^eWS$* of $G_{8.27}$- $G_{8.30}$

Matrices **LC** and **L^{1}W** (see Chap. Topological Matrices) degenerate in the set $G_{8.27}$ - $G_{8.30}$ which suggests that these graphs are very similar. Despite the fact that matrices **L^{e}W** are not more degenerated at $e > 2$, they only differ in the entries corresponding to the remote vertices in graphs. Similar behavior shows the matrix **LDS**. The X-type indices show little differences, with respect to these graphs. Better results were obtained by using the C-type indices, when calculating the Manhattan distance, D_M :[53]

$$D_{Mh} = \sum_e \sum_i \left\| C(L^e M)_{i1} - C(L^e M)_{i2} \right| / (C(L^e M)_{i1} + C(L^e M)_{i2}) \right] \tag{8.44}$$

The results are listed in Table 8.11.

Table 8.11. Manhattan Distance D_M for the Graphs $G_{8.27}$ - $G_{8.30}$.

(a) D_M Calculated with $C(LDS)_i$ 10^3 $(dsp = 20)$ Values.			
graph	$G_{8.27}$	$G_{8.28}$	$G_{8.29}$	$G_{8.30}$
$G_{8.27}$	0	3.3760	5.5463	1.2138
$G_{8.28}$		0	2.8633	3.4901
$G_{8.29}$			0	5.8184
$G_{8.30}$				0
(b) D_M Calculated with $C(L^e WS)_i$ 10^2 $(dsp = 20; e = 2\text{-}6)$ Mean Values.			
graph	$G_{8.27}$	$G_{8.28}$	$G_{8.29}$	$G_{8.30}$
$G_{8.27}$	0	11.8007	13.8506	11.0244
$G_{8.28}$		0	6.6727	10.0055
$G_{8.29}$			0	11.0654
$G_{8.30}$				0

The correlating arrays (a) and (b) (Table 8.11) show that the structure $G_{8.27}$ is *closer* to $G_{8.30}$ and $G_{8.28}$ to $G_{8.29}$, the last pair being the closest among the whole set. This result is confirmed by the sequences ^eWS (Table 8.11(b)): they differ only at elongation $e = 9$ thus demonstrating that all four structures are very similar. For the pair $G_{8.28}$: $G_{8.29}$ the above sequences differ only for $e = 13$.

8.6 PATH, TERMINAL PATH AND CYCLE COMPLEXITY

In trees, path count superimposes on distance count: any two vertices in a connected graph are joined by a path which is the shortest one and unique. In cycle-containing graphs, more than one path may exist between two vertices. As the number of paths increases as the complexity of structure increases.

In the above section, the *Distance Degree Sequence, DDS*, was considered in comparing and ordering graphs. It was shown that graphs having degenerate (i.e., identical) *DDS*, provide degenerate topological indices based on distances in graph. It is conceivable that there exist graphs with degenerate *All Path Sequence, APS*. The idea may be extended to the detour degree sequence, ΔDS, all *Shortest Path Sequence, SPS*, all *Longest Path Sequence, LPS*, as well as to the *Terminal Path Sequences, TPS*, in graph.

All these six sequences give information on the graph complexity. They could become *criteria of similarity*, in comparing rather than ordering structures within a set of molecules. None of them is unique for a certain structure, that is why they could not be criteria of isomorphism. In the following several selected structures are characterized by the above sequences and similarity aspects are discussed.

8.6.1. Graphs with Degenerate Sequences: *APS, TPS, DDS* and/or ΔDS

The graphs in Figure 8.11 were published by Diudea et al.[93] and Dobrynin et al.[94] The pair G_{31} : G_{32} shows degenerate *APS, DDS* and ΔDS but different *TPS*. Cluj indices calculated on it are degenerated excepting the corresponding reciprocal ones (proving the degeneracy came out only at the operational level). Szeged indices solve the degeneracy but only the path-calculated indices. Far more useful proved to be the Schultz-type indices, as it can be seen in the bottom of Table 8.12.

The pair G_{33} : G_{34} is reported to have not only the same *APS* but even the same path sequence matrix.[94] However, the two graphs show different *DDS*. This is reflected in the different values of Harary index, H_{De}, despite the degeneracy of the Wiener index (i.e., the sum of all distances in graph). This pair also shows degenerate ΔDS, *TPS*, Cluj indices and the classical Szeged index, $Sz = IE(SZD)$. Among the simple indices, only the hyper-Szeged index, $IP(SZD)$, solves this pair. The Schultz-type indices $I(SCH_{(M1,A,M3)})$, are again more discriminating ones (boldface, in Table 8.12).

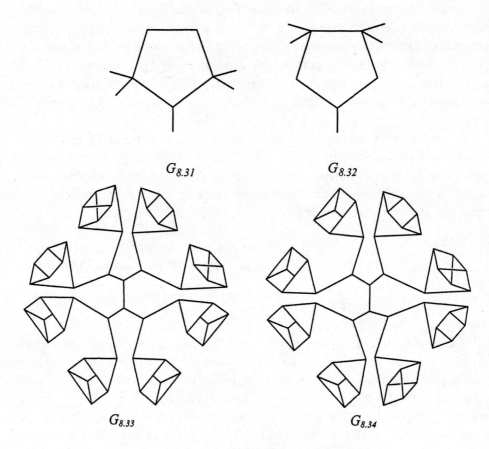

$G_{8.31}$ $G_{8.32}$

$G_{8.33}$ $G_{8.34}$

Figure 8.11. Graphs with degenerate *APS* sequence.

The sequences of the graphs of Figure 8.11 are as follows:

$G_{8.31}$:

TPS:	**5.14.18.22.22.8.0.0.0**
APS:	10.17.19.19.14.4.0.0.0
DDS:	10.17.14.4.0.0.0.0.0
ΔDS:	5.2.5.15.14.4.0.0.0

$G_{8.32}$:

TPS:	**5.14.18.22.24.8.0.0.0**
APS:	10.17.19.19.14.4.0.0.0
DDS:	10.17.14.4.0.0.0.0.0
ΔDS:	5.2.5.15.14.4.0.0.0

$G_{8.33}$

TPS: 0.0.0.48.208.424.176.352.704.1408.2784.4912.8016.11216.13504.12992.6272.0...0
APS:
93.186.348.576.848.1164.1680.2848.4672.3208.4080.4152.2544.3304.3200.2976.3136.0.
..0
DDS: 93.118.84.132.200.274.316.358.248.68.0...0
ΔDS: 13.18.20.24.44.140.48.96.96.192.192.0.0.144.0.288.576.0...0

$G_{8.34}$

TPS: 0.0.0.48.208.424.176.352.704.1408.2784.4912.8016.11216.13504.12992.6272.0...0
APS:
93.186.348.576.848.1164.1680.2848.4672.3208.4080.4152.2544.3304.3200.2976.3136.0.
..0
DDS: 93.118.84.132.200.274.316.354.260.56.4.0...0
ΔDS: 13.18.20.24.44.140.48.96.96.192.192.0.0.144.0.288.576.0...0

Table 8.12. Distance- and Path-Based Indices for the Graphs of Figure 8.11

I	$G_{8.31}$	$G_{8.32}$	$G_{8.33}$	$G_{8.34}$
W	102	102	11741	11741
H_{De}	24.1667	24.1667	**422.9150**	**422.9120**
w	178	178	23681	23681
$H_{\Delta e}$	14.8833	14.8833	195.1464	195.1464
$IP(CFD)$	333	333	124897	124897
$IP(RCFD)$	**14.5401**	**14.5813**	60.91107	60.91107
$IE(CFD)$	121	121	16917	16917
$IP(CF\Delta)$	147	147	33491	33491
$IP(RCF\Delta)$	**24.3000**	**24.5000**	1182.963	1182.963
$IE(CF\Delta)$	64	64	7821	7821
$IP(CJD)$	297	297	120901	120901
$IP(RCJD)$	**16.1472**	**16.1885**	62.90328	62.90328
$IE(CJD)$	121	121	16917	16917
$IP(CJ\Delta)$	147	147	33491	33491
$IP(RCJ\Delta)$	**24.3000**	**24.5000**	1182.963	1182.963
$IE(CJ\Delta)$	64	64	7821	7821
$IP(SZD)$	**549**	**537**	**922067**	**929875**
$IE(SZD)$	121	121	16917	16917
$IP(SZ\Delta)$	**631**	**647**	892955	892955
$IE(SZ\Delta)$	121	121	9781	9781
$IP(SCH_{(A.A.CFD)})$	**3255**	**3279**	1504350	1504350
$IP(SCH_{(A.A.CJD)})$	**2843**	**2847**	1428474	1428474
$IP(SCH_{(De.A.CFD)})$	**177969**	**178041**	**20548161320**	**20553577720**
$IP(SCH_{(De.A.CJD)})$	**133837**	**134529**	**19810608456**	**19812619992**

8.6.2. Cycle Complexity

8.6.2.1. Cycles in Graphs

By visiting the *TP* structure of a graph it is possible to count the cycles in that graph. The procedure works on a *List of vertex neighborhood*, (VN: Array[0..|V|,0..14]of Integer) according to the construction $C_{8.1}$:

Searching for Cycles:

```
For i:=1,|V| do
    For each tp ∈ TP_G(i) do {each terminal path of vertex i}
    For j:=1, VN[tp[0],0] do {all neighbors of the last vertex}
        If VN[tp[0],j]<> tp[tp[0]-1] then {vertex different from the last one}
            For k:=1 to tp[0]-2 execute {at least 3 vertices in a cycle}
            If tp[k]=VN[tp[0],j] then {there exist cycles}
            The vertex sequence tp[k],...,tp[tp[0]] means a cycle
                EndIf;
            EndFor;
        EndIf;
        EndFor;
    EndFor;
EndFor;
```
$(C_{8.1})$

The above algorithm searches for cycles at the end of *TP* of a graph. It counts all cycles, according to the observation that: "for any cycle there exists a terminal path that ends in that cycle".

The list of cycles, provided by the algorithm, may be ordered cf. the cycle length and then only the distinct cycles are listed, in increasing ring size order. A *sequence of cycle* matrix **SCy** associated with the graph is thus constructed:

$$[\mathbf{SCy}]_{i,j} = \text{No. of } j\text{-membered cycles traversing vertex } i \tag{8.45}$$

A global *cycle sequence*, *CyS*, is finally provided:

$$CyS_j = (1/j)\sum_i [\mathbf{SCy}]_{ij}; \ j = 3,...,N \tag{8.46}$$

The procedure is exemplified on the graph representing 2 Azabicyclo [2, 2, 1 – hept-5-en-3-one] (Figure 8.12):

$G_{8.35}$

(1) Unsorted List of Cycles:

4	5	3	6	7			4	2	1	3	6	7
4	7	6	3	5			4	7	6	3	1	2
1	3	6	7	4	2		3	6	7	4	2	1
3	6	7	4	5			6	3	5	4	7	
1	3	5	4	2			3	5	4	2	1	
2	4	5	3	1			3	1	2	4	5	
4	5	3	6	7			6	3	1	2	4	7
4	7	6	3	5			4	2	1	3	5	
2	4	7	6	3	1		4	5	3	1	2	
3	6	7	4	5			4	5	3	1	2	
3	1	2	4	5			4	2	1	3	5	
3	5	4	2	1			3	1	2	4	5	
3	6	7	4	5			3	5	4	2	1	
3	6	7	4	2	1		6	3	5	4	7	
4	7	6	3	1	2		3	5	4	2	1	
4	7	6	3	5			3	1	2	4	5	
4	5	3	6	7			6	3	1	2	4	7
4	5	3	1	2			4	2	1	3	5	
4	2	1	3	5			4	5	3	1	2	
4	2	1	3	6	7							

(2) List of Cycles Ordered by Length:

4	5	3	6	7		4	5	3	1	2	
4	7	6	3	5		4	2	1	3	5	
3	6	7	4	5		3	1	2	4	5	
1	3	5	4	2		3	5	4	2	1	
2	4	5	3	1		6	3	5	4	7	
4	5	3	6	7		3	5	4	2	1	
4	7	6	3	5		3	1	2	4	5	
3	6	7	4	5.		4	2	1	3	5	
3	1	2	4	5		4	5	3	1	2	
3	5	4	2	1		1	3	6	7	4	2
3	6	7	4	5		2	4	7	6	3	1
4	7	6	3	5		3	6	7	4	2	1
4	5	3	6	7		4	7	6	3	1	2
4	5	3	1	2		4	2 ·	1	3	6	7
4	2	1	3	5		4	2	1	3	6	7
6	3	5	4	7		4	7	6	3	1	2
3	5	4	2	1		3	6	7	4	2	1
3	1	2	4	5		6	3	1	2	4	7
4	2	1	3	5		6	3	1·	2	4	7
4	5	3	1	2							

(3) List of Distinct Cycles:

4	5	3	6	7	
1	3	5	4	2	
1	3	6	7	4	2

(4) Sequence of Cycle Matrix, **SCy**:

0	0	0	0	1	1	0	0
0	0	0	0	1	1	0	0
0	0	0	0	2	1	0	0
0	0	0	0	2	1	0	0
0	0	0	0	2	0	0	0
0	0	0	0	1	1	0	0
0	0	0	0	1	1	0	0
0	0	0	0	0	0	0	0

(5) Cycle Sequence, *CyS*:   0.0.0.0.2.1.0.0

Figure 8.12. Cycle counting of 2 Azabicyclo [2, 2, 1 – hept-5-en-3-one], $G_{8.35}$

Cycle counting as given by the above algorithm is an *exact solution of the ring perception problem*, very similar (but not identical) to the algorithm proposed by Balducci and Pearlman.[95]

Cycle counting can be used as a *cycle complexity* criterion, *CyC*: the *increasing lexicographic order* of *CyS* shows the graph with *the larger number of smallest rings*, which is the most complex and symmetrical among a set of isomeric graphs. A *CyS* can be used as a first fingerprint for a cycle-containing graph.

8.6.2.2. Cubic Graphs with Degenerate Sequences SPS, LPS, DDS *and/or* ΔDS *but Different Cycle-Count*

Figure 8.13. presents a collection of 14 regular cubic graphs (i.e. graphs having the degree 3 for all of their vertices) with $N = 12$ and degenerate sequences *SPS, LPS, DDS* and/or *ΔDS* but different *TPS, APS* and *CyS*. All these graphs show different cycle sequences. The ordering of the graphs $G_{8.mn}$, given by increasing lexicographic ordered *CyS*, is shown in the following array:

CyS **Lexicographic Ordering** for the 14 Graphs of Figure 8.13

8.37.   0.0.0.1.6.12.10.11.22.20.8.1
8.36.   0.0.0.1.8.6.12.21.12.18.12.1
8.43.   0.0.0.3.0.20.0.24.0.24.0.1
8.42.   0.0.0.3.4.8.12.12.20.14.8.1
8.41.   0.0.0.3.4.8.12.13.22.14.12.1
8.40.   0.0.0.3.5.5.13.18.15.19.7.1
8.44.   0.0.0.4.0.16.0.29.0.36.0.1
8.47.   0.0.0.4.2.8.14.9.26.10.12.1

8.48. 0.0.0.4.4.2.16.17.16.14.12.1
8.49. 0.0.0.4.4.4.12.16.20.16.12.1
8.45. 0.0.0.6.0.6.12.6.36.6.12.1
8.46. 0.0.0.6.0.8.0.36.0.36.0.1
8.39. 0.0.0.0.8.12.8.12.24.20.8.1
8.38. 0.0.0.0.9.9.9.18.18.18.12.1

All of them are Hamiltonian circuits: they can be drawn on a circle. Cycle sequence can be also used as a clustering criterion. Only three of the graphs in Figure 8.13 show *all even-membered cycles* ($G_{8.43}$; $G_{8.44}$ and $G_{8.46}$). Note also that $G_{8.46}$ and $G_{8.48}$ are polyhedra (see also[96]).

Half of the graphs in Figure 8.13 are *full Hamiltonian detour* graphs, $FH\Delta$ - (see Chap. Cluj Indices) - property that could be another similarity criterion.

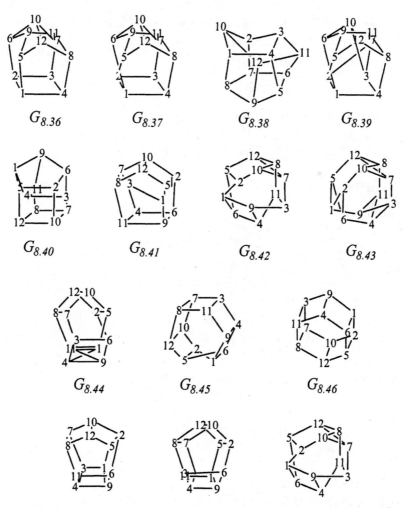

Figure 8.13. Graphs with degenerate sequences *SPS*, *LPS*, *DDS* and/or *ΔDS* but different *TPS*, *APS*, and *CyS*

A simple similarity view indicates some clustering of these graphs, as shown below:

SPS: (8.37; 8.49); (8.40; 8.48)
LPS: (8.47; 8.49)
DDS: (8.36; 8.37);(8.38; 8.39);(8.40; 8.41; 8.42; 8.43)
ΔDS: (8.37; 8.42); (8.43; 8.44; 8.46)
$DDS\&\Delta DS$:   (8.47; 8.48; 8.49)
$FH\Delta$:   (8.36; 8.39; 8.41; 8.45; 8.47; 8.48; 8.49)

The sequences and vertex orbits (identified according to the length of all terminal paths starting from the vertex i, $L(TP_G(i))$) for the graphs of Figure 8.13 are as follows:

$G_{8.36}$: $FH\Delta$

TPS: 0.0.0.0.0.16.88.184.384.736.464
APS: 18.36.72.140.236.388.560.676.704.600.232
SPS: 18.36.28.0.0.0.0.0.0.0.0.0
LPS: 0.0.0.0.0.0.0.0.0.0.0.232
DDS: 18.34.14.0.0.0.0.0.0.0.0.0
ΔDS: 0.0.0.0.0.0.0.0.0.0.0.66
CyS: 0.0.0.1.8.6.12.21.12.18.12.1
Vertex Orbits: {1,2,3,4}; {5,6,7,8}; {9,10,11,12}

$G_{8.37}$:

TPS: 0.0.0.0.0.0.8.56.216.468.648.488
APS: 18.36.72.140.246.382.546.698.738.568.244
SPS: 18.36.38.0.0.0.0.0.0.0.0.0
LPS: 0.0.0.0.0.0.0.0.0.0.20.244
DDS: 18.34.14.0.0.0.0.0.0.0.0.0
ΔDS: 0.0.0.0.0.0.0.0.0.0.2.64
CyS: 0.0.0.1.6.12.10.11.22.20.8.1
Vertex Orbits: {1,2,3,4}; {5,6,7,8}; {9,11}; {10, 12}

$G_{8.38}$:

TPS: 0.0.0.0.0.0.72.216.396.702.504
APS: 18.36.72.144.243.387.567.693.711.603.252
SPS: 18.36.27.0.0.0.0.0.0.0.0.0
LPS: 0.0.0.0.0.0.0.0.0.45.252
DDS: 18.36.12.0.0.0.0.0.0.0.0.0
ΔDS: 0.0.0.0.0.0.0.0.0.3.63
CyS: 0.0.0.0.9.9.9.18.18.18.12.1

Vertex Orbits: {1,2,3,5,6,8,9,10,11}; {4,7,12}

$G_{8.39}$: $FH\Delta$

TPS: 0.0.0.0.0.0.48.240.432.688.496
APS: 18.36.72.144.248.384.560.704.728.592.248
SPS: 18.36.32.0.0.0.0.0.0.0.0.0
LPS: 0.0.0.0.0.0.0.0.0.0.0.248
DDS: 18.36.12.0.0.0.0.0.0.0.0.0
ΔDS: 0.0.0.0.0.0.0.0.0.0.0.66
CyS: 0.0.0.0.8.12.8.12.24.20.8.1
Vertex Orbits: {1,2,6,7,8,10,11,12}; {3,4,5,9}

$G_{8.40}$:

TPS: 0.0.0.0.4.28.88.210.420.564.496
APS: 18.36.72.132.227.375.543.673.690.530.248
SPS: 18.36.36.0.0.0.0.0.0.0.0.0
LPS: 0.0.0.0.0.0.0.0.0.42.248
DDS: 18.30.18.0.0.0.0.0.0.0.0.0
ΔDS: 0.0.0.0.0.0.0.0.0.0.4.62
CyS: 0.0.0.3.5.5.13.18.15.19.7.1
Vertex Orbits: {1,11}; {2,5,7,8}; {3,6,}; {4,9}; {10,12}

$G_{8.41}$: $FH\Delta$

TPS: 0.0.0.0.4.20.80.228.420.580.492
APS: 18.36.72.132.232.372.536.684.698.536.246
SPS: 18.36.41.0.0.0.0.0.0.0.0.0
LPS: 0.0.0.0.0.0.0.0.0.0.246
DDS: 18.30.18.0.0.0.0.0.0.0.0.0
ΔDS: 0.0.0.0.0.0.0.0.0.0.0.66
CyS: 0.0.0.3.4.8.12.13.22.14.12.1
Vertex Orbits: {1,2,3,5,7,8}; {4,9}; {6,11}; {10,12}

$G_{8.42}$:

TPS: 0.0.0.0.0.16.64.232.400.544.544
APS: 18.36.72.132.232.372.536.692.696.544.272
SPS: 18.36.42.0.0.0.0.0.0.0.0.0
LPS: 0.0.0.0.0.0.0.0.0.16.272
DDS 18.30.18.0.0.0.0.0.0.0.0.0
ΔDS: 0.0.0.0.0.0.0.0.0.0.2.64

CyS: 0.0.0.3.4.8.12.12.20.14.8.1
Vertex Orbits: {1,2,3,4,5,6,9,11}; {7,8,10,12}

$G_{8.43}$: H(12,5,-5)* = P(6,1,3)**

TPS: 0.0.0.0.0.0.96.96.672.384.576
APS: 18.36.72.132.252.360.552.648.816.480.288
SPS: 18.36.60.0.0.0.0.0.0.0.0.0
LPS: 0.0.0.0.0.0.0.0.0.480.288
DDS: 18.30.18.0.0.0.0.0.0.0.0.0
ΔDS: 0.0.0.0.0.0.0.0.0.0.30.36
CyS: 0.0.0.3.0.20.0.24.0.24.0.1
Vertex Orbits: {1,2,3,4,5,6,7,8,9,10,11,12}

$G_{8.44}$:

TPS: 0.0.0.0.8.8.128.128.608.392.536
APS: 18.36.72.128.240.352.556.632.768.464.268
SPS: 18.36.56.16.0.0.0.0.0.0.0.0
LPS: 0.0.0.0.0.0.0.0.0.464.268
DDS: 18.28.18.2.0.0.0.0.0.0.0.0
ΔDS: 0.0.0.0.0.0.0.0.0.0.30.36
CyS: 0.0.0.4.0.16.0.29.0.36.0.1
Vertex Orbits: {1,2,3,5,6,7,8,11}; {4,9,10,12}

$G_{8.45}$: *FHΔ*

TPS: 0.0.0.0.24.24.96.288.408.480.456
APS: 18.36.72.120.216.348.516.708.648.468.228
SPS: 18.36.48.0.0.0.0.0.0.0.0.0
LPS: 0.0.0.0.0.0.0.0.0.0.228
DDS: 18.24.24.0.0.0.0.0.0.0.0.0
ΔDS: 0.0.0.0.0.0.0.0.0.0.0.66
CyS: 0.0.0.6.0.6.12.6.36.6.12.1
Vertex Orbits: {1,2,3,4,5,6,7,8,9,10,11,12}

$G_{8.46}$: H(12,3,-3)*

TPS: 0.0.0.0.24.24.144.144.576.408.456
APS: 18.36.72.120.216.336.564.624.720.432.228
SPS: 18.36.48.48.0.0.0.0.0.0.0.0
LPS: 0.0.0.0.0.0.0.0.0.432.228
DDS: 18.24.18.6.0.0.0.0.0.0.0.0

ΔDS: 0.0.0.0.0.0.0.0.0.0.30.36
CyS: 0.0.0.6.0.8.0.36.0.36.0.1
Vertex Orbits: {1,2,3,4,5,6,7,8,9,10,11,12}

$G_{8.47}$: $FH\Delta$

TPS: 0.0.0.0.8.32.80.252.428.492.504
APS: 18.36.72.128.230.370.522.694.680.498.252
SPS: 18.36.46.0.0.0.0.0.0.0.0.0
LPS: 0.0.0.0.0.0.0.0.0.0.0.252
DDS: 18.28.20.0.0.0.0.0.0.0.0.0
ΔDS: 0.0.0.0.0.0.0.0.0.0.0.66
CyS: 0.0.0.4.2.8.14.9.26.10.12.1
Vertex Orbits: {1,3,6,11}; {2,5,7,8}, {4,9}, {10,12}

$G_{8.48}$: $FH\Delta$; H(12, 3,6,3)*

TPS: 0.0.0.0.8.56.80.232.416.520.472
APS: 18.36.72.128.220.376.536.672.664.496.236
SPS: 18.36.36.0.0.0.0.0.0.0.0.0
LPS: 0.0.0.0.0.0.0.0.0.0.0.236
DDS: 18.28.20.0.0.0.0.0.0.0.0.0
ΔDS: 0.0.0.0.0.0.0.0.0.0.0.66
CyS: 0.0.0.4.4.2.16.17.16.14.12.1
Vertex Orbits: {1,2,3,5,6,7,8,11}; {4,9,10,12}

$G_{8.49}$: $FH\Delta$

TPS: 0.0.0.0.8.24.88.224.416.520.504
APS: 18.36.72.128.220.364.540.676.680.512.252
SPS: 18.36.38.0.0.0.0.0.0.0.0.0
LPS: 0.0.0.0.0.0.0.0.0.0.0.252
DDS: 18.28.20.0.0.0.0.0.0.0.0.0
ΔDS: 0.0.0.0.0.0.0.0.0.0.0.66
CyS: 0.0.0.4.4.4.12.16.20.16.12.1
Vertex Orbits: {1,2,3,4,5,6,9,11}; {7,8,10,12}
* Hamiltonian circuit symbol (see Sect.8.7)
** Petersen generalized graph (see Sect.8.7)

Table 8.13. Distance-and Path-Based Indices for the Graphs of Figure 8.13

I G	$G_{8.36}$	$G_{8.37}$	$G_{8.38}$	$G_{8.39}$	$G_{8.40}$	$G_{8.41}$	$G_{8.42}$	$G_{8.43}$	$G_{8.44}$	$G_{8.45}$	$G_{8.46}$	$G_{8.47}$	$G_{8.48}$	$G_{8.49}$
W	128	128	126	126	132	132	132	132	136	138	144	134	134	134
w	726	724	723	726	722	726	724	696	696	726	696	726	726	726
$IP(CFD)$	1282	1332	1218	1286	1381	1385	1376	1476	1582	1434	1578	1396	1402	1410
$IE(CFD)$	358	456	369	418	403	452	476	648	640	450	648	472	374	416
$IP(CF\Delta)$	66	72	75	66	78	66	72	156	156	66	156	66	66	66
$IE(CF\Delta)$	18	18	18	18	18	18	18	18	18	18	18	18	18	18
$IP(CJD)$	1152	1168	1134	1146	1202	1210	1242	1338	1360	1242	1452	1238	1222	1226
$IE(CJD)$	358	456	369	418	403	452	476	648	640	450	648	472	374	416
$IP(CJ\Delta)$	66	72	75	66	78	66	72	156	156	66	156	66	66	66
$IE(CJ\Delta)$	18	18	18	18	18	18	18	18	18	18	18	18	18	18
$IP(SZD)$	1258	1294	1218	1246	1358	1376	1408	1536	1572	1458	1696	1418	1394	1406
$IE(SZD)$	358	456	369	418	403	452	476	648	640	450	648	472	374	416
$IP(SZ\Delta)$	66	110	120	66	152	66	110	1326	1326	66	1326	66	66	66
$IE(SZ\Delta)$	18	34	36	18	46	18	32	648	648	18	648	18	18	18

From Figure 8.13 and Table 8.13 it can be seen that sequences *SPS*, *LPS*, and particularly *DDS* and *ΔDS* induce a clustering among the set of these structures. The consequence is the degeneracy of indices based on distances and detours, respectively. The cluster of full Hamiltonian detour graphs, *FHΔ*, show a minimal value for the detour-based indices: $IP(M\Delta)$; $M = CJ$, CF and SZ: $\min IP(M\Delta) = \binom{N}{2}$. Also, the corresponding edge-computed indices show a minimal value: $\min IE(M\Delta) = |E| = 3N/2$ in cubic graphs. The distance-based Cluj and Szeged hyper indices are all different: $IP(CJD) \neq IP(CFD) \neq IP(SZD)$.

8.6.3. Families of Graphs with Degenerate Sequences and Rearrangements

8.6.3.1. Spiro-Graphs with Degenerate Sequences

Figure 8.14 illustrates a set of spiro-graphs (i.e. graphs having two simple cycles incident in a single collapsed atom). These graphs represent the *spiro*-copy of the graphs $G_{8.27}$-$G_{8.30}$. The two families show degenerate sequences, *TPS*, *APS*, *DDS*, *ΔDS*, and *CyS* for the spiro-family being presented below. It is obvious that a calculation of some chemical distance by using such sequences is impossible. Obviously, the four spiro-structures are very similar. Moreover, a whole list of TI based on these sequences are degenerated (Table 8.14 includes only the distance- and detour-based indices which are degenerated).

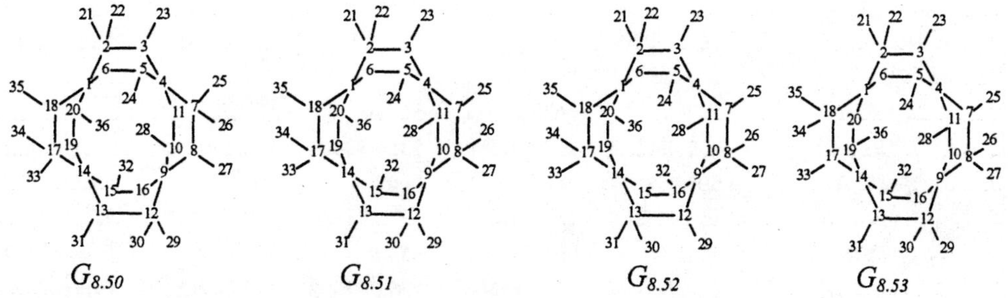

$G_{8.50}$ $\quad\quad$ $G_{8.51}$ $\quad\quad$ $G_{8.52}$ $\quad\quad$ $G_{8.53}$

Figure 8.14. Spiro-graphs with degenerate TPS, APS, DDS, ΔDS, and CyS sequences

The degenerate sequences for the spiro-graphs of Figure 8.14 are as follows:

TPS: $\quad$ 16.40.80.136.200.244.416.496.536.912.1160.1200.1728.1328.448.0...0
APS: $\quad$ 40.76.128.188.264.356.488.592.768.1056.1248.1360.1408.832.224.0...0
DDS: $\quad$ 40.76.116.132.120.82.48.16.0...0
ΔDS: $\quad$ 16.4.0.0.0.2.0.0.4.32.72.104.176.164.56.0...0
CyS: $\quad$ 0.0.0.0.0.4.0.0.0.0.0.15.0...0

Table 8.14. Topological Indices for the Spiro-Graphs of Figure 8.14.

	Index	$G_{8.50}$	$G_{8.51}$	$G_{8.52}$	$G_{8.53}$	Similarity
1	**W**	2624	2624	2624	2624	degenerated
2	H_{De}	196.190476	196.190476	196.190476	196.190476	degenerated
3	w	7856	7856	7856	7856	degenerated
4.	$H_{\Delta e}$	66.175980	66.175980	66.175980	66.175980	degenerated
5	$IP(CFD)$	56732	**56736**	**56740**	56732	$G_{8.50}$ - $G_{8.53}$
6	**$IP(RCFD)$**	**132.450543**	**132.441500**	**132.432463**	**132.450531**	**discriminated**
7	$IP(CF\Delta)$	6064	6064	6064	6064	degenerated
8	$IP(RCF\Delta)$	259.478160	**259.367049**	259.478160	**259.033715**	$G_{8.50}$ - $G_{8.52}$
9	$IP(CJD)$	45408	45408	45408	45408	degenerated
10	**$IP(RCJD)$**	**139.544074**	**139.544098**	**139.544095**	**139.544126**	**discriminated**
11	$IP(CJ\Delta)$	5752	5752	5752	5752	degenerated
12	$IP(RCJ\Delta)$	259.858843	**259.747732**	259.858843	**259.414398**	$G_{8.50}$ - $G_{8.52}$
13	$IP(SCH_{(A,A,CFD)})$	717320	**717516**	**717712**	717320	$G_{8.50}$ - $G_{8.53}$
14	$IP(SCH_{(A,A,CF\Delta)})$	79984	79984	79984	79984	degenerated
15	$IP(SCH_{(A,A,CJD)})$	**498708**	498712	498712	**498716**	$G_{8.51}$ - $G_{8.52}$
16	$IP(SCH_{(A,A,CJ\Delta)})$	74216	74216	74216	74216	degenerated
17	$IP(SCH_{(A,A,SZD)})$	**795024**	795032	795032	**795040**	$G_{8.51}$ - $G_{8.52}$
18	$IP(SCH_{(A,A,SZ\Delta)})$	682600	**682612**	682600	**682648**	$G_{8.50}$ - $G_{8.52}$
19	**$IP(SCH_{(De,A,CFD)})$**	**1842206288**	**1843228300**	**1844250520**	**1842206400**	**discriminated**
20	$IP(SCH_{(De,A,CF\Delta)})$	**140244912**	**140244732**	140244672	140244672	$G_{8.52}$ - $G_{8.53}$
21	**$IP(SCH_{(De,A,CJD)})$**	**1021397364**	**1021397421**	**1021396876**	**1021398568**	**discriminated**
22	$IP(SCH_{(De,A,CJ\Delta)})$	**135360880**	**135360700**	135360640	135360640	$G_{8.52}$ - $G_{8.53}$
23	**$IP(SCH_{(\Delta e,A,CFD)})$**	**17815556248**	**17825108344**	**17834666560**	**17815549848**	**discriminated**
24	$IP(SCH_{(\Delta e,A,CF\Delta)})$	1322993536	**1322992844**	1322993536	**1322990768**	$G_{8.50}$ - $G_{8.52}$
25	**$IP(SCH_{(\Delta e,A,CJD)})$**	**10099587348**	**10099580053**	**10099581748**	**10099569368**	**discriminated**
26	$IP(SCH_{(\Delta e,A,CJ\Delta)})$	1271920544	**1271919852**	1271920544	**1271917776**	$G_{8.50}$ - $G_{8.52}$
27	**$IP(SCH_{(De,A,SZD)})$**	**3169720572**	**3169684668**	**3169689676**	**3169638748**	**discriminated**
28	**$IP(SCH_{(De,A,SZ\Delta)})$**	**2484822748**	**2484806956**	**2484852940**	**2484699196**	**discriminated**
29	**$IP(SCH_{(\Delta e,A,SZD)})$**	**28437624492**	**28437008460**	**28437077836**	**28436253676**	**discriminated**
30	$IP(SCH_{(\Delta e,A,SZ\Delta)})$	22416183548	**22415626044**	22416183548	**22413953532**	$G_{8.50}$ - $G_{8.52}$

A very interesting behavior is shown the Cluj and Szeged indices, both as basic indices and as Schultz-type composite indices (Table 8.14). Indices induce different *clustering* within this set (indicated in the last column of Table 8.14). Only the reciprocal Cluj-distance indices, among the basic indices, discriminated the whole set (boldface values). Among the composite indices, those constructed on distance were more discriminating than those based on detours or adjacency. Since different indices induce different clustering, the occurrence of one or another cluster may be used in drawing the similarity in a set of structures, anyhow, very related.

Despite the degeneracy of *TPS*, the length of all terminal paths starting from the vertex i, $L(TP_G(i))$, (as *LTP* descriptor, in Table 8.15) succeeded in separating the *orbits of equivalent vertices* in all these structures.

Table 8.15. Vertex Orbits *VO*'s, of the Graphs of Figure 8.14 and Their *LTP* Values.

$VO(G_{8.50})$	$LTP(G_{8.50})$	$VO(G_{8.51})$	$LTP(G_{8.51})$	$VO(G_{8.52})$	$LTP(G_{8.52})$	$VO(G_{8.53})$	$LTP(G_{8.53})$
{1,4,9,14}	1817	{4} {14} {1} {9}	1724 1789 1845 1910	{4} {1,9} {14}	1752 1817 1882	{4,14} {1,9}	1696 1938
{2,7,12,17} {3,8,13,18} {5,10,15,20} {6,11,16,19}	2598 2701 2914 2932	{2} {8} {17} {12} {3} {18} {7} {13} {5} {6} {20} {11} {10} {19} {15} {16}	2578 2590 2598 2610 2689 2701 2709 2721 2902 2912 2914 2922 2924 2932 2934 2944	{2,8} {13,17} {3,7} {12,18} {5,11} {6,10} {16,20} {15,19}	2578 2610 2689 2721 2902 2912 2934 2944	{2,8,12,18} {3,7,13,17} {5,11,15,19} {6,10,16,20}	2590 2709 2922 2924
{21,22,25,26,29,30,33,34} {23,27,31,35} {24,28,32,36}	2840 2952 3182	{21,22} {26,27} {33,34} {29,30} {36} {23} {35} {25} {31} {24} {28} {32}	2818 2832 2840 2854 2914 2938 2952 2960 2974 3168 3190 3204	{21,22,26,27} {30,31,33,34} {23,25} {29,35} {24,28} {32,36}	2818 2854 2938 2974 3168 3204	{21,22,26,27,29,30,34,35} {23,25,31,33} {24,28,32,36}	2832 2960 3190

8.6.3.2. Spiro-Graphs with Degenerate Rearrangements

Again the terminal paths proved to be useful descriptors in separating the vertex orbits and again the spiro-graphs (Figure 8.15) show interesting properties. Thus, the graph $G_{8.54\text{-}a}$ may be viewed as a knot in 3D optimized geometry ($G_{8.54\text{-}b}$). In a 3D configuration, $G_{8.55\_a}$ looks like $G_{8.55\_d}$ (a true catenand).

By crossing two edges, say $G_{8.54-a}$ {(2,5); (9,12)} results in $G_{8.55}$(a) {(5,20); (2,19)}. This last graph, by a further crossing process: $G_{8.55\_a}$ {(7,9); (16,15)}

$\longrightarrow$ $G_{8.55\_b}$ {(7,15); (9,16)} lead to the isomorphic graph $G_{8.55\_b}$. The renumbering of $G_{8.55\_b}$ offers a labeling (as in $G_{8.55\_c}$) that preserves the connectivity in $G_{8.55\_a}$. Thus $G_{8.55\_c}$ is automorphic with $G_{8.55\_a}$. In other words, the crossing process $G_{8.55a}$ $\longrightarrow$ $G_{8.55\_b}$ represents a *degenerate rearrangement*.

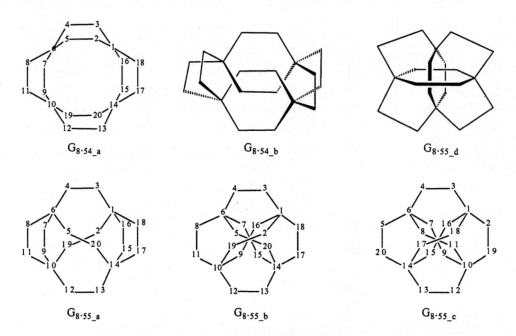

Figure 8.15. Spiro-graphs and a degenerate rearrangement $G_{8.55\_a}$ {(7,9);(16,15)} $G_{8.55\_b}$ {(7,15);(9,16)}

The sequences and vertex orbits (identified according to the length of all terminal paths starting from the vertex i, $L(TP_G(i))$) for the spiro-graphs of Figure 8.15 are as follows:

$G_{8.54}$

TPS:	0.0.0.0.16.16.16.16.48.48.160.224.224.0.0.0.0.0.0
APS:	24.40.56.72.120.144.160.240.320.336.480.448.224.0...0
DDS:	24.40.44.32.32.18.0.0.0.0.0.0.0.0.0.0.0.0.0.0
ΔDS:	0.0.0.0.0.2.0.0.4.32.40.56.56.0.0.0.0.0.0
CyS:	0.0.0.0.0.4.0.0.0.0.0.16.0.0.0.0.0.0.0.0
Vertex Orbits:	{1,6,10,14}; (2,3,4,5,7,8,9,11,12,13,15,16,17,18,19,20}

$G_{8.55}$

TPS:	0.0.0.0.8.8.8.56.72.72.120.216.216.0.0.0.0.0.0
APS:	24.40.56.72.120.156.188.300.340.324.420.432.216
DDS:	24.40.50.52.24.0.0.0.0.0.0.0.0.0.0.0.0.0.0
ΔDS:	0.0.0.0.0.0.0.0.6.16.56.28.84.0.0.0.0.0.0
CyS:	0.0.0.0.0.2.0.0.8.0.0.9.0.0.0.0.0.0.0.0
Vertex Orbits:	{1,6,10,14};{2,3,4,5,12,13,19,20}; {7,8,9,11,15,16,17,18}

8.6.3.3. A Family of FHΔ Cubic Graphs

Figure 8.16 illustrates a collection of cubic graphs, (in Schlegel projection, 8.16 (a) and as 3D optimized structures, 8.16 (b)) whose point molecular symmetry is C_{3v} (the first three) and C_1 (the last two). Note that the structure $G_{8.57}$, was published by Diudea et. al.[97] in the $G_{8.57-c}$ representation (Figure 8.17). Also note that $G_{8.56}$ is a polyhedron (see $G_{8.56-b}$ and also[96]).

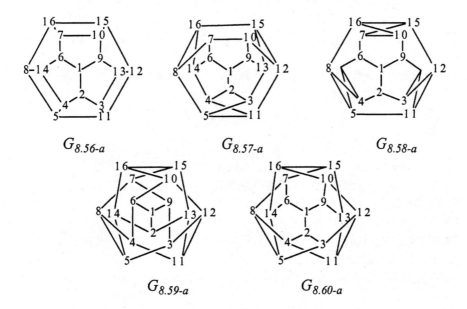

$G_{8.56-a}$ $G_{8.57-a}$ $G_{8.58-a}$

$G_{8.59-a}$ $G_{8.60-a}$

Figure 8.16.a. A family of *FHΔ* cubic graphs: Schlegel projection

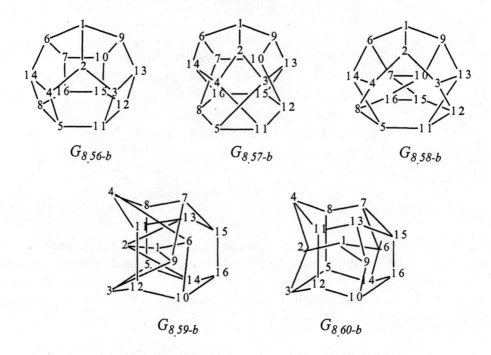

$G_{8.56\text{-}b}$ $G_{8.57\text{-}b}$ $G_{8.58\text{-}b}$

$G_{8.59\text{-}b}$ $G_{8.60\text{-}b}$

Figure 8.16.b. A family of $FH\Delta$ cubic graphs: 3D optimized structures

The sequences and vertex orbits (identified according to the length of all terminal paths starting from the vertex i, $L(TP_G(i))$) for the spiro-graphs of Figure 8.16 are as follows:

$G_{8.56}$

TPS:	0.0.0.0.0.36.84.156.396.708.1332.2166.2364.2544.1212
APS:	24.48.96.180.318.576.975.1569.277.973.1519.1573.844.1878.606
SPS:	24.48.54.42.0.0.0.0.0.0.0.0.0.0.0.0
LPS:	0.0.0.0.0.0.0.0.0.0.0.0.0.0.0.606
DDS:	24.42.39.15.0.0.0.0.0.0.0.0.0.0.0.0
ΔDS:	**0.0.0.0.0.0.0.0.0.0.0.0.0.0.0.120**
CyS:	0.0.0.3.6.1.9.18.31.48.39.46.54.30.16.1

$G_{8.57}$

TPS:	0.0.0.0.0.0.36.108.216.696.1560.2220.2820.3132.1308
APS:	24.48.96.192.354.636.1086.1668.352.1288.1936.1784.1336.172.654
SPS:	24.48.66.0.0.0.0.0.0.0.0.0.0.0.0.0
LPS:	0.0.0.0.0.0.0.0.0.0.0.0.0.0.0.654
DDS:	**24.48.48.0.0.0.0.0.0.0.0.0.0.0.0.0**
ΔDS:	**0.0.0.0.0.0.0.0.0.0.0.0.0.0.0.120**
CyS:	0.0.0.0.6.7.12.27.28.24.54.76.54.36.16.1

$G_{8.58}$

TPS:	0.0.0.0.0.12.36.108.384.654.1404.2370.2328.2808.1380
APS:	24.48.96.180.333.621.1011.1557.280.1066.1780.1724.1018.430.690
SPS:	24.48.69.0.0.0.0.0.0.0.0.0.0.0.0.0
LPS:	0.0.0.0.0.0.0.0.0.0.0.0.0.0.0.690
DDS:	24.42.54.0.0.0.0.0.0.0.0.0.0.0.0.0
ΔDS:	***0.0.0.0.0.0.0.0.0.0.0.0.0.0.0.120***
CyS:	0.0.0.3.3.1.21.21.16.42.42.52.60.24.16.1

$G_{8.59}$

TPS:	0.0.0.0.0.0.0.48.240.816.1572.2346.2796.2910.1548
APS:	24.48.96.192.369.657.1071.1683.430.1354.817.1151.1387.181.774
SPS:	24.48.81.0.0.0.0.0.0.0.0.0.0.0.0.0
LPS:	0.0.0.0.0.0.0.0.0.0.0.0.0.0.0.774
DDS:	***24.48.48.0.0.0.0.0.0.0.0.0.0.0.0.0***
ΔDS:	***0.0.0.0.0.0.0.0.0.0.0.0.0.0.0.120***
CyS:	0.0.0.0.3.11.21.15.18.39.66.69.48.33.16.1

$G_{8.60}$

TPS:	0.0.0.0.0.0.12.132.276.576.1344.2436.2928.2748.1512
APS:	24.48.96.192.354.612.1056.1716.370.1168.1822.1258.1366.210.756
SPS:	24.48.66.48.0.0.0.0.0.0.0.0.0.0.0.0
LPS:	0.0.0.0.0.0.0.0.0.0.0.0.0.0.0.756
DDS:	24.48.36.12.0.0.0.0.0.0.0.0.0.0.0.0
ΔDS:	***0.0.0.0.0.0.0.0.0.0.0.0.0.0.0.120***
CyS:	0.0.0.0.6.11.6.15.48.33.30.69.72.36.16.1

Vertex Orbits: {1}; {2,6,9}; {3,4,7,10,13,14}; {5,8,11,12,15,16}

All graphs shown in Figure 8.16 are well discriminated both by sequences (see above) and topological (2D) indices (Table 8.16): $IP(CJD) \neq IP(CFD) \neq IP(SZD)$. All these graphs have the same vertex orbit structure (see above). The pair $G_{8.57}$: $G_{8.59}$ shows degenerate *DDS* and, of course the corresponding Wiener and Harary degenerate indices. For this pair, the *SPS* is not degenerated. This family represents a cluster among the cubic cages with 16 vertices and *girth*[46] (i.e. the smallest circuit in a cage) ≥4: they all are *FHΔ* graphs, with degenerate *ΔDS* and degenerate indices based on detours. The *LPS* fully discriminates among these structures.

As shown above, the $FH\Delta$ graphs show a minimal value for the detour-based indices: $IP(M\Delta)$; $M = CJ$, CF and SZ, $\min IP(M\Delta) = \binom{N}{2}$ and the corresponding edge-computed indices show a minimal value, in these cubic graphs, $\min IE(M\Delta) = |E| = 3N/2$.

Table 8.16. Topological Indices for the Graphs of Figure 8.16.

I	$G_{8.56}$	$G_{8.57}$	$G_{8.58}$	$G_{8.59}$	$G_{8.60}$
W	285	264	270	264	276
H_{De}	61.7500	64.0000	63.0000	64.0000	63.0000
w	1800	1800	1800	1800	1800
$H_{\Delta e}$	7.9999	7.9999	7.9999	7.9999	7.9999
$IP(CFD)$	4692	4059	4035	4050	4476
$IE(CFD)$	942	804	981	981	1149
$IP(CF\Delta)$	120	120	120	120	120
$IE(CF\Delta)$	24	24	24	24	24
$IP(CJD)$	4194	3558	3408	3576	4146
$IE(CJD)$	942	804	762	981	1149
$IP(CJ\Delta)$	120	120	120	120	120
$IE(CJ\Delta)$	24	24	24	24	24
$IP(SZD)$	4848	4098	4002	4104	4734
$IE(SZD)$	942	804	762	981	1149
$IP(SZ\Delta)$	120	120	120	120	120
$IE(SZ\Delta)$	24	24	24	24	24

As they represent a family is supported by the structure of their edge orbits of automorphism (Table 8.17 - as given by *MOLORD* algorithm).

It can be seen that, for the first three graphs, $G_{8.56}$ to $G_{8.58}$, the orbits denoted by A to E are identical . Only the orbit F is different. These three graphs may be called *basic* members of the family. The two remaining graphs may be viewed as *derivative* members since $G_{8.59}$ shows a combination of the F orbits of $G_{8.57}$ and $G_{8.58}$ with a new orbit E, while $G_{8.60}$ represent a different combination of the same F orbits but preserving the E orbit of the basic members of family. The structure of edge orbits was confirmed by calculating the Wiener index of rank 3 on the distance matrix of their line graphs, $^3W_{D(L1)}$. Any other mixing of the edge orbits (i.e. changing in their connectivity) provides graphs no more belonging to the family of $FH\Delta$ graphs with girth ≥ 4.

Table 8.17. Edge Orbits of Automorphism of the Graphs of Figure 8.16.

G	A	B	C	D	E	F
$G_{8.56}$	(1,2), (1,6), (1,9)	(3,13), (4,14), (7,10)	(5,8), (11,12),(15,16)	(5,11), (8,16), (12,15)	(2,3), (2,4), (6,7), (6,14), (9,10), (9,13)	(3,11), (4,5), (7,16), (8,14), (10,15), (12,13)
$G_{8.57}$	(1,2), (1,6), (1,9)	(3,13), (4,14), (7,10)	(5,8), (11,12),(15,16)	(5,11), (8,16), (12,15)	(2,3), (2,4), (6,7), (6,14), (9,10), (9,13)	(3,5), (4,11), (7,8), (14,16), (10,12), (13,15),
$G_{8.58}$	(1,2), (1,6), (1,9)	(3,13), (4,14), (7,10)	(5,8), (11,12),(15,16)	(5,11), (8,16), (12,15)	(2,3), (2,4), (6,7), (6,14), (9,10), (9,13)	(3,12), (11,13), (4,8), (5,14), (7,15), (10,16)
$G_{8.59}$	(1,2), (1,6), (1,9)		(5,8), (11,12),(15,16)	(3,5), (4,11), (7,8), (14,16) (10,12), (13,15)	(2,13), (3,9), (2,14), (4,6), (6,10), (7,9)	(3,12), (11,13), (4,8), (5,14), (7,15), (10,16)
$G_{8.60}$	(1,2), (1,6), (1,9)		(5,8), (11,12),(15,16)		(2,3), (2,4), (6,7), (6,14), (9,10), (9,13)	(3,12), (11,13), (4,8), (5,14), (7,15), (10,16) (3,5), (4,11), (7,8), (14,16), (10,12), (13,15)

Another nice property is encountered in $G_{8.57}$: the degenerate rearrangements (Figure 8.17). Note that it is the unique member of this family showing such a property. The crossing process herein considered was *monocrossing* (i.e. a pair of edges interchange one of the two endpoints while the other one remain as an already existing edge belonging to a different orbit – see below) and *triplecrossing* (i.e. three pairs of edges are interchanged as above mentioned). The trivial full crossing (possible in all basic members of family) was not considered.

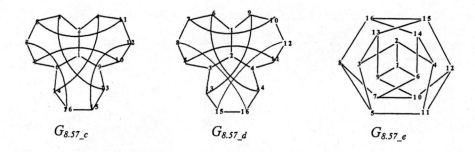

$$G_{8.57\_c} \qquad\qquad G_{8.57\_d} \qquad\qquad G_{8.57\_e}$$

Figure 8.17. Degenerate rearrangements of $G_{8.57}$

The degenerate rearrangements of $G_{8.57}$ are as follows:

(1) Monocrossing: $G_{8.57}$ (D/C) $\longrightarrow G_{8.57}$
　　　(the boldface pairs are edges belonging to the C orbit)

(a) $(5,11) \rightarrow (8,11)$　(b) $(5,11) \rightarrow (5,12)$　(c) $(12,15) \rightarrow (12,16)$
　　$(8,16) \rightarrow (5,16)$　　　$(12,15) \rightarrow (11,15)$　　$(8,16) \rightarrow (8,15)$

(2) Triplecrossing:

(a) $G_{8.57}$ (F/B) $\longrightarrow G_{8.57}$
　　(the boldface pairs are edges belonging to the B orbit)

$$(3,5) \rightarrow (5,13) \qquad (4,11) \rightarrow (11,14) \qquad (7,8) \rightarrow (8,10)$$
$$(13,15) \rightarrow (3,15) \qquad (14,16) \rightarrow (4,16) \qquad (10,12) \rightarrow (7,12)$$

(b) $G_{8.57}$ (E/B) $\longrightarrow G_{8.57}$
　　(the boldface pairs are edges belonging to the B orbit)

$$(2,3) \rightarrow (2,13) \qquad (2,4) \rightarrow (2,14) \qquad (6,7) \rightarrow (6,10)$$
$$(9,13) \rightarrow (3,9) \qquad (6,14) \rightarrow (4,6) \qquad (9,10) \rightarrow (7,9)$$

The graph $G_{8.57\text{-}d}$ (Figure 8.17) represents the monocrossing rearrangement (1, c) and $G_{8.57\text{-}e}$ denotes the triplecrossing rearrangement (2, b).

When the Manhattan distance, D_M, was evaluated by using the calculated sequences, the dissimilarity (increasing) ordering was as follows:

Table 8.18. D_M of Structures of Figure 8.16, by *TPS*

	$G_{8.57}$	$G_{8.58}$	$G_{8.59}$	$G_{8.60}$
$G_{8.56}$	873	465	1023	867
$G_{8.57}$	0	708	432	678
$G_{8.58}$		0	672	582
$G_{8.59}$			0	510

Increasing dissimilarity ordering is: $(G_{8.57} - G_{8.59})$, $(G_{8.56} - G_{8.58})$, $(G_{8.59} - G_{8.60})$, $(G_{8.58} - G_{8.60})$, $(G_{8.58} - G_{8.59})$, **$(G_{8.57} - G_{8.60})$, $(G_{8.57} - G_{8.58})$, $(G_{8.56} - G_{8.60})$, $(G_{8.56} - G_{8.57})$, $(G_{8.56} - G_{8.59})$**.

Table 8.19. D_M of Structures of Figure 8.16, by *APS*

	$G_{8.57}$	$G_{8.58}$	$G_{8.59}$	$G_{8.60}$
$G_{8.56}$	3582	2322	4420	3558
$G_{8.57}$	0	1356	2142	1050
$G_{8.58}$		0	2946	1580
$G_{8.59}$			0	1534

Increasing dissimilarity ordering is: $(G_{8.57} - G_{8.60})$, $(G_{8.57} - G_{8.58})$, $(G_{8.59} - G_{8.60})$, $(G_{8.58} - G_{8.60})$, $(G_{8.57} - G_{8.59})$, **$(G_{8.56} - G_{8.58})$, $(G_{8.58} - G_{8.59})$, $(G_{8.56} - G_{8.60})$, $(G_{8.56} - G_{8.57})$, $(G_{8.56} - G_{8.59})$**.

Table 8.20 D_M of Structures of Figure 8.16, by *SPS*

	$G_{8.57}$	$G_{8.58}$	$G_{8.59}$	$G_{8.60}$
$G_{8.56}$	54	57	69	18
$G_{8.57}$	0	3	15	48
$G_{8.58}$		0	12	51
$G_{8.59}$			0	63

Increasing dissimilarity ordering is: $(G_{8.57} - G_{8.58})$, $(G_{8.58} - G_{8.59})$, $(G_{8.57} - G_{8.59})$, $(G_{8.56} - G_{8.60})$, $(G_{8.57} - G_{8.60})$, **$(G_{8.58} - G_{8.60})$, $(G_{8.56} - G_{8.57})$, $(G_{8.56} - G_{8.58})$, $(G_{8.59} - G_{8.60})$, $(G_{8.56} - G_{8.59})$**.

Table 8.21. D_M of Structures of Figure 8.16, by *DDS*

	$G_{8.57}$	$G_{8.58}$	$G_{8.59}$	$G_{8.60}$
$G_{8.56}$	30	30	30	12
$G_{8.57}$	0	12	0	24
$G_{8.58}$		0	12	36
$G_{8.59}$			0	24

Increasing dissimilarity ordering is: $(G_{8.57} = G_{8.59})$, $(G_{8.57} - G_{8.58})$, $(G_{8.58} - G_{8.59})$, $(G_{8.56} - G_{8.60})$, $(G_{8.57} - G_{8.60})$, $(G_{8.59} - G_{8.60})$, $(G_{8.56} - G_{8.57})$, $(G_{8.56} - G_{8.58})$, $(G_{8.56} - G_{8.59})$, $(G_{8.58} - G_{8.60})$.

Table 8.22. D_M of Structures of Figure 8.16, by CyS

	$G_{8.57}$	$G_{8.58}$	$G_{8.59}$	$G_{8.60}$
$G_{8.56}$	99	60	112	107
$G_{8.57}$	0	111	81	100
$G_{8.58}$		0	86	131
$G_{8.59}$			0	117

Increasing dissimilarity ordering is: $(G_{8.56} - G_{8.58})$, $(G_{8.57} - G_{8.59})$, $(G_{8.58} - G_{8.59})$, $(G_{8.56} - G_{8.57})$, $(G_{8.57} - G_{8.60})$, $(G_{8.56} - G_{8.60})$, $(G_{8.57} - G_{8.58})$, $(G_{8.56} - G_{8.59})$, $(G_{8.59} - G_{8.60})$, $(G_{8.58} - G_{8.60})$.

By following the occurrence of graphs within the above pair ordering, the most dissimilar three graphs according to each sequence are: *TPS* $(G_{8.56}, G_{8.57}, G_{8.60})$; *APS* $(G_{8.56}, G_{8.58}, G_{8.59})$; *SPS* $(G_{8.56}, G_{8.59}, G_{8.60})$; *DDS* $(G_{8.56}, G_{8.58}, G_{8.60})$ and *CyS* $(G_{8.58}, G_{8.59}, G_{8.60})$, with the most dissimilar three graphs cf. to all five criteria: $G_{8.56}$, $G_{8.59}$ and $G_{8.60}$. Conversely, the most similar pair is $G_{8.57} - G_{8.58}$. In a larger set of structures such an analysis would be, of course, more reliable.

For other aspects about symmetry and similarity in molecular graphs, the reader can consult refs.[1, 98-102]

8.7. HIGHLY SYMMETRIC STRUCTURES

A molecular structure having all substructures of a given dimension (i.e. the number of its edges $|e|$) equivalent is called a structure S_e transitive. Thus, a transitive structure shows a single orbit of the fragments of dimension $|e|$.[103]

The present section refers to the topological symmetry of some geometrical structures, irrespectively they were already synthesized or are only *paper* molecules.

In the last two decades, the synthesists have made considerable efforts for *building*, at molecular level, highly symmetric geometric structures, in the hope that the Euclidean symmetry must induce unexpected molecular properties. Platonic polyhedra:[96, 104] tetrahedron, cube, prism and dodecahedron have been synthesized. In the last years, the fullerenes (polyhedra having faces of five and six atoms) have opened a wide field of research. Many articles deal with the synthesis and functionalization of fullerenes, but also with related theoretical aspects (quantum chemical or topological calculations).

Dendrimers, hyper-branched structures, with spherical shape and strictly tailored constitution represent another new field of interest for the scientists also referred to as *supramolecules*. They can be functionalized and used for simulating enzymatic reactions (i.e. *host-guest* reactions[105]).

As a tool for the symmetry perception the *MOLORD* algorithm[43, 53, 106] was chosen. The Layer matrices on which the indices are computed, are given (when needed) in line form.

8.7.1. Cube Orbits of Automorphism

Cube is a polyhedron having 8 vertices and 12 edges all equivalent (i.e. characterized by the same graph-theoretical parameter). Moreover, subgraphs of two edges are topologically indistinguishable. Only the fragments of three edges (and larger) can be separated. Thus, the cube is a structure S_0, S_1 and S_2 transitive. Figure 8.18 shows the cube orbits of automorphism, with respect to three edge fragments: $a\{8\}$; $b\{24\}$ and $c\{24\}$ (the number of equivalent fragments given in brackets). Representative fragments are depicted by bold line.

Values of the indices $C(LDS)_i$ and $X(LDS)_i$ are written under the corresponding structures. The ordering is here less important.

In opposition to the cube, the Möbius cube (Figure 8.18) is only S_0 transitive. Its edges ($n = 1$) show two orbits : $a\{8\}$ and $b\{4\}$. Fragments of two edges ($n = 2$) show two orbits: $a\{8\}$ and $b\{4\}$. Fragments of three edge ($n = 3$) show 6 orbits: $a\{8\}$; $b\{8\}$; $c\{16\}$; $d\{8\}$; $e\{8\}$ and $f\{8\}$.

The values of indices for the fragments of three edges were derived from the L_4 line graph, since the index $X(LDS)_i$ does not discriminate between the fragments of type e and f. These fragments each represent only four distinct (i.e. real) fragments in L_0. Of course, other layer matrices and other descriptors may be used for better discriminating of fragments.

8.7.2. Homeomorphic Transforms of Tetrahedron

An insertion of vertices of degree two on the edges of a graph is called a homeomorphic transform.[46] In molecular graphs such a transformation can be achieved by various fragments: -CH$_2$- (methylene), -CH$_2$-CH$_2$- (ethylene) etc. and it results in the lowering of the strain energy of small rings. Figure 8.19 illustrates some possible homeomorphic transforms of tetrahedron (another Platonic solid[104]), which, completed by additional connections, could lead to highly symmetric structures. Similar *reactions* are suggested in Figure 8.20, actually starting from the Schlegel projection of tetrahedron.

Cube; $n = 3$; $m = 3$.

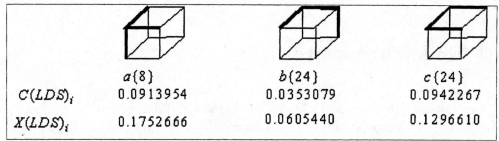

	$a\{8\}$	$b\{24\}$	$c\{24\}$
$C(LDS)_i$	0.0913954	0.0353079	0.0942267
$X(LDS)_i$	0.1752666	0.0605440	0.1296610

Möbius cube;

$n = 1$; $m = 1$. $\qquad\qquad\qquad\qquad$ $n = 2$; $m = 2$.

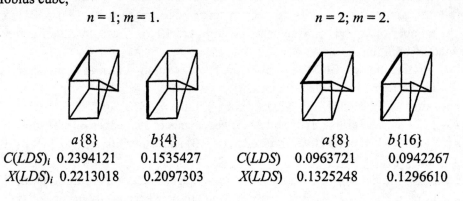

	$a\{8\}$	$b\{4\}$		$a\{8\}$	$b\{16\}$
$C(LDS)_i$	0.2394121	0.1535427	$C(LDS)$	0.0963721	0.0942267
$X(LDS)_i$	0.2213018	0.2097303	$X(LDS)$	0.1325248	0.1296610

$n = 3$; $m = 4$.

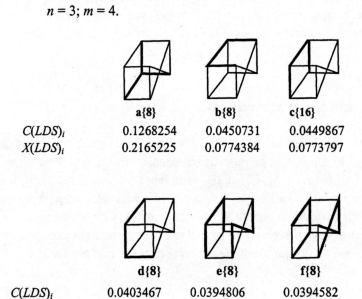

	$a\{8\}$	$b\{8\}$	$c\{16\}$
$C(LDS)_i$	0.1268254	0.0450731	0.0449867
$X(LDS)_i$	0.2165225	0.0774384	0.0773797

	$d\{8\}$	$e\{8\}$	$f\{8\}$
$C(LDS)_i$	0.0403467	0.0394806	0.0394582
$X(LDS)_i$	0.0765613	0.0781188	0.0781098

Figure 8.18. Fragments of cube and Möbius cube and local values $SI_m(i_n)$
cf. *MOLORD* algorithm (fragment occurrence in brackets).

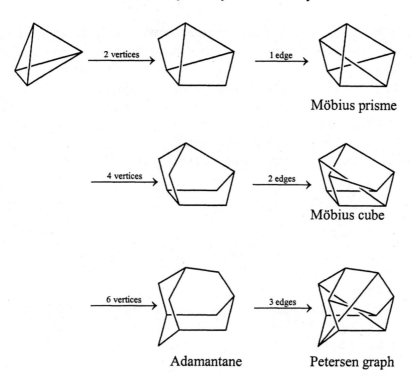

2 vertices → 1 edge →

Möbius prisme

4 vertices → 2 edges →

Möbius cube

6 vertices → 3 edges →

Adamantane   Petersen graph

Figure 8.19. Homeomorphic transforms of tetrahedron.

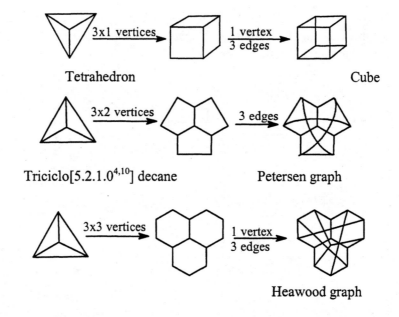

Tetrahedron   3x1 vertices →   1 vertex / 3 edges →   Cube

Triciclo[5.2.1.0^{4,10}] decane   3x2 vertices →   3 edges →   Petersen graph

3x3 vertices →   1 vertex / 3 edges →   Heawood graph

Figure 8.20. Homeomorphic transforms of tetrahedron (Schlegel projections).

Note that some of the *intermediates* in the above figures are real chemical compounds. Among these, adamantane is considered the *stabilomere* in the series of C_{10} cyclic hydrocarbons.[107] As a molecular graph, adamantane shows two vertex orbits: $a\{4\}$ and $b\{6\}$ but its edges are all equivalent (see its line graph L_1, in Figure 8.21), the graph being S_1 transitive. On the other hand, adamantane is a bipartite graph, so it is not surprising that its edges are equivalent whereas its vertices are not. The equivalence of edges (i.e. covalent bonds) in six member rings (practically without tension), condensed by following the tetrahedron faces (see Figure 8.19) explains the exceptional stability of adamantane.

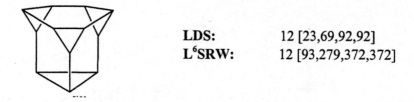

LDS: 12 [23,69,92,92]
L⁶SRW: 12 [93,279,372,372]

Figure 8.21. Line graph L_1 of adamantane and its matrices **LDS** and **L^{e}SRW**.

8. 7. 3. Other Routes for Some Highly Symmetric Structures

Successive transforms of the Möbius cube (Figure 8.22) could lead to the well known, symmetric, graphs: Petersen[108] and Heawood,[109] respectively. Their actual pictorial representation is Möbius cube patterned.

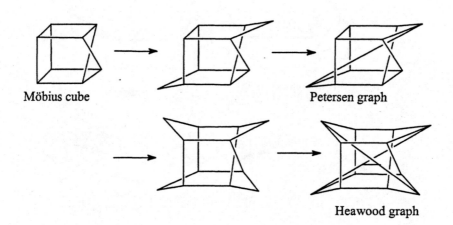

Figure 8.22. Homeomorphic transforms of Möbius cube.

Furthermore, the Heawood graph can be derived from the cube and the diamantane, a hydrocarbure obtained by *condensing* two adamantane units, [110] as shown in Figure 8.23.

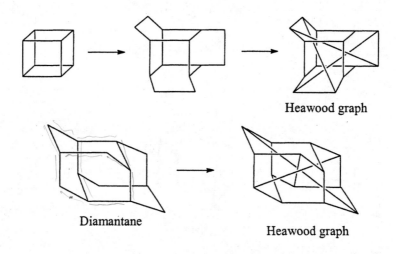

Heawood graph

Diamantane

Heawood graph

Figure 8.23. *Synthesis* of Heawood graph.

From Figures 8.19-8.23 it is obvious that the *retrosynthesys* of structures of the Peterson and Heawood graphs type could follow various ways and various "intermediates".

Finally, the célebre Desargues-Levi graph,[111] used as a reaction graph,[96] is presented. Diudea[47] proposed its derivation from a tetramantane. Figure 8.24 shows this *synthesis* and the Desargues-Levi graph designed by Randić, as two interlocked adamantanes.

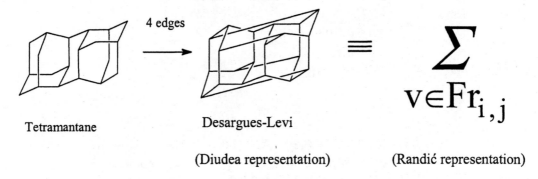

4 edges

$$\equiv \sum_{v \in Fr_{i,j}}$$

Tetramantane

Desargues-Levi

(Diudea representation)

(Randić representation)

Figure 8.24. *Synthesis* of Desargues-Levi graph from tetramantane.

Petersen graph and two of generalized Petersen graphs, along with the Heawood graph, characterized according to their sequences and vertex orbits are presented in Figure 8.25.

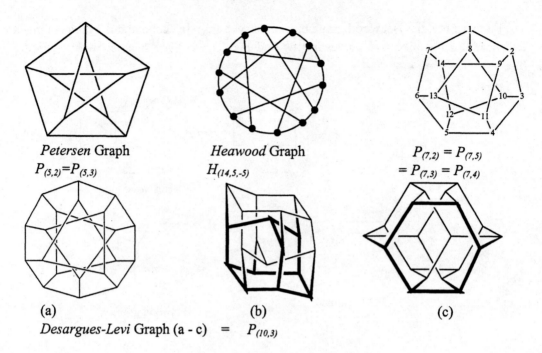

Petersen Graph
$P_{(5,2)}=P_{(5,3)}$

Heawood Graph
$H_{(14,5,-5)}$

$P_{(7,2)} = P_{(7,5)}$
$= P_{(7,3)} = P_{(7,4)}$

(a)   (b)   (c)

Desargues-Levi Graph (a - c) = $P_{(10,3)}$

Figure 8.25. Highly symmetric graphs.

The sequences of the graphs shown in Figure 8.25 are as follows:

Petersen Graph = $P_{(5,2)}=P_{(5,3)}$

TPS:	0.0.0.0.0.0.120.360.240
APS:	15.30.60.120.180.240.300.300.120
DDS:	15.30.0.0.0.0.0.0.0.0
ΔDS:	0.0.0.0.0.0.0.0.15.30
CyS:	0.0.0.0.12.10.0.15.10.0
Vertex Orbits:	{all vertices}

Heawood Graph = $H_{(14,5,-5)}$

TPS:	0.0.0.0.0.0.0.0.672.672.1680.1008.1008
APS:	21.42.84.168.336.504.840.1176.1680.1680.56.1008.504
DDS:	21.42.28.0.0.0.0.0.0.0.0.0.0.0
ΔDS:	0.0.0.0.0.0.0.0.0.0.0.0.42.49
CyS:	0.0.0.0.0.28.0.21.0.84.0.28.0.1
Vertex Orbits:	{all vertices}

Generalized Petersen Graph, $P_{(7,2)} = P_{(7,5)} = P_{(7,3)} = P_{(7,4)}$
$$H_{(14,6,-4,5,7,-5,4,-6,-5,4,-4,-7,4,-4,5)}$$

TPS:	0.0.0.0.0.0.56.84.392.826.1120.1512.840
APS:	21.42.84.168.301.525.826.1148.1582.868.1596.1176.420
DDS:	21.42.28.0.0.0.0.0.0.0.0.0.0.0
ΔDS:	0.0.0.0.0.0.0.0.0.0.0.0.0.91
CyS:	0.0.0.0.7.7.16.21.14.35.42.28.14.1
Vertex Orbits:	{1,2,3,4,5,6,7}; {8,9,10,11,12,13,14}

Desargues-Levi Graph = $P_{(10,3)}$

TPS:

0.0.0.0.0.0.0.0.240.240.1440.1200.6240.6480.14160.10080.17760.7440.4560

APS:

30.60.120.240.480.840.1560.80.2000.1600.80.520.680.2000.1720.2160.2080.88
0.2280

DDS:	30.60.60.30.10.0.0.0.0.0.0.0.0.0.0.0.0.0.0.0
ΔDS:	0.0.0.0.0.0.0.0.0.0.0.0.0.0.0.0.0.0.90.100
CyS:	0.0.0.0.0.20.0.30.0.132.0.150.0.420.0.300.0.100.0.1
Vertex Orbits:	{all vertices}

A *generalized Petersen graph*,[112, 113] denoted $P_{(n,j)}$, is a cyclic comb graph, composed of a cycle C_n and n branches of unit length. By joining all the terminal vertices of the comb graph with their (clockwise) j^{th} neighbors one obtains the $P_{(n,j)}$ graph. In this notation, the original Petersen graph is $P_{(5,2)} = P_{(5,3)}$. It is a S_5 transitive graph, showing various geometric symmetries, function of its pictorial representation.[114] For example, the representation in Figure 8.25 the apparent symmetry is D_{5h}. It is neither a *FHΔ* graph, nor a Hamiltonian circuit graph (see below).

A *Hamiltonian wheel graph*, denoted $H_{(n,j)}$,[113] is constructed by periodic joining of n points of a cycle graph, C_n, clockwise or anticlockwise. In this notation, the Heawood graph is $H_{(14, 5,-5)}$. Some Petersen generalized graphs, such as $P_{(7,2)}$ (= $P_{(7,5)}$= $P_{(7,3)}$= $P_{(7,4)}$), are at the same time Hamiltonian wheel graphs (possess N-membered circuits - see *CyS*, in Figure 8.25), such as they may be symbolized like the Heawood graph. However, in many cases, as in the case of $P_{(7,2)}$, such a symbol is cumbersome. This graph is the single *FHΔ* graph in Figure 8.25.

The Desargues-Levi Graph[111] (Figure 8.25, a - c) is another example of generalized Petersen graph, symbolized as $P_{(10,3)}$. It is also an S_5 transitive graph. Its cycles are all even-membered ones.

For these graphs, Table 8.23 includes the values of most important topological indices ussed in this book.

Table 8.23. Topological Indices of Some Highly Symmetric Graphs

I	Petersen	Heawood	$P(7,2)$	Desargues-Levi
W	75	189	189	500
H_{De}	30.0000	51.3334	51.3334	89.5000
w	390	1141	1183	3520
$H_{\Delta e}$	5.2083	7.2692	7.0000	10.2632
$IP(CFD)$	405	2415	2499	11740
$IE(CFD)$	135	1029	602	3000
$IP(CF\Delta)$	90	217	91	460
$IE(CF\Delta)$	60	60	21	30
$IP(CJD)$	405	2415	2177	11410
$IE(CJD)$	135	1029	602	3000
$IP(CJ\Delta)$	90	217	91	460
$IE(CJ\Delta)$		60	21	30
$IP(SZD)$	405	2779	2429	13240
$IE(SZD)$	135	1029	602	3000
$IP(SZ\Delta)$	405	2443	91	10090
$IE(SZ\Delta)$	135	1029	21	3000

REFERENCES

1. Balasubramanian, K. Computer Generation of Nuclear Equivalence Classes Based on the Three-Dimensional Molecular Structure. *J. Chem. Inf. Comput. Sci.* **1995**, *35*, 243-250; Computational Strategies for the Generation of Equivalence Classes of Hadamard Matrices. *Ibid.* **1995**, *35*, 581-589; Computer Perception of Molecular Symmetry. *Ibid.* **1995**, *35*, 761-770.

2. Gutman, I.; Polansky, O.E. *Mathematical Concepts in Organic Chemistry.* Springer, Berlin, 1986, Chap. 9, pp. 108-116.

3. Polansky, O. E. in: *Chemical Graph Theory. Introduction and Fundamentals.* eds. Bonchev, D.; Rouvray, D.H. Abacus Press/Gordon & Breach, New York, 1991, Chap. 2, pp. 41-96.

4. Razinger, M.; Balasubramanian, K.; Munk, M.E. Graph Automorphism Perception Algorithms in Computer-Enhanced Structure Elucidation. *J. Chem. Inf. Comput. Sci.* **1993**, *33*, 197-201.

5. Bangov, I.P. Graph Isomorphism: A Consequence of the Vertex Equivalence *J. Chem. Inf. Comput. Sci.* **1994**, *34*, 318-324.

6. Klin, M.H.; Zefirov, N.S. Group Theoretical Approach to the Investigation of Reaction Graphs for Highly Degenerate Rearrangements of Chemical Compounds 2 Fundamental Concepts. *Commun. Math. Comput. Chem. (MATCH)*, **1991**, *26*, 171-190.

7. Balasubramanian, K. Computer Generation of Automorphism Graphs of Weighted Graphs. *J. Chem. Inf. Comput. Sci.* **1994**, *34*, 1146-1150.

8. Rosen, R. in: Johnson, M. A.; Maggiora, G. M. Eds. *Concepts and Applications of Molecular Similarity*, Wiley, New York, 1990, Chap. 12, pp. 369-382.

9. Ivanciuc, O. Canonical Numbering and Constitutional Symmetry, in: *The Enciclopedia of Computational Chemistry*, Eds: Schleyer, P. v.; Allinger, N.L.; Clark, T.; Gasteiger, J.; Kollman, P.A.; Schaefer III, H.F.; Schreiner, P.R. John Wiley&Sons, Chichester, 1998, pp.176-183.

10. Balasubramanian, K. Computational Techniques for the Automorphism Groups of Graphs. J. Chem. Inf. Comput. Sci. **1994**, 34, 621-626.

11. Bohanec, S.; Perdih, M. Symmetry of Chemical Structures: A Novel Method of Graph Automorphism Group Determination. *J. Chem. Inf. Comput. Sci.* **1993**, *33*, 719-726.

12. Shelley, C.A.; Munk, M.E. Signal Number Prediction in Carbon-13 Nuclear Magnetic Resonance Spectrometry, *Anal. Chem.* **1978**, *50*, 1522-1527.

13. Morgan, H. The generation of a unique machine description for chemical structures. A technique developed at Chemical Abstracts Service, *J. Chem. Doc.* **1965**, *5*, 107-113.

14. Randić, M. On the Recognition of Identical Graphs Representing Molecular Topology. *J. Chem. Phys.* **1974**, *60*, 3920-3928.

15. Randić, M. On Unique Numbering of Atoms and Unique Codes for Molecular Graphs. J. Chem. Inf. Comput. Sci. **1975**, 15, 105-108.

16. Randić, M. On Discerning Symmetry Properties of Graphs. *Chem. Phys. Lett.* **1976**, *42*, 283-287.

17. Shelley, C.A.; Munk, M.E. Computer Perception of Topological Symmetry, *J. Chem. Inf. Comput. Sci.* **1977**, *17*, 110-113.

18. Jochum, C.; Gasteiger, J. Canonical Numbering and Constitutional Symmetry. *J. Chem. Inf. Comput. Sci.* **1977**, *17*, 113-117.

19. Jochum, C.; Gasteiger, J. On the Misinterpretation of Our Algorithm for the Perception of Constitutional Symmetry. *J. Chem. Inf. Comput. Sci.* **1979**, *19*, 49-50.

20. Moreau, G. A Topological Code for Molecular Structures. A Modified Morgan Algorithm. *Nouv. J. Chim.* **1980**, *4*, 17-22.

21. Randić, M.; Brissey, G.M.; Wilkins, C.L. Computer Perception of Topological Symmetry via Canonical Numbering of Atoms. *J. Chem. Inf. Comput. Sci.* **1981**, *21*, 52-59.

22. Golender, V. E.; Drboglav, V.; Rosenblit, A. B.J. Graph Potentials Method and Its Application for Chemical Information Processing. *J. Chem. Inf. Comput. Sci.* **1981**, *21*, 196-204.

23. Hendrickson, J.B.; Toczko, A.G. Unique Numbering and Cataloguing of Molecular Structures. *J. Chem. Inf. Comput. Sci.* **1983**, *23*, 171-177.

24. Balaban, A. T.; Mekenyan, O.; Bonchev, D. Unique Description of Chemical Structures Based on Hierarchically Ordered Extended Connectivities (HOC Procedures). I. Algorithms for Finding graph Orbits and Cannonical Numbering of Atoms, *J. Comput. Chem.* **1985**, *6*, 538-551; Unique Description of Chemical Structures Based on Hierarchically Ordered Extended Connectivities (HOC Procedures). II. Mathematical Proofs for the HOC Algorithm, *Ibid.* **1985**, *6*, 552-561.

25. Fujita, S. Description of Organic Reactions Based on Imaginary Transition Structures. 1. Introduction of New Concepts. *J. Chem. Inf. Comput. Sci.* **1986**, *26*, 205-212.

26. Filip, P.A.; Balaban, T.-S.; Balaban, A. T. A New Approach for Devising Local Graph Invariants: Derived Topological Indices with Low Degeneracy and Good Correlation Ability, *J. Math. Chem.* **1987**, *1*, 61-83.

27. Bersohn, M. A Matrix Method for Partitioning the Atoms of a Molecule Into Equivalence Classes. *Comput. Chem.* **1987**, *11*, 67-72.

28. Fujita, S. Canonical Numbering and Coding of Reaction Center Graphs and Reduced Reaction Center Graphs Abstracted from Imaginary Transition Structures. A Novel Approach to the Linear Coding of Reaction Types *J. Chem. Inf. Comput. Sci.* **1988**, *28*, 137-142.

29. Gasteiger, J.; Ihlenfeldt, W. D.; Rose, P.; Wanke, R. Computer-Assisted Reaction Prediction and Synthesis Design *Anal. Chim. Acta*, **1990**, *235*, 65-75.

30. Liu, X.; Balasubramanian, K.; Munk, M.E. Computational Techniques for Vertex Partitioning of Graphs. *J. Chem. Inf. Comput. Sci.* **1990**, *30*, 263-269.

31. Liu, X.; Klein, D. J. The Graph Isomorphism Problem. *J. Comput. Chem.* **1991**, *12*, 1243-1251.

32. Rücker, G.; Rücker, C. Computer Perception of Constitutional (Topological) Symmetry: TOPSYM, a Fast Algorithm for Partitioning Atoms and Pairwise Relations among Atoms into Equivalence Classes *J. Chem. Inf. Comput. Sci.* **1990**, *30*, 187-191.

33. Rücker, G.; Rücker, C. On Using the Adjacency Matrix Power Method for Perception of Symmetry and for Isomorphism Testing of Highly Intricate Graphs. *J. Chem. Inf. Comput. Sci.* **1991**, *31*, 123-126; Isocodal and Isospectral Points, Edges, and Pairs in Graphs and How To Cope with Them in Computerized Symmetry Recognition. *Ibid.* **1991**,*31*,422-427.

34. Figueras, J. Morgan Revisited, *J. Chem. Inf. Comput. Sci.* **1992**, *32*, 153-157.

35. Kvasnička, V.; Pospichal, J. Maximal Common Subgraphs of Molecular Graphs, *Reports in Molecular Theory*, **1990**, *1*, 99-106.

36. Gasteiger, J.; Hanebeck, W.; Schulz, K.-P. Prediction of Mass Spectra from Structural Information. *J. Chem. Inf. Comput. Sci.* **1992**, *32*, 264-271.

37. Ihlenfeldt, W.D.; Gasteiger, J. *J. Comput. Chem.* **1994**, *15*, 793-813.

38. Kvasnička, V.; Pospichal, J. Fast Evaluation of Chemical Distance by Tabu Search Algorithm. *J. Chem. Inf. Comput. Sci.* **1994**, *34*, 1109-1112.

39. Wipke, W. T.; Dyott, T.M. Stereochemically Unique Naming Algorithm. *J. Am. Chem. Soc.* **1974**, *96*, 4834-4842.

40. Mekenyan, O.; Balaban, A.T.; Bonchev, D. Unique Description of Chemical Structures Based on Hierarchically Ordered Extended Connectivities (HOC Procedures). VI. Condensed Benzenoid Hydrocarbons and Their 1H-NMR Chemical Shifts. *J. Magn. Reson.* **1985**, *63*, 1-13.

41. Balasubramanian, K. Graph Theoretical Perception of Molecular Symmetry. *Chem. Phys. Lett.* **1995**, *232*, 415-423.

42. Balasubramanian, K. Computer Perception of NMR Symmetry. *J. Magn. Reson.* **1995**, *A112*, 182-190.

43. Diudea, M.V.; Horvath, D.; Bonchev, D. MOLORD Algorithm and Real Number Subgraph Invariants. *Croat. Chem. Acta*, **1995**, *68*, 131-148.

44. Trinajstić, N. *Chemical Graph Theory* : CRC Press: Boca Raton, FL, 1983, Vol.2, Chap. 4.

45. Bertz, S.H. Branching in Graphs and Molecules, *Discr. Appl. Math.* **1988**, *19*, 65-83.

46. Harary, F. *Graph Theory*, Addison - Wesley, Reading, M.A., 1969.

47. Diudea, M.V. *Multiple Line Graphs*, MATH/CHEM/COMP'94 Conference, Dubrovnik, Croatia.

48. Diudea, M.V.; Horvath, D.; Kacso', I.E; Minailiuc, O. M.; Parv, B. Centricities in Molecular Graphs. The MOLCEN Algorithm, *J. Math. Chem.* **1992**, *11*, 259-270.

49. Bonchev, D.; Balaban, A.T.; Randić, M. The Graph Center Concept for Polycyclic Graphs, *Int. J. Quantum Chem.* **1981**, *19*, 61-82.

50. Bonchev, D.; Mekenyan, O.; Balaban, A.T. Iterative Procedure for the Generalized Graph Center in Polycyclic Graphs, *J. Chem. Inf. Comput. Sci.* **1989**, *29*, 91-97.

51. Dobrynin, A. Degeneracy of some matrix invariants and derived topological indices, *J. Math. Chem.* **1993**, *14*, 175 - 184.

52. Ivanciuc, O.; Balaban, T.S.; Balaban, A.T. Chemical Graphs with DegenerateTopological Indices Based of Information on Distance, *J. Math. Chem.* **1993**, *12*, 21-31.

53. Diudea, M.V. Layer Matrices in Molecular Graphs, *J. Chem. Inf. Comput. Sci.* **1994**, *34*, 1064-1071.

54. Balaban, A. T.; Motoc, I. Correlations between Octane Number and Topological Indices of Alkanes, *Commun. Math. Comput. Chem. (MATCH)*, **1979**, *5*, 197-218.

55. Diudea, M.V.; Ivanciuc, O. *Molecular Topology*, COMPREX, Cluj, Romania, 1995, (in Romanian).

56. Kier, L.B.; Hall, L.H. *Molecular Connectivity in Chemistry and Drug Research*, Acad. Press, 1976.

57. Randić, M. On Characterization of Molecular Branching, *J. Am. Chem. Soc.* **1975**, *97*, 6609-6615.

58. Balaban, A. T. Topological Index J for Heteroatom-Containing Molecules Taking into Account Periodicities of Element Properties, *Commun. Math. Comput. Chem. (MATCH)*, **1986**, *21*, 115-122.

59. Diudea, M.V.; Silaghi-Dumitrescu, I. Valence Group Electronegativity as a Vertex Discriminator. *Rev. Roum. Chim.* **1989**, *34*, 1175-1182.

60. Diudea, M.V.; Kacso', I.E; Topan, M.I. A QSPR/QSAR Study by Using New Valence Group Carbon - Related Electronegativities, *Rev. Roum. Chim.* **1996**, *41*, 141-157.

61. Allread, A. L.; Rochow, E. G. *J. Inorg. Nucl. Chem.* **1958**, *5*, 264.

62. Diudea, M. V.; Topan, M.; Graovac, A. Layer Matrices of Walk Degrees, *J. Chem. Inf. Comput. Sci.* **1994**, *34*, 1072-1078.

63. Diudea, M.V. Walk Numbers ^eW_M : Wiener Numbers of Higher Rank, *J. Chem. Inf. Comput. Sci.* **1996**, *36*, 535-540.

64. Rouvray, D. H. The Evolution of the Concept of Molecular Similarity. In: Johnson, M. A.;Maggiora, G. M. Eds. *Concepts and Applications of Molecular Similarity*, Wiley, New York, 1990, Chap. 2, pp. 15-42.

65. Maggiora, G. M.; Johnson, M. A. Introduction to Similarity in Chemistry. In: Johnson, M. A.; Maggiora, G. M. Eds. *Concepts and Applications of Molecular Similarity*, Wiley, New York, 1990, Chap. 1, pp. 1-13.

66. Balaban, A. T. Topological and Stereochemical Molecular Descriptors for Databases Useful in QSAR, Similarity/Dissimilarity and Drug Design, *SAR and QSAR in Environmental Research*, **1998**, *8*, 1-21.

67. Mezey, P. G. Three-Dimensional Topological Aspects of Molecular Similarity. In: Johnson, M.A.; Maggiora, G. M. Eds. *Concepts and Applications of Molecular Similarity*, Wiley, New York, 1990, Chap. 11, pp. 321-368. 68. Tremblay, J. P.; Manohar, R. *Discrete Mathematical Structures with Applications to Computer Science*, McGraw-Hill, New-York, 1975, p. 186.

69. Randić, M. Design of Molecules with Desired Properties. A Molecular Similarity Approach to Property Optimization. In: Johnson, M. A.; Maggiora, G. M. Eds. *Concepts and Applications of Molecular Similarity*, Wiley, New York, 1990, Chap. 5, pp. 77-145.

70. Tsai, C. -c.; Johnson, M. A.; Nicholson, V.; Naim, M. Eds., *Graph Theory and Topology in Chemistry*, Elsevier, Amsterdam, 1987, p. 231.

71. Balaban, A. T.; Chiriac, A.; Motoc, I.; Simon, Z. *Steric Fit in QSAR* (Lecture Notes in Chemistry, Vol. 15), Springer, Berlin, 1980, Chap. 6.

72. Dugundji, J.; Ugi, I. An Algebraic Model of Constitutional Chemistry as a Basis for Chemical Computer Programs, *Top. Curr. Chem.* **1973**, *39*, 19-64.

73. Ugi, I.; Wochner, M.A.; Fontain, E.; Bauer, J.; Gruber, B.; Karl, R. Chemical Similarity, Chemical Distance, and Computer Assisted Formalized Reasoning by Analogy, n: Maggiora, G. M. Eds. *Concepts and Applications of Molecular Similarity*, Wiley, New York, 1990, Chap. 9, pp. 239-288.

74. Balaban, A.T.; Ciubotariu, D.;Ivanciuc,O. Design of Topological Indices part 2. Distance Measure Connectivity Indices, *Commun.Math.Comput.Chem.(MATCH)*, **1990**, *25*, 41-70.

75. Balaban, A. T.; Bonchev, D.; Seitz, W.A. Topological /Chemical Distances and Graph Centers in Molecular Graphs with Multiple Bonds, *J. Molec. Structure (Theochem.)*, **1993**, *99*, 253-260.

76. Basak, S.C.; Magnusson, V.R.; Niemi, G.J.; Regal, R.R., Determining Structural Similarity of Chemicals Using Graph-Theoretic Indices, *Discr. Appl. Math.* **1988**, *19*, 17-44.

77. Gower, J. C. *J. Classification*, **1986**, *3*, 5.

78. Johnson, M. A. A Review and Examination of the Mathematical Spaces Underlying Molecular Similarity Analysis. *J. Math. Chem.* **1989**, *3*, 117-145.

79. Willett, P. Algorithms for the Calculation of Similarity in Chemical Structure Databases in: Johnson, M. A.; Maggiora, G. M. Eds. *Concepts and Applications of Molecular Similarity*, Wiley, New York, 1990, Chap. 3, pp. 43-63.

80. Willett, P. *Similarity and Clustering in Chemical Information Systems*, Research Studies Press, Letchworth, England, 1987, Chap. 2.

81. Mezey, P. G. Global and Local Relative Convexity and Oriented Relative Convexity; Application to Molecular Shapes in External Fields. *J. Math. Chem.* **1988**, *2*, 325-346.

82. Mezey, P.G. Descriptors of Molecular Shape in 3D, in: Balaban, A. T. Ed., *From Chemical Topology to Three-Dimensional Geometry*, Plenum Press, New York and London, 1997, Chap. 2, pp. 25-42.

83. Randic, M.; Razinger, M. Molecular Topographic Indices, *J. Chem. Inf. Comput. Sci.* **1995**, *35*, 140-147.

84. Randic, M.; Razinger, M. On Characterization of 3D Molecular Structure, in: Balaban, A. T. Ed., *From Chemical Topology to Three-Dimensional Geometry*, Plenum Press, New York and London, 1997, Chap. 6, pp. 159-236.

85. Simon, Z.; Chiriac, A.; Holban, S. Steric Fit in QSAR. IV. MTD – Receptor Site Mapping, *Preprint Univ. Timisoara*, **1980**, *4*, 1-50.

86. Motoc, I. Biological Receptor Maps.1. Steric Maps. The SIBIS Method, *Quant. Struct. Act. Relat.* **1984**, *3*, 43-47.

87. Milne, G. W. A.; Nicklaus, M. C.; Driscoll, J. S.; Wang, S.; Zaharevitz, D. National Cancer Institute Drug Information System 3D Database, *J. Chem. Inf. Comput. Sci.* **1994**, *34*, 1219-1224.

88. Martin, Y.; Burres, M. G.; Willett, P. in *Computational Chemistry*, (Eds. Lipkowitz, K. B.; Boyd, D. B.) VCH Publishers, New York, 1990, p. 213.

89. Diudea, M.V.; Horvath, D.; Graovac, A. 3-D Distance Matrices and Related Topological Indices, *J. Chem. Inf. Comput. Sci.* **1995**, *35*, 129-135.

90. Randić, M.; Wilkins, C.L. Graph-Theoretical Ordering of Structures as a Basis for Systematic Search for Regularities in Molecular Data, *J. Phys. Chem.* **1979**, *83*, 1525-1540.

91. Bonchev, D.; Kier, L.B.; Mekenyan, O. Self-Returning Walks and Fractional Charges of Atoms in Molecules, *Int. J. Quantum Chem.* **1993**, *46*, 635-649.

92. Balaban, A. T. Local versus Global (i.e. Atomic versus Molecular) Numerical Modeling of Molecular Graphs, *J. Chem. Inf. Comput. Sci.* **1994**, *34*, 398-402.

93. Kiss, A. A.; Katona, G.; Diudea, M.V. Szeged and Cluj Matrices within the Matrix Operator $W_{(M1,M2,M3)}$, *Coll. Sci. Papers Fac. Sci. Kragujevac*, **1997**, *19*, 95-107.

94. Dobrynin, A. Cubic Graphs with 62 Vertices Having the Same Path Layer Matrix, *J. Graph Theory*, **1993**, *17*, 1-4.

95. Balducci, R.; Pearlman, R. S. Efficient Exact Solution Of The Ring Perception Problem, *J. Chem. Inf. Comput. Sci.* **1994**, *34*, 822-831.

96. King, R. B. Applications of Topology and Graph Theory in Understanding Inorganic Molecules, in Balaban, A. T. Ed., *From Chemical Topology to Three-Dimensional Geometry*, Plenum Press, New York and London, 1997, Chap. 10, pp. 343-414.

97. Diudea, M.V.; Parv, B.; Gutman, I. Detour-Cluj Matrix and Derived Invariants, *J. Chem. Inf. Comput. Sci.* **1997**, *37*, 1101-1108.

98. Basak, S.C.; Bertelsen, S.; Grunwald, G.D. Application of Graph Theoretical Parameters in Quantifying Molecular Similarity and Structure-Activity Relationships, *J. Chem. Inf. Comput. Sci.* **1994**, *34*, 270-276.

99. Basak, S.C.; Grunwald, G.D. Molecular Similarity and Estimation of Molecular Properties, *J. Chem. Inf. Comput. Sci.* **1995**, *35*, 366-372.

100. Balaban, A. T. Ed., *From Chemical Topology to Three-Dimensional Geometry*, Plenum Press, New York and London, 1997

101. Hu, C.Y.; Xu, L. Algorithm for Computer Perception of Topological Symmetry, *Anal. Chim. Acta,* **1994**, *295*, 127-134.

102. Mezey, P.G. Iterated Similarity Sequences and Shape ID Numbers for Molecules, *J. Chem. Inf. Comput. Sci.* **1994**, *34*, 244-247

103. King, R.B. Chirality Polynomials, *J. Math. Chem.* **1988**, *2*, 89-115.

104. Warrington, J., Ed., *Plato's Timaeus*, Dent, London, 1965.

105. Tomalia, D.A.; Naylor, A.M.; Goddard III, W.A. Starburst Dendrimers: Molecular Level Control of Size, Shape, Surface Chemistry, Topology and Flexibility from Atoms to Macroscopic Matter, *Angew. Chem., Int. Ed. Engl.* **1990**, *29*, 138-175.

106. Diudea, M.V.; Minailiuc, O.M.; Balaban, A.T. Regressive Vertex Degrees (New Graph Invariants) and Derived Topological Indices, *J. Comput. Chem.* **1991**, *12*, 527-535.

107. Engler, E.M; Farcasiu, M.; Sevin, A.; Cense, J.M; Schleyer, P.v.R. On the Mechanism of Adamantane Rearrangements, *J. Am. Chem. Soc.* **1973**, *95*, 5769-5771.

108. Petersen, J. Die Theorie der regulären Graphen, *Acta Math.* **1891**, *15*, 193-220.

109. Heawood, P. J. Map Colour Theorems, *Quart. J. Math.* Oxford Ser. **1890**, *24*, 332-338.

110. Balaban, A. T.; Schleyer, P.v.R. Systematic Classification and Nomenclature of Diamond Hydrocarbons I. Graph-Theoretical Enumeration of Polymantanes, *Tetrahedron,* **1978**, *34*, 3599-3609.

111. Randić, M. Symmetry Properties of Graphs of Interest in Chemistry.II. Desargues-Levi Graph, *Int. J. Quantum Chem.* **1979**, *15*, 663-682.

112. Hosoya, H. Factorization and Recursion Relations of the Matching and Characteristic Polynomials of Periodic Polymer Networks, *J. Math. Chem.* **1991**, *7*, 289-305.

113. Hosoya, H.; Okuma, Y.; Tsukano, Y.; Nakada, Y. Multilayered Cyclic Fence Graphs: Novel Cubic Graphs Related to the Graphite Network, *J. Chem. Inf. Comput. Sci.* **1995**, *35*, 351-356.

114. Randić, M. A Systematic Study of Symmetry Properties of Graphs. I. Petersen Graph, *Croat. Chem. Acta,* **1977**, *49*, 643-655.

Chapter 9

ELEMENTS OF STATISTICS

The design of molecular structures with desired physico-chemical or biological properties is the major target of the molecular topology. An insight of a set of molecules could reveal the crucial factors involved in the structure-property relationship.

This is performed by the aid of some molecular descriptors (e.g. topological indices) and/or the *regression analysis*, within various models (i.e. algorithms). The results of this analysis have a *diagnostic* meaning (e.g. the partitioning of a molecular property into fragmental contributions to a *computed* global property) and a *prognostic* one (e.g. the estimation of a molecular property from some fragmental mathematical or physico-chemical properties). The topological characterization of the chemical structures allows their classification according to some similarity criteria. The regression analysis is based on some basic *statistics*.

9.1. ELEMENTARY CONCEPTS

Elementary statistical concepts providing the necessary foundations[1] for more specific expertise in any area of statistical data analysis are briefly discussed.

Because of space limitations, the reader is invited to consult more detailed textbooks[2,3,4,5].

9.1.1 Mean values

Let X be a string of n values $X_1, X_2, ..., X_n$. The following main indicators are most used:

Arithmetic mean AM(X) is the number calculated by:

$$AM(X) = \frac{\sum_{i=1}^{n} X_i}{n} \qquad (9.1)$$

Geometric mean GM(X) is obtained by:

$$GM(X) = \sqrt[n]{\prod_{i=1}^{n} X_i} \underset{X_i > 0}{=} \exp(AM(\ln(X))) \tag{9.2}$$

Note that for $n = 2k$, k integer, the expression for GM can be indeterminate if the product ΠX_i is negative.

Harmonic mean HM(X) is the number given by:

$$HM(X) = \frac{n}{\sum_{i=1}^{n} \frac{1}{X_i}} = 1/AM(1/X) \tag{9.3}$$

Eulerian Mean EM(X) is calculated as:

$$EM(X) = \sqrt{\frac{\sum_{i=1}^{n} X_i^2}{n}} = \sqrt{AM(X^2)} \tag{9.4}$$

Median value m(X) is the number given by:

$$m(X) = \begin{cases} \left(X_{\pi(\frac{n}{2})} + X_{\pi(\frac{n}{2}+1)} \right) \Big/ 2, & for \quad n \quad even \\ X_{\pi(\frac{n+1}{2})}, & for \quad n \quad odd \end{cases} \quad , \pi \text{ such that } X_\pi \text{ is ordered} \tag{9.5}$$

9.1.2. Indicators of Spreading

Hereafter, $M(X)$ denotes any mean value (9.1-9.5).
Dispersion D is the number given by:

$$D_M(X) = EM(X - M(X)) \tag{9.6}$$

and is a measure of spreading of X values around the mean value $M(X)$. The subscript M is the label for the type of mean around the statistical indicator considered. If the label is missing, the arithmetic mean AM is assumed.

Standard deviation s is the number calculated as:

$$s_M(X) = \sqrt{n/(n-1)} \, D_M(X) \tag{9.7}$$

Absolute mean deviation am *is the quantity:*

$$am_M(X) = AM(abs(X-M(X))) \tag{9.8}$$

It is called *variance* (s^2, am^2, D^2) the square of any deviation (s, am, D).

9.2. CORRELATIONS

Correlation is a measure of the relation between two or more variables. The correlation coefficient is a measure of linear dependencies of two or more series of data and is not dependent on the measurement scales of series.[6] Correlation coefficients range from -1.00 to +1.00. The value of -1.00 or +1.00 represents a perfect linear correlation while a value of 0.00 represents a lack of linear correlation.

The most widely used correlation coefficient is that of Pearson, r, also called linear or product-moment correlation[7].

9.2.1. Pearson Correlation, r

Let X and Y be two series of data.
The quantity defined by:
$$\mu(X,Y) = AM(XY) - AM(X)AM(Y) \tag{9.9}$$
is called the *second degree moment* or *covariance* or *correlation* of the two data through the numeric series before considered.

The *Pearson correlation coefficient, r* is given by:
$$r(X,Y) = \frac{\mu(X,Y)}{\sqrt{\mu(X,X)\cdot\mu(Y,Y)}} \tag{9.10}$$
The quantity $\mu(X, X)$ provides the same values as the *square dispersion* of X:

$$D^2(X) = \mu(X,X) \tag{9.11}$$

Among all these quantities, the correlation coefficient is more often used for the statistical characterization of the correlation between two series of data.

The higher is $\mu(X, Y)$ the greater is the functional dependence between X and Y, and r becomes higher too. When $r = 1$ the correlation reaches the maximum, and X and Y become directly proportional.

The smaller is $\mu(X, Y)$, the stronger is the functional dependence between X and Y, but this time Y decreases with increasing X. When $r = -1$, the correlation is at the minimum value, X and Y are changing in an inversely proportional manner. The above relations are true. of course. for a linear correlation.

9.2.2. Rank Correlation. Spearman ρ and Kendall τ

The rank correlation is used especially when the series of inputs do not have rigorous values, being affected by systematic errors. In such a case, the only useful parameter is the position of measurement in the ordered string (file) of these ones.

We now introduce the notion of rank: the *rank* is the position of a measured value in the string of the measured values ordered in an increasing manner. Consider the series $X_1, X_2, \ldots, X_n$ and the permutation

$$\pi: \{1,\ldots,n\} \to \{1,\ldots, n\}: \ X_{\pi(i)} \le X_{\pi(i+1)} , \ 1 \le i \le n\text{-}1 \tag{9.12}$$

that put into increasing order the measurements, namely the *rank* of X_i is $X_{\pi(i)}$ (see also the same π in eq 9.5).

Let be the series $X_1, X_2, \ldots, X_n$ and $Y_1, Y_2, \ldots, Y_n$ and (according to 9.12) π_1, π_2 permutations that put in order X and Y, respectively:

$$X_{\pi_1(i)} \le X_{\pi_1(i+1)}, \ Y_{\pi_2(i)} \le Y_{\pi_2(i+1)}, \ 1 \le i \le n\text{-}1 \tag{9.13}$$

and let be $d_k = \pi_1(k) - \pi_2(k), \ 1 \le k \le n$, and also $d = \Sigma|d_k|$. $\tag{9.14}$

If $d = 0$ then the considered series are on the same order and there is a perfect correspondence of ranks.

Taking into account that:

$$AM(\pi_1) = AM(\pi_2) = \frac{n+1}{2} \tag{9.15}$$

the *Spearman correlation coefficient ρ (correlation of rank)*, is obtained by performing the Pearson r calculations for π_1 and π_2 variables:

$$\rho(X,Y) = r(\pi_1,\pi_2) = 1 - \frac{6\sum_{i=1}^{n} d_i^2}{n(n^2 - 1)} \tag{9.16}$$

Detailed discussions upon the *Spearman ρ* statistic can be found in refs [8, 4, 9, 10].

In order to define the *Kendall correlation coefficient* we need to introduce the functions K_1 and K_2 according to:

$$K_1(i) = \left|\{k|\pi_2(k) < \pi_2(i),\pi_1(k) < \pi_1(i),k < i\}\right| \tag{9.17}$$

that is the number of ranks in Y smaller than the rank i from Y and in the series of X the ranks from 1 to i; $\ K_2(i) = \left|\{k|\pi_2(k) > \pi_2(i),\pi_1(k) < \pi_1(i),k < i\}\right|$ $\tag{9.18}$

that is the number of ranks from Y larger than the rank i from Y and in the series of X, the ranks from 1 to i.

The quantities

$$P_i = 1-\pi_2(i)+K_1(i); \quad Q_i = n-\pi_2(i)-K_2(i); \quad S_i = P_i + Q_i; \quad S = (S_i)_{1 \le i \le n} \tag{9.19}$$

once calculated, the *Kendall correlation coefficient* τ is obtained as:

$$\tau(X,Y) = \frac{2}{n-1} AM(S) \tag{9.20}$$

Note that: (i) $k = 1$ when both series are in the same order $\pi_1 = \pi_2$; (ii) $k = -1$ when both series are in the opposite order $\pi_1 \circ \pi_2 = 1_N$. The rank correlation is successfully used at *Genetic Programming*[11,12].

Kendall τ and Spearman ρ statistics are comparable in terms of their statistical power. However, the two statistics are usually not identical in magnitude because their underlying logic, as well as their computational formulas are very different. Siegel and Castellan [13] express the relationship of the two measures in terms of the inequality:

$$-1 \le 3\cdot\tau - 2\cdot\rho \le 1 \tag{9.21}$$

More importantly, they imply different interpretations: While *Spearman* ρ can be thought as the regular *Pearson* product-moment correlation coefficient as computed from ranks, *Kendall* τ rather represents a probability. Specifically, it is the difference between the probability that the observed data are in the same order for the two variable vs the probability that the observed data are in different orders for the two variables. For details see the refs. [13,14,15]

9.2.3. Correlations in Non-Homogeneous Groups

A lack of homogeneity in the sample from which a correlation was calculated can be another factor that biases the value of the correlation. Imagine a case where a correlation coefficient is calculated from data points coming from two different experimental groups but this fact is ignored when the correlation is calculated. Let us assume that the experimental manipulation in one of the groups increased the values of both correlated variables and thus the data from each group form a *distinctive cluster* in the scatterplot.

In such cases, a high correlation may result that is entirely due to the arrangement of the two groups, but which does not represent the *true* relation between the two variables. If you suspect the influence of such a phenomenon on your correlations and know how to identify such *subsets* of data, try to run the correlations separately in each subset of observations.

9.3. REGRESSION MODELS

Regardless of their type, two or more variables are related if in a sample of observations the values of those variables are distributed in a consistent manner. In other words, variables are related if their values systematically correspond to each other for these observations.

The general purpose of *multiple regression* (the term was first used by Pearson[16], 1908) is to learn more about the relationship between several *independent* (or predictor) variables and a *dependent* (or criterion) variable.

In general, multiple regression allows the researcher to ask (and hopefully answer) the general question *what is the best predictor of ....*

The most frequently used multiple regression is *multiple linear regression* because this type of regression offers maximum capability in prediction.[17] First of all, it is assumed that the *relationship between variables is linear*. In practice this assumption can virtually never be confirmed; fortunately, multiple regression procedures are not greatly affected by minor deviations from the linearity. However, it is prudent to always look at bivariate scatterplot of the variables of interest. If curvature in the relationships is evident, one may consider either transforming the variables, or explicitly allowing for nonlinear components.

Once this so-called *regression line* has been determined, the analyst can now easily construct a graph of the expected (predicted) values and the actual values of dependent variable. Thus, the researcher is able to determine which position is below the regression line, above the regression line, or at the regression line.

9.3.1 Loss Function in Regression Models

The *loss function* (the term loss was first used by Wald [18] in 1939) is the function that is *minimized* in the process of fitting a model, and it represents a selected measure of the discrepancy between the observed data and data *predicted* by the fitted function.

For example, in many traditional linear model techniques, the loss function (commonly known as *least squares*) is the sum of squared deviations from the fitted line. One of the properties of that common loss function is that it is very sensitive to outliers.

A common alternative to the least squares loss function is to *maximize* the likelihood or log-likelihood function.

Let Y be a string of measured data and $\quad$ a string of predicted Y values. The loss function is of the form:

$$loss(Y, \;) = \sum_i f(Y - \hat{Y}) \quad (9.22)$$

where f is a positive function ($f : \mathcal{R} \rightarrow \mathcal{R}_+$).

Model parameters are determined by minimizing the loss function

$$loss(Y, \;) = min. \tag{9.23}$$

Minimization of Risk. Least Squares Method

A well known estimation model for parameters is based[19] on the *minimization of risk* defined as mean of square loss function, (promoted by Kolmogorov[20]) best known as the *least squares* method. Expression of loss function is

$$f(z) = z^2 \qquad (9.24)$$

Many papers[21,22,23] have described different approaches of the estimation model based on the loss function. Most used are presented in the following:

Fisher[24] introduced the *maximum likelihood method* given by[25]

$$f(z) = 1 - e^{-z^2/2} \qquad (9.25)$$

Newman and Waad proposed the *minimax method* given by a function

$$f(z) = |z| \qquad (9.26)$$

Bayes (1750), was first that introduced maximum aposteriory probability method by

$$f(z) = \begin{cases} 0, & z < D(Z)/2 \\ 1, & z \geq D(Z)/2 \end{cases}, \text{ where } D \text{ is the dispersion (see eqs 9.6, 9.11)} \qquad (9.27)$$

In many variants of the least squares, weighted loss functions are used

$$f_l(z) = w \cdot f(z) \qquad (9.28)$$

where w is a weight dependent on values of dependent variable Y, independent variable(s) X or predicted variable .

A widely used weighted function is (see [26] p. 168)

$$loss = loss(Y, \, , X) = \sum_i \frac{(Y_i - \hat{Y}_i)^2}{X_i^2} \qquad (9.29)$$

This method will yield more stable estimates of the regression parameters (for more details, see [26]).

An interesting model is obtained if expression of regression model is written in implicit form

$$g(Y, \,) = \varepsilon \qquad (9.30)$$

when the loss function becomes

$$loss(Y, \,) = \sum_i f(g(Y, \hat{Y})) \qquad (9.31)$$

This kind of model is useful when both the predicted variable and the predictor variable are affected by measurement errors.[27]

9.3.2. Simple Linear Model

Let X be an independent variable and Y a dependent variable ($Y = Y(X)$). The linear model assumes that X and Y are linked in a dependence of the form

$$\hat{Y} = b_0 + b_1 X; \quad Y = \hat{Y} + \varepsilon \tag{9.32}$$

where ε is the residue of the estimate of Y.

The loss function for the model is defined as in eq 9.24-9.28. The parameters b_1 and b_0 are determined by eq 9.23.

For the most of the cases, the loss function is the *minimization of risk* and the values for parameters are:

$$b_1 = \frac{AM(XY) - AM(X)AM(Y)}{AM(X^2) - AM^2(X)}; \quad b_0 = AM(Y) - b_1 AM(X) \tag{9.33}$$

9.3.3. Multiple Linear Model

Let Y be a dependent variable, and independent variables $X^1, ..., X^p$ where $p < n$, n being the number of experiments ($Y_1, Y_2, ..., Y_n$). The model for *multiple linear regression* is

$$\hat{Y} = b_0 + b_1 X^1 + b_2 X^2 + ... + b_p X^p; \quad Y = \hat{Y} + \varepsilon \tag{9.34}$$

The coefficients can be obtained by applying eqs 9.22- 9.24 (for other cases, see eqs 9.25-9.28) when results a system of linear equations

$$\frac{\partial\left(AM\left((BX - Y)^2\right)\right)}{\partial B} = 0, \quad \text{where } B^T = [b_0, b_1, ..., b_p] \tag{9.35}$$

with solution (if exists):

$$B = CZ^{-1}; \quad C^T = [AM(X^k Y)]_{0 \le k \le p} \text{ and } Z = [AM(X^{k+i})]_{0 \le k,i \le p} \tag{9.36}$$

As a regression power measure the Pearson r_p is used

$$r_p(X^1, X^2, ...,X^p, Y) = r(\hat{Y}, Y) \tag{9.37}$$

or multiple r, namely r_M

$$r_M(X^1, X^2, ...,X^p, Y) = \sqrt{\frac{AM\left((\hat{Y} - AM(Y))^2\right)}{AM\left((Y - AM(Y))^2\right)}} \tag{9.38}$$

9.3.4. Other Regression Models

In pharmacology, the following model is often used to describe the effects of different dose levels of a drug

$$Y = b_0 - b_0 \frac{1}{1 + (X / b_2)^{b_1}}$$ (9.39)

In this model, X is the dose level ($X \geq 1$) and Y is the responsiveness, in terms of the percent of maximum possible responsiveness.

The parameter b_0 denotes the expected response at the level of dose saturation while b_2 is the concentration that produces a half-maximal response; the parameter b_1 determines the slope of the function.

For specific problems, non-linear regression models are used.[28, 29, 30, 31, 32]

9.4. REDUCTION TO LINEAR MODELS

According to the concept of linear dependence, a regression equation is linear if the functional dependence between the considered variables can be linearized. Transforming the independent variables can be achieved following the procedures described in ref.[33] at p. 560. The estimation of the u parameters for this procedure is not iterative in nature, but is accomplished by expanding the terms of the regression model for the transformed predictor variables in a first-order Taylor series. For example, the following regression equations

$$Y = a \cdot log(X) + b; \quad Y = a\{1/X\} + b; \quad Y = a \cdot e^X + b; \quad log(1/Y) = a \cdot X + b;$$ (9.40)

can be linearized and the dependence can be associated with the linear model
$$Y = a \cdot Z + b$$ (9.41)

where the new independent variable z is obtained by substitutions[27, 28]
$$Z = log(X); \quad Z = 1/X; \quad Z = e^X \ or \ Z = e^Y$$ (9.42)

We can minimize the residual sums of squares for the regression model, after transforming the dependent variable via $Z = Y^u$ ($u \neq 0$), or, best known $Z = log(Y)$ (see $log P$ calculations and correlating studies)[33,34]. Note that this kind of substitutions require that all values of Y be greater than zero. For additional details see refs.[35,36,33,37]

Another extension of linear regression model can be obtained when the error factor influences both variables involved in the regression. In this case, the formulas for the validation of regression parameters have another form.[27]

9.5. FACTOR ANALYSIS AND PCA

Thurstone first introduced the term *factor analysis*.[38] The factor analysis is applied in connection with a variant of multiple linear regression, which applies successively the simple linear model to the non-explicated data[39]

Step 1. $\hat{Y}^1 = a_1 + b_1 X^1;$ $Y^1 = Y - \hat{Y}^1$

Step 2. $\hat{Y}^2 = a_2 + b_2 X^2;$ $Y^2 = Y^1 - \hat{Y}^2$

Step p. $\hat{Y}^p = a_p + b_p X^p;$ $Y^p = Y^{p-1} - \hat{Y}^p; \; \varepsilon = Y^p$ (9.43)

which, in terms of multiple linear regression is:

Step Σ. $\hat{Y}^p = (a_1 + a_2 + ... + a_p) + b_1 X^1 + b_2 X^2 + ... + b_p X^p + \varepsilon;$

          $Y^p = Y - \hat{Y}^p$ (9.44)

Note that this technique of multiple linear regression leaves unchanged the values b_i ($1 \leq i < p$). The values b_i are *invariants* at the application of any additional step $k > p$ of regression. This technique is referred to as the *Principal Component Analysis PCA*.[40,41]

The main applications of factor analysis techniques is to reduce the number of variables p and to detect structure in the relationships between variables, that to classify variables[42,43]. Therefore, factor analysis is applied as a data reduction or structure detection method[44,45]. Many excellent books on factor analysis already exist.[46,47,48,49] The interpretation of secondary factors in hierarchical factor analysis, as an alternative to traditional oblique rotational strategies, is explained in detail in ref.[50] At the *heart* of factor analysis is the *problem of regression coefficients b_p*, evaluated usually via *LS* (least squares) procedures. In most of the cases, this problem is solved via the *Householder* method.[51,52,53]

9.6. DOMINANT COMPONENT ANALYSIS, DCA

It is a variant of linear multiple regression and/or *PCA*. The method starts with the observation that in regression equations (9.34) and (9.44) the descriptors $X_1, X_2, ..., X_p$ are intercorrelated.

DCA approach proposes a method of orthogonalization of independent variables involved in the regression equation. In this way, a new set of non-correlated descriptors is created. Note that in this type of multi-linear regression the best correlation score makes the selection of the next descriptor from the set of descriptors. The algorithm of *DCA* is:

Step 1. Make linear regressions: $\hat{Y}^i = a_i \cdot X^i + b_i;$ $Y^i = Y - \hat{Y}^i; \; i = 1, ..., p;$

Let $k1$: $r(Y, X^{k1}) = max \; \{r(Y, X^i), i = 1, ..., p\};$

Make 1-variate regressions: $\hat{X}^i = a_i \cdot X^{k1} + b_i;$ $W^i = X^i - \hat{X}^i; \; i \neq k1.$

Step 2. Make linear regressions: $\hat{Y}^i = A_i \cdot W^i + B_i;$ $Y^i = Y^{k1} - \hat{Y}^i; \; I \neq K1;$

Let $k2$: $r(Y^{k1}, W^{k2}) = max \; \{r(Y^{k1}, W^i), i \neq k1\};$

Make 2-variate regressions: $^i = a_i \cdot X^{k1} + b_i X^{k2} + c_i;$ $W^i = X^i - {}^i;$ $i \neq k1, k2.$
Step 3. Make linear regressions: $^l = A_l \cdot W^i + B_i;$ $Y^i = Y^{k2} - {}^l; I \neq K1, K2;$
Let $k3$: $r(Y^{k2}, W^{k3}) = max \{r(Y^{k2}, W^i), i \neq k1, k2\};$
Make 3-variate regressions: $^i = a_i \cdot X^{k1} + b_i X^{k2} + c_i X^{k3};$ $W^i = X^i - {}^i;$ $i \neq k1, k2.$
...

$$(9.45)$$

The orthogonal descriptors are X^{k1} (Step 1), W^{k2} (Step 2), W^{k3} (Step 3), etc. Coefficients in the regression equation

$$Y = \alpha_1 X^{k1} + \alpha_2 W^{k2} + \alpha_3 W^{k3} + ... \qquad (9.46)$$

are obtained through substitutions in the algorithm equations (Step 1, ...) or making multiple linear regression (eq 9.46).

The method was first reported by Randić[54] and further in refs.[55,56,57]

$$***$$

In more general terms, there are three types of multiple regression: *standard regression, forward stepwise regression* and *backward stepwise regression.*[58,59]

In *standard regression* all variables will be entered into the regression equation in one single step. This is the most frequently used case, which is also described in (9.34-9.36).

In *forward stepwise regression* the independent variables will be individually added or deleted from the model at each step of the regression, depending on the choice based on the statistical significance of the regression equation, until the best model is obtained. This is the case both in PCA and DCA, also described in eqs 9.43, 9.44 and 9.45, 9.46.

In *backward stepwise regression* the independent variables will be removed from the regression equation one at a time, depending on the researcher choice, until the best regression model is obtained. This last procedure is more flexible, it could be made at an equation of the form (9.34-9.36), (9.43, 9.44) and (9.45, 9.46).

For the cases when the independent variables are highly intercorrelated, and stable estimates for the regression coefficients cannot be obtained via ordinary least squares methods, the *ridge regression analysis* [60, 61,25] is used.

9.7. TESTS FOR VALIDATION

We can test differences between groups (independent samples), differences between variables (dependent samples), and relationships between variables. For regression equations, tests are called *significance tests*.

9.7.1 Differences Between Independent Groups

Usually, when we have two samples that we want to compare concerning their mean value for some variable of interest, we would use the t-test for independent samples; alternatives for this test are the *Wald-Wolfowitz* runs test, the *Mann-Whitney U* test, and the *Kolmogorov-Smirnov two-sample test*.

9.7.1.1 The t-Test for Independent Samples

The t-test is the most commonly used method to evaluate the *differences in mean values* between two groups. Theoretically, the t-test can be used even if the sample size is very small (< 10).

The *normality assumption* can be evaluated by looking at the distribution of the data or by performing a normality test. The *equality of variances assumption* can be verified by the F test, or by using the *Levene test*. If these conditions are not met, then the differences in means between two groups can be evaluated by using one of the alternatives to the t-test.

The p-level included in t-test represents the probability of error involved in accepting the research hypothesis about the existence of a difference. Technically speaking, this is the probability of error associated with the rejecting of the hypothesis of no difference between the two group populations when, in fact, the hypothesis is true.

Some researchers suggest that if the difference is in the predicted direction, you can consider only one half (*one tail*) of the probability distribution and thus divide the standard p-level reported with a t-test by two (a *two-tailed* probability).

9.7.1.2 Wald-Wolfowitz Runs Test

This test assumes that the variable under consideration is continuous, and that it was measured on at least an ordinal scale (i.e. rank order).

The *Wald-Wolfowitz runs* test assesses the hypothesis that two independent samples were drawn from two populations that differ in some respect, i.e. not just *with respect to the mean*, but also *with respect to the general shape* of the distribution. The null hypothesis is that the two samples were drawn from the same population. In this respect, this test is different from the parametric t-test, which strictly tests for differences in locations (means) of two samples.

9.7.1.3 Mann-Whitney U Test

The *Mann-Whitney U* test is a nonparametric alternative to the *t*-test for independent samples. The procedure expects the data to be arranged in the same way as for the *t*-test for independent samples.

Specifically, the data file should contain a coding variable (independent variable) with at least two distinct codes that uniquely identify the group membership of each case in the data.

The *Mann-Whitney U* test assumes that the variable under consideration was measured on at least an ordinal (rank order) scale. The interpretation of the test is essentially identical to the interpretation of the result of a *t*-test for independent samples, except that the *U* test is computed based on rank sums rather than means (it is a measure of *differences in average ranks*). The *U* test is the most powerful (or sensitive) alternative to the *t*-test; in fact, in some instances it may offer even a greater power to reject the null hypothesis than the *t*-test.

With samples larger than 20, the sampling distribution of the *U* statistics rapidly approaches the normal distribution[62]. Hence, the *U* statistics (adjusted for ties) will be accompanied by a *z* value (normal distribution variate value), and the respective *p* value.

9.7.1.4 Kolmogorov-Smirnov Test

The *Kolmogorov-Smirnov* test assesses the hypothesis that two samples were drawn from different populations. Unlike the parametric *t*-test for independent samples or the Mann-Whitney *U* test, which test for differences in the location of two samples (differences in means, differences in average ranks, respectively), the *Kolmogorov-Smirnov* test is also sensitive to *differences in the general shapes* of the distributions in the two samples, i.e. to *differences in dispersion, skewness,* etc.

9.7.2 Differences Between Dependent Groups

If we want to compare two variables measured in the same sample we would use the t-test for dependent samples. Alternatives to this test are the *Sign test* and *Wilcoxon's matched pairs* test. If the variables of interest are dichotomous in nature (i.e. pass vs. no pass) then *McNemar's Chi-square* test is appropriate.

9.7.2.1. The t-test for Dependent Samples

The *t*-test for dependent samples helps us to take advantage of one specific type of design in which an important source of *within-group variation* (or so-called, error) can be easily identified and excluded from the analysis.

Specifically, if two groups of observations (that are to be compared) are based on the same sample which was tested twice (e.g. before and after a treatment), then a considerable part of the within-group variation in both groups of scores can be attributed to the initial individual differences between samples. Note that, in a sense, this fact is not much different than in cases when the two groups are entirely independent (see the t-test for independent samples), where individual differences also contribute to the error variance. Note that in the case of independent samples, we cannot do anything about it because we cannot identify (or *subtract*) the variation due to individual differences in subjects. However, if the same sample was tested twice, then we can easily identify (or subtract) this variation.

Specifically, instead of treating each group separately, and analyzing raw scores, we can look only at the differences between the two measures (e.g. *pre-test* and *post test*) in each sample.

By subtracting the first score from the second one for each sample and then analyzing only those *pure (paired) differences*, we will exclude the entire part of the variation in our data set that results from unequal base levels of individual subjects. This is precisely what is being done in the t-test for dependent samples, and, as compared to the t-test for independent samples, it always produces *better* results (i.e. it is always more sensitive).

Paired Differences

Let Y^1 and Y^2 be two variables, which estimate the same measured property. Then, let be

$$D = Y^1 - Y^2 \tag{9.47}$$

Variable D (*paired differences*) provides the mean $AM(D)$ and next the standard deviation is obtained as

$$s(D) = \sqrt{\frac{n}{n-1}\left(AM(D^2) - AM^2(D)\right)} \tag{9.48}$$

In this case, the associate (calculated from experimental data) t-value will be

$$t = \frac{AM(D)}{s(D)}\sqrt{n} = \frac{\sqrt{n-1}}{\sqrt{\dfrac{AM(D^2)}{AM^2(D)} - 1}} \tag{9.49}$$

9.7.2.2 Sign Test

The *sign test* is an alternative to the t-test for dependent samples. The test is applicable in situations when the researcher has two measures (under two conditions) for each subject and wants to establish that the two measurements (or conditions) are different. Each variable in the first list will be compared to each variable in the second list.

The only assumption required by this test is that the underlying distribution of the variable of interest is continuous; no assumptions about the nature or shape of the underlying distribution are required. The test simply computes the number of times (across subjects) that the value of the first variable (Y^1) is larger than that of the second variable (Y^2). Under the null hypothesis (stating that the two variables are not different from each other) we expect this to be the case about 50% of the time. Based on the binomial distribution we can compute a z value for the observed number of cases where $Y^1 > Y^2$, and compute the associated tail probability for that z value.

Wilcoxon Matched Pairs Test

This procedure assumes that the variables under consideration were measured on a scale that allows the rank ordering of observations based on each variable and that allows rank ordering of the differences between variables (this type of scale is sometimes referred to as an ordered metric scale[63]. Thus, the required assumptions for this test are more stringent than those for the *Sign* test.

However, if they are met, that is, if the *magnitudes of differences* (e.g., different ratings by the same individual) contain meaningful information, then this test is more powerful than the *Sign* test.

In fact, if the assumptions for the parametric *t*-test for dependent samples (interval scale) are met, then this test is almost as powerful as the *t*-test.

9.7.2.4 McNemar Chi-square

This test is applicable in situations where the frequencies in the table in form

$$\begin{pmatrix} A & B \\ C & D \end{pmatrix} \tag{9.50}$$

represent dependent samples. Two *Chi-square values* can be computed: *A/D* and *B/C*. The *Chi-square A/D* tests the hypothesis that the frequencies in cells A and D are identical. The *Chi-square B/C* tests the hypothesis that the frequencies in cells B and C are identical.

9.7.3 Relationships between variables

To express a relationship between two variables one usually computes the correlation coefficient r. Equivalents to the standard correlation coefficient are *Spearman* ρ and *Kendall* τ.

If the two variables of interest are categorical in nature, appropriate statistics for testing the relationship between the two variables are the *Chi-square test*, the *Phi square* coefficient, and the *Fisher exact* test.

9.7.3.1 Chi-square Test of Goodness of Fit

Chi-square test is computed using the observed Y's frequencies and expected $\hat{Y}$'s frequencies. Categories where the expected frequency is less than 5 are collapsed to form larger categories. If this test is significant, we reject the hypothesis that the observed data follow the hypothesized distribution. The *degrees of freedom* for the *Chi-square* test are computed as:

df = number of categories - number of parameters − 1 $\hspace{2cm}$ (9.51)

where the number of categories refers to the number of categories in the frequency table where the expected frequencies are greater than 5 and number of parameters refers to the number of parameters defining the respective theoretical distribution.

Yates Correction

The approximation of the *Chi-square* statistic in small 2x2 tables (9.50) can be improved by reducing the absolute value of differences between expected and observed frequencies by 0.5 before squaring (*Yates' correction*). This correction, which makes the estimation more conservative, is usually applied when the table contains only small observed frequencies, so that some expected frequencies become less than 10. For details see refs. [64,15,4,5,65]

9.7.3.2 Phi-Square Coefficient

The Phi-square is a measure of correlation between the two categorical variables in the table of data. Its value can range from 0 (no relation between factors; Chi-square = 0.0) to 1 (perfect relation between the two factors in the table). For more details concerning this statistic see [13] at p. 232.

9.7.3.3 Fisher Exact Test

Given the marginal frequencies in the data table, and assuming that in the population the two factors in the table are not related, how likely is it to obtain cell frequencies as uneven or worse than the ones that were observed?

For small n, this probability can be computed exactly by counting all possible tables that can be constructed based on the marginal frequencies. This is the underlying rationale for the *Fisher exact* test.

It computes the exact probability under the null hypothesis of obtaining the current distribution of frequencies across cells, or one that is more uneven.

9.7.4. Statistical Significance Tests

The standard formulas can be used for calculating the variance of the error, s_e, F-value associated with the multiple r, and t-values associated with the regression coefficients (e.g., see [26,42,43,47,48,66,67,68]).

9.7.4.1 Variance of the Error, s_e

Let p be the number of independent variables in equation of regression (see eq 9.34). Variance of error ε is estimated by s_e in formula

$$s_e^2 = \frac{n}{n-p-1} AM\left((Y-\hat{Y})^2\right)$$

(9.52)

9.7.4.2 F-Value Associated with the Multiple r_M

Let Y be a string of values and an estimation for Y. F-value is given by

$$F = (n-p-1)\frac{AM\left((\hat{Y}-AM(Y))^2\right)}{AM\left((Y-\hat{Y})^2\right)}$$

(9.53)

9.7.4.2 The t-Value for the Slope

The estimator $s_{b_k}^2$ of error in calculus of b_k coefficient is calculated by using eq 9.52 and further

$$s_{b_k}^2 = \frac{1}{n}\frac{s_e^2}{AM\left((X^k-AM(X^k))^2\right)} = \frac{1}{n-p-1}\frac{AM\left((Y-\hat{Y})^2\right)}{AM\left((X^k-AM(X^k))^2\right)}$$

(9.54)

The t-value for b_k is

$$t_{b_k} = \frac{abs(b_k)}{s_{b_k}}$$

(9.55)

9.7.4.3 Confidence Interval

Let α be the probability of error involved in accepting our research hypothesis that b_k is coefficient of X^k.

Theoretical value for t, t^* is of the form $t^*(\alpha, n\text{-}p\text{-}1)$ that is obtained through inversion of the function

$$\alpha = St(x,d) = \frac{\Gamma\left(\frac{d+1}{2}\right)}{\Gamma\left(\frac{d}{2}\right)\sqrt{\pi d}} \cdot \frac{1}{\sqrt{\left(1+\frac{x^2}{d}\right)^{d+1}}}$$

(9.56)

when is obtained:

$$t^*(\alpha, d) = x \text{ which obeys } St^{-1}(x,d) = \alpha.$$

(9.57)

with d being the degrees of freedom $(n\text{-}p\text{-}1)$, x is a real number and α a probability. The hypothesis that b_k is the coefficient of X^k is accepted if

$$t_{b_k} > t^*(\alpha, n\text{-}p\text{-}1) \tag{9.58}$$

With the value for $t^*(\alpha/2, n\text{-}p\text{-}1)$ and s_{b_k} we can calculate *confidence interval* for b_k

$$b_k \pm t^*(\alpha/2, n\text{-}p\text{-}1) \cdot s_{b_k} \tag{9.59}$$

and the *confidence interval* for   values (see eq 9.34)

$$= b_0 + \sum_{k=1}^{p} \left(b_k \pm t^*(\frac{\alpha}{2}, n - p - 1) \cdot s_{b_k} \right) \tag{9.60}$$

REFERENCES

1.   1 Nisbett, R. E., Fong, G. F., Lehman, D. R., Cheng, P. W. Teaching Reasoning. *Science*, **1987**, *238*, 625-631.

2.   2 Kachigan, S. K. Statistical Analysis: An Interdisciplinary Introduction to Univariate and Multivariate Methods, New York, Redius Press, **1986**.

3.   3 Runyon, R. P., Haber, A. *Fundamentals of Behavioral Statistics*. Reading, MA, Addison-Wesley, **1976**.

4.   4 Hays, W. L. *Statistics*, New York, CBS College Publishing, **1988**.

5.   5 Kendall, M., Stuart, A. *The Advanced Theory of Statistics* (Vol. 2), New York, Hafner, 1979.

6.   6 Galton, F. Co-relations and Their Measurement, *Proceedings of the Royal Society of London*, **1888**, *45*, 135-145.

7.   7 Pearson, K. Regression, Heredity, and Panmixia, *Philosophical Transactions of the Royal Society of London*, **1896**, *Ser. A*, *187*, 253-318.

8.   8 Gibbons, J. D. *Nonparametric Methods for Quantitative Analysis*, New York, Holt, Rinehart and Winston, **1976**.

9.   9 Olds, E. G. The 5% Significance Levels for Sums of Squares of Rank Differences and a Correction, *Annals of Mathematical Statistics*, **1949**, *20*, 117-118.

10.   10 Hotelling, H., Pabst, M. R. Rank Correlation and Tests of Significance Involving no Assumption of Normality, *Annals of Mathematical Statistics*, **1936**, *7*, 29-43.

11.   11 Gilbert R., Goodacre R., Woodward A.M., Kell D.B. Using Genetic Programming in Analysis of Chromatographic Data for a Series of Bromo-Alkenes. *Anal. Chem.* **1997**, *69*, 4381-4389.

12.   12 Goldberger G. M.; Duncan O. D. *Structural Equation Models in the Social Sciences*, New York, Seminar Press, **1989**.

13.   13 Siegel, S., Castellan, N. J. *Nonparametric Statistics for the Behavioral Sciences*, New York, McGraw-Hill, **1988**.

14.   14 Kendall, M. G. *Rank Correlation Methods*, London, Griffin, **1975**.

15.   15 Everitt, B. S. *The Analysis of Contingency Tables*, London, Chapman and Hall, **1977**.

16.   16 Pearson, K. On the Generalized Probable Error in Multiple Normal Correlation, *Biometrika*. **1908**. *6*. 59-68.

17. 17 Cristopher S. J. W., Wrigglerworth R., Bevan S., Campbell E. A., Dray A., James I. F., Masdin K. J., Perkins M. N., Winter J. On the Development of Novel Analgesic Agents, *J. Med. Chem.* **1993**, *36*, 2381.

18. 18 Wald, A. Contributions to the Theory of Statistical Estimation and Testing Hypotheses, *Annals of Mathematical Statistics*, **1939**, *10*, 299-326.

19. 19 Moritz H. *Advanced Physical Geodesy*, Herbert Wichman Verlag, **1980**.

20. 20 Kolmogorov, A. Confidence Limits for an Unknown Distribution Function. *Annals of Mathematical Statistics*, **1941**, *12*, 461-463.

21. 21 Bjerhammar A. *Theory of Errors on Generalized Matrix Inverses*; Amsterdam-London-New York, Elvister, **1973**.

22. 22 *** Matematiceskie Osnovî Kibernetiki, Kiev, *Visşaia Şkola*, **1977**.

23. 23 Tiron M. *Errors Theory and Least Squares Method*, Bucharest, Ed. Tehnica **1972**.

24. 24 Fisher, R. A. On the Interpretation of Chi-square From Contingency Tables, and the Calculation of p. *Journal of the Royal Statistical Society*, **1922**, *85*, 87-94.

25. 25 Schmidt, P., Muller, E. N. The Problem of Multicollinearity in a Multistage Causal Alienation Model: A Comparison of Ordinary Least Squares, Maximum-Likelihood and Ridge Estimators. *Quality and Quantity*, **1978**, *12*, 267-297.

26. 26 Neter, J., Wasserman, W., Kutner, M. H. Applied Linear Statistical Models: Regression, Analysis of Variance, and Experimental Designs, Homewood, **1985**.

27. 27 Sârbu C., Jäntschi L. Statistical Evaluation and Validation of Analytical Method by Comparative Studies (I. Validation of Analytical Method Using Regression Analysis). *Rev. Roum. Chim.* **1998**, *49*, 19-24.

28. 28 Jennrich, R. I., Sampson, P. F. Application of Stepwise Regression to Non-Linear Estimation. *Technometrics*, **1968**, *10*, 63-72.

29. 29 Ostrom, C. W. *Time Series Analysis: Regression Techniques*, Beverly Hills, CA: Sage Publications, **1978**.

30. 30 Ryan, T. P. *Modern Regression Methods*, New York, Wiley, **1997**.

31. 31 Seber, G. A. F., Wild, C. J. *Nonlinear Regression*. New York, Wiley, **1989**.

32. 32 Bates, D. M., Watts, D. G. *Nonlinear Regression Analysis and Its Applications*. New York, Wiley, **1988**.

33. 33 Mason, R. L., Gunst, R. F., Hess, J. L. Statistical Design and Analysis of Experiments with Applications to Engineering and Science, New York, Wiley, **1989**.

34. 34 Maddala, G. S. *Econometrics*, New York, McGraw-Hill, **1977**, p. 315.

35. 35 Box, G. E. P., Cox, D. R. An Analysis of Transformations. *Journal of the Royal Statistical Society*, **1964**, *26*, 211-253.

36. 36 Box, G. E. P., Tidwell, P. W. Transformation of the Independent Variables. *Technometrics*, **1962**, *4*, 531-550.

37. 37 Snee, R. D. An Alternative Approach to Fitting Models When Re-Expression of the Response is Useful. *Journal of Quality Technology*, **1986**, *18*, 211-225.

38. 38 Thurstone, L. L. Multiple Factor Analysis. *Psychological Review*, **1931**, *38*, 406-427.

39. 39 Thurstone, L. L. *Multiple Factor Analysis*. Chicago, University of Chicago Press, 1947.
40. 40 Viswanadhan V. N., Mueller G. A., Basak S. C., Wienstein J. N. A New QSAR Algorithm Combining Principal Component Analysis with a Neural Network: Application to Calcium Channel Antagonists. *Network Science*, **1996**, *Jan*, http://www.netschi.org/Science/Compchem/feature07.html.
41. 41 Westerhuis, J. A., Kourti T., MacGregor J. F. Analysis of Multiblock and Hierarchical PCA and PLS Models, *J. Chemometrics*, **1998**, *12*, 301-321.
42. 42 Stevens, J. Applied Multivariate Statistics for the Social Sciences, Hillsdale, NJ, Erlbaum, **1986**.
43. 43 Cooley, W. W., Lohnes, P. R. *Multivariate Data Analysis*. New York, Wiley, **1971**.
44. 44 Harman, H. H. *Modern Factor Analysis*. Chicago, University of Chicago Press, **1967**.
45. 45 Kim, J. O., Mueller, C. W. *Introduction to Factor Analysis: What It Is and How to Do It*. Beverly Hills, CA, Sage Publications, **1978**; Kim, J. O., Mueller, C. W. *Factor analysis: Statistical Methods and Practical Issues*. Beverly Hills, CA, Sage Publications, **1978**.
46. 46 Lawley, D. N., Maxwell, A. E. *Factor Analysis as a Statistical Method*, New York, American Elsevier, **1971**; Lawley, D. N., Maxwell, A. E. *Factor Analysis as a Statistical Method* (2nd. ed.). London, Butterworth Company, **1971**.
47. 47 Lindeman, R. H., Merenda, P. F., Gold, R. *Introduction to Bivariate and Multivariate Analysis*, New York, Scott, Foresman, and Co, **1980**.
48. 48 Morrison, D. *Multivariate Statistical Methods*. New York, McGraw-Hill, **1967**.
49. 49 Mulaik, S. A. *The Foundations of Factor Analysis*. New York, McGraw Hill, **1972**.
50. 50 Wherry, R. J. *Contributions to Correlational Analysis*, New York, Academic Press, **1984**.
51. 51 Golub, G. H., Van Loan, C. F. *Matrix Computations*, Baltimore, Johns Hopkins University Press, **1983**.
52. 52 Jacobs, D. A. H. (Ed.). *The State of the Art in Numerical Analysis*. London, Academic Press, **1977**.
53. 53 Ralston, A., Wilf, H.S. (Eds.), *Mathematical Methods for Digital Computers* (Vol. II), New York, Wiley, **1967**.
54. 54 Randić M. Search for Optimal Molecular Descriptors. *Croat. Chem, Acta*, **1991**, *64*, 43-54.
55. 55 Randić M. Resolution of Ambiguities in Structure-Property Studies by Use of Orthogonal Descriptors. *J. Chem. Inf. Comput. Sci.* **1991**, *31*, 311-320.
56. 56 Randić M. Orthogonal Molecular Descriptors. *New J. Chem.* **1991**, *15*, 517-525.
57. 57 Randić M. Correlation of Enthalpy of Octants with Orthogonal Connectivity Indices. *J. of Molecular Structure (Theochem)*, **1991**, *233*, 45-59.
58. 58 Jennrich, R. I. Stepwise Regression. In Enslein, K.; Ralston, A.; Wilf, H. S. *Statistical Methods for Digital Computers*, New York, Wiley, **1977**.

59. 59 Jennrich, R. I., Sampson, P. F. Application of Stepwise Regression to Non-Linear Estimation, *Technometrics*, **1968**, *10*, 63-72.

60. 60 Hoerl, A. E. Application of Ridge Analysis to Regression Problems. *Chemical Engineering Progress*, **1962**, *58*, 54-59.

61. 61 Rozeboom, W. W. Ridge Regression: Bonanza or Beguilement? *Psychological Bulletin*, **1979**, *86*, 242-249.

62. 62 Siegel, A. E. Film-Mediated Fantasy Aggression and Strength of Aggressive Drive. *Child Development*, **1956**, *27*, 365-378.

63. 63 Coombs, C. H. Psychological Scaling Without a Unit of Measurement. *Psychological Rev.* **1950**, *57*, 145-158.

64. 64 Conover, W. J. Some Reasons For Not Using the Yates Continuity Correction on 2 x 2 Contingency Tables. *J. Am. Stat. Assoc.* **1974**, *69*, 374-376.

65. 65 Mantel, N. Comment and Suggestion on the Yates Continuity Correction. *J. Am. Stat. Assoc.* **1974**, *69*, 378-380.

66. 66 Darlington, R. B. *Regression and Linear Models*. New York, McGraw-Hill, **1990**.

67. 67 Pedhazur, E. J. *Multiple Regression in Behavioral Research*. New York, Holt, Rinehart, and Winston, **1973**.

68. 68 Younger, M. S. *A First Course in Linear Regression*. (2nd ed.), Boston, Duxbury Press, **1985**.

INDEX